studies in physical and theoretical chemistry 65

STRUCTURE AND REACTIVITY IN REVERSE MICELLES

studies in physical and theoretical chemistry 65

STRUCTURE AND REACTIVITY IN REVERSE MICELLES

Edited by

M.P. Pileni

*Laboratoire de Structure
et Réactivité aux Interfaces,
Université Pierre et Marie Curie,
F-75005 Paris, France*

ELSEVIER
Amsterdam — Oxford — New York — Tokyo   1989

CHEMISTRY

ELSEVIER SCIENCE PUBLISHERS B.V.
Sara Burgerhartstraat 25
P.O. Box 211, 1000 AE Amsterdam, The Netherlands

Distributors for the United States and Canada:

ELSEVIER SCIENCE PUBLISHING COMPANY INC.
655, Avenue of the Americas
New York, NY 10010, U.S.A.

ISBN 0-444-88166-2

studies in physical and theoretical chemistry

Other titles in this series

STRUCTURE AND REACTIVITY IN REVERSE MICELLES

FOREWORD

Man has been using water, oil and surfactant mixtures - that is to say "microemulsions"- for several thousands of years, and, as for many crafts, it was without any understanding of the origin of their properties. It has only been a little more than ten years since scientists have felt mature enough and interested enough to apply their skills to these complex physico-chemical systems. Maturity is indeed a requisite to tackle these ternary fluids where small modifications in the concentrations of constituants can bring about dramatic changes in the properties of the solutions for which a myriad of structures can exist. All the panoply of modern physical chemistry, its concepts as well as its experimental techniques, is not too large for the beautiful questions posed by these systems. The motivation behind the increasing interest in microemulsions is not merely the ambition to understand them, there is also the hope to master them, to forge new tools for human activity. Bio-systems, the connection between their impressive abilities and the properties of physico-chemically heterogeneous systems often akin to microemulsions, have exerted their fascination here. The hope of robbing them of some of their secrets for practical purposes (synthesis of efficient pharmaceuticals, selective chemistry, energy storage, etc.) is a frequent background to a number of scientific studies in this field.

In view of the huge variety of microemulsion systems - whether it be from the point of view of chemical composition or the structural phase - it is a useful approach to restrict oneself to reversed micelles, as is the case in this book. Structural and chemical properties are today more amenable to physical chemistry than many other microemulsion phases and also their topology puts us in the presence of this stable and confined aqueous volume where peculiar chemical properties are to be looked for. However, studying even this restricted class of systems is still is a formidable task to undertake, and studies of reversed micelles must define concepts and methods before gathering measurements for practical applications. That explains, for example why most today refer to AOT as a surfactant, a prototype system now well mastered from the point of view of reliable preparations.

This book first presents concepts and results pertaining to the study of structural properties, multiple experimental techniques must be used and specific interpretative methods must to be developed. Basic chemical properties (identification of dynamic processes, behaviour of injected electrons, etc.), that allow gross characterization of reversed micelles as reactive media are then tackled. In a different approach, more macroscopical in nature, reactions sensitive to the specificity of the environment are studied, photochemistry, magnetic field effects, metal particle aggregate formation, polymerization, etc. ; they all provide hints for possible future applications. With no surprise, in view of the amphiphilic character of reversed micelles, it is finally with regard to the physical chemistry of proteins that the book directs itself. Bio-mimetic reasoning is the basis here ; if Nature makes use of such mixed systems, is it not by modeling and copying it that one will understand it and ultimately become powerful enough to make ourselves able to use these secrets for our own purposes ? As this book shows, the task is demanding and everything is ready for this new scientific venture.

Paul RIGNY

CONTENTS

X

LIST OF AUTHORS

ANTONETTI A.
E.N.S.T.A./Optique Appliquée
Batterie de l'Yvette
91129 PALAISEAU (FRANCE)

BARDEZ E. (Mrs)
C.N.A.M./ Chimie générale
292, rue Saint-Martin
75003 PARIS (FRANCE)

BRUS L.E.
A.T.T. Bell Laboratories
600 Mountain Ave.
MURRAY HILL NJ 07974 (USA)

CANDAU F. (Mrs)
Institut Charles Sadron
C.R.M./E.A.H.P.
6, rue Boussingault
67083 STRASBOURG Cedex (FRANCE)

CARVER M.
Institut Charles Sadron
C.R.M./E.A.H.P.
6, rue Boussingault
67083 STRASBOURG Cedex (FRANCE)

DE SCHRYVER F.C.
K.U. Leuven Chemistry Department
Celestijnenlaan 200 F
3030 LEUVEN (BELGIQUE)

FLETCHER P.D.I.
School of Chemistry
University of Hull
HULL, HU67RX (U.K.)

GAUDUEL Y.
E.N.S.T.A./Optique Appliquée
Batterie de l'Yvette
91129 PALAISEAU (FRANCE)

HATTON T.A.
Department of Geophysical Engineering
Massachusetts Institute of Technology
CAMBRIDGE, MA 02139 (USA)

HILHORST R. (Mrs)
Wageningen Agricultural University
De Dreijen 11/6703 BC WAGENINGEN (PAYS BAS)

KHAN-LODHI A.N.
School of Chemical Sciences
University of East Anglia
NORWICH NR4 7TJ (U.K.)

LANGEVIN D. (Mrs)
E.N.S./Spectroscopie Hertzienne
24, rue Lhomond
75005 PARIS (FRANCE)

LEODIDIS E.B.
Department of Geophysical Engineering
Massachusetts Institute of Technology
CAMBRIDGE, MA 02139 (USA)

LLOR A.
DLPC-SCM
CEN Saclay
91191 GIF SUR YVETTE (FRANCE)

LUISI P.
E.T.H. Zürich, Institut für Polymere
Universitätsrasse 6
CH 8092 ZURICH (SUISSE)

MARTINEK K.
Institut of Organic Chemistry and Biochemistry
Fleming Sq 2
CS 166-10 PRAHA 6 (CZECHOSLOVAKIA)

NICOT C.
Université René Descartes
C.N.R.S./UA 64.01
45, rue des Saint Pères
75006 PARIS (FRANCE)

NINHAM B.W.
Department of Applied Mathematics
Research School of Physical Sciences
Australian National University
CANBERRA, ACT 2601 (AUSTRALIA)

PARROTT D.
School of Chemistry
University of Hull
HULL, HU67RX (U.K.)

PILENI M.P. (Mrs)
Université Pierre et Marie Curie
Laboratoire "Structure et Réactivité aux Interfaces"
11, rue Pierre et Marie Curie
75005 PARIS (FRANCE)

and

DPC-SCM
CEN Saclay
91191 GIF SUR YVETTE (FRANCE)

POMMERET S.
E.N.S.T.A./Optique Appliquée
Batterie de l'Yvette
91129 PALAISEAU (FRANCE)

RIGNY P.
Directeur/DESICP
CEN Saclay
91191 GIF SUR YVETTE

ROBINSON B.H.
School of Chemical Sciences
University of East Anglia
NORWICH NR4 7TJ (U.K.)

STEIGERWALD M.L.
A.T.T. Bell Laboratories
600 Mountain Ave.
MURRAY HILL NJ 07974 (USA)

STEINER U.E.
Fakultat fur Chemie der Universitat
Universitatsrasse 10
7750 KONSTANZ 1 (FRG)

TOWEY T.
School of Chemical Sciences
University of East Anglia
NORWICH NR4 7TJ (U.K.)

VALEUR B.
C.N.A.M./ Chimie générale
292, rue Saint-Martin
75003 PARIS (FRANCE)

VAN DER AUWERAER M.
K.U. Leuven Chemistry Department
Celestijnenlaan 200 F
3030 LEUVEN (BELGIQUE)

VOORTMANS G.
K.U. Leuven Chemistry Department
Celestijnenlaan 200 F
3030 LEUVEN (BELGIQUE)

WAKS M.
Université René Descartes
C.N.R.S./UA 64.01
45, rue des Saint Pères
75006 PARIS (FRANCE)

ZEMB T.
DLPC-SCM
CEN Saclay
91191 GIF SUR YVETTE (FRANCE)

INTRODUCTION

A large number of books on colloids have been published in recent years. However, none of these works deal specifically with reverse micelles, unless we include the report of the proceedings of a conference organized by P.L. Luisi in 1982.

In this book our aim has been to describe the results of the studies carried out on reverse micelles, in terms of both structure and reactivity. In fact, in recent years a number of groups have attempted to study the variation in micellar structure with the addition of water. Studies using techniques of neutron, X-ray and light diffusion have shown that the size of the reverse micelles increases in proportion to the quantity of water added. It is therefore an easy matter to compare the reverse micelle to a microreactor, the size of which can be modified simply by changing the quantity of water. Numerous chemical and photochemical reactions have therefore been studied. These have shown that the location of the species plays a fundamental role in the efficacy of the reaction. Many kinetic studies have been undertaken in order to increase our knowledge of the way in which the velocity constants in such microreactors vary. In order to obtain more quantitative results, certain groups have tried to ascertain whether the addition of reagents is liable to result in significant perturbation of the micellar structure. When the reagents comprise small molecules it appears that the perturbations observed are minimal. When they involve proteins, the results often appear contradictory, but are not however in total disagreement. It would appear to be established that for cases of high water content the whole system reorganizes itself to form monodispersed population aggregates. With low values of water content the problems seem more delicate but the data are not, in my opinion, contradictory. As these problems are not wholly resolved, the interpretations do not always concur. This is true of the two groups which have been working for some years on these questions, those of P.L. Luisi and K. Martinek. In order to allow each point of view to be presented, several pages are devoted to the ideas of Professor Luisi while those of Professor Martinek are presented in the last chapter on enzymatic reactions. The question is a very important one as it must account for the effects of structure on reactivity and vice versa. We must therefore understand why, in most cases, the enzymes are more active for small quantities of water than for large micelles. In order to use these processes, a greater comprehension of the phenomena is necessary in order to exploit to the full the resources of these systems which have so much to offer.

This is why this book presents various approaches currently being studied in the domain of biotechnology.

A large number of applications should be possible, using reverse micelles as microreactors which encourage the formation of small aggregates of variable size. This is of interest for semiconductors and also for metallic particles and small latex systems. In the first two cases, the proximity of the species in the reverse micelles and the fact that the small aggregates present a large catalytic surface mean that catalytic yields can increase by several orders of magnitude. In the third case, numerous applications (drug transport, etc.) should be possible.

I would like to conclude this introduction by acknowledging the constant assistance and expert advice so generously given to me by C. Troyanowsky, without whom this book could not have been written. I would also like to thank the Chemistry Department of the Centre National de la Recherche Scientifique, and the Département de Physico-Chimie of the Centre d'Etudes Nucléaires de Saclay, for their help. Finally my thanks are due to Mme A. Labarre who undertook overall coordination.

M.P. PILENI

IN SEARCH OF MICROSTRUCTURE

B. W. NINHAM

1. INTRODUCTION

"As is well known to students of statistical physics and chemistry, Boltzmann's central thrust was to deduce the macroscopic properties of material systems, especially gases, from constituent atomic and molecular models. Unfortunately, in his time an influential component of the scientific community, the so called "energeticist" School led by Ostwald, Duhem, Helm and Mach, rejected evidence for the existence of atoms and molecules and any physical theory based on them. They preferred to introduce generalised thermodynamic arguments which they called "energetics".

This comment, taken from the last paper[1] of Elliott Montroll, is a paradigm for the modern debate on properties of surfactant solutions that self assemble from amphiphilic molecules, be they micelles, reverse micelles, vesicles, microemulsions, cubic, hexagonal or other phases. Only several years ago some of the most eminent leaders in the field of association colloid science clung to what we might call a pseudo-energeticist view. It was admitted that micelles, reverse micelles (spheres), hexagonal phases (cylinders) and lamellar phases (planes of bilayers) existed. Typical micelles are spherical aggregates of about 50 surfactant molecules that associate spontaneously in water to form if we like the analogue of atoms. Typical reverse micelles are droplets of water (radius ~ 5nm) surrounded by about 500 surfactant molecules dispersed in oil. For such systems what we visualise as microstructure is fairly clear. But for microemulsions, mixtures of oil, water, surfactant with or without "cosurfactant", there was a distinction made, and the only possible path to understanding had to follow a path via thermodynamics. Vesicles were held to be always metastable, i.e. non existent as equilibrium structures. So too were emulsions. Phase diagrams and universality as exemplified by behaviour near critical points was all, and the notion of microstructure within such thermodynamically stable solutions was held to be nonsense or at best indeterminate. It is worth recalling that Hartley's picture[2] of a spherical micelle - as opposed to the older idea that micelle forming surfactants were fragments of lamellar liquid crystalline phases - had to wait until the ideas of Tanford[3] on self-assembly gained general acceptance. Until recently people made a distinction between reverse micelles and microemulsions, and even maintained that ionic microemulsions could not exist without co-surfactants. By now however at least the idea of Scriven[4], that middle phases - optically isotropic surfactant phases in equilibrium with excess oil and water - could be bicontinuous oil swollen bilayers forming minimal surfaces, is acceptable. And cubic phases[5] exist as well as other more exotic bicontinuous crystalline phases, e.g. tetragonal and orthorhombic[6]. Yet still the energeticist school remains predominant, amongst physicists. Drawing on parallels with theories of phase transitions, the invocation of a Hamiltonian based on truncation of a Landau-Ginsberg form was seen to be essential in explaining spectra obtained by neutron or low angle X-ray scattering, at least one route to understanding an apparently inexplicable morass[7]. Such explanations are as tautological as thermodynamics itself, in the sense that they say nothing of microstructure. From one point of view this, the energeticist approach, is reasonable. Even the conceptualisation, let alone the definition of a micelle, is an open question[8], so that what one means by a thermodynamically stable solution is indeed still unclear. In those circumstances appeal to extensions of the Gibbsian framework is mandatory in order to explain this microstructural diversity.

From another viewpoint it is a curious anomaly as biology has taught us that microstructure, equilibrium or non equilibrium, stable or metastable, certainly exists. It must be understood and be predictable if we are ever to optimise and exploit the capacity of nature to contain reactants, effect rapid transport, to catalyse, or to make new colloidal particles and materials. Once microstructure is understood, a rich field of new scientific endeavour comes into sight. If microstructure can be prescribed one can tune up curvature, and water structure, and connectivity and, for example, use microemulsions (or reverse micelles) optimally for molecular templates of prodigious surface area, and/or for new materials either by polymerisation of the oil, or by precipitation in optimal molecular containers.

This introduction attempts to delineate directions and trends in our understanding of association colloid science - still imperfect - that bear on parallels with Boltzmann's view. The (kinetic) aggregates, the micelles and reverse micelles and vesicles, that form in dilute surfactant solutions play the role of atoms.

The paradigm is deliberately provocative. This is because there has been a curious disjunction and absence of communication between workers in the field that deserves to be challenged. On the one hand those (chemists) concerned, and rightly excited by the prospects for micellar catalysis, membrane mimetic chemistry, surfactant aggregates for templating fine particles so eloquently advocated by Fendler[9] have virtually ignored microstructure. On the other hand condensed matter physicists concerned with random and chaotic media have been reluctant to broaden their conceptual framework to admit that surfactant assemblies have size, shape and intra and inter aggregate interactions. Again, the pioneers in surfactant solution chemistry have for too long clung to a characterisation by phase behaviour alone; that is, thermodynamics, to the exclusion of a consideration of microstructure within phases.

2.   CHARACTERISATION OF SELF ASSEMBLY

Clearly there is a problem. Unlike the atoms or molecules that make up simple gases and liquids, the pseudo-atoms that form surfactant aggregates are dynamic and depending on concentration, an infinite hierarchy is possible. The possibility that all the bewildering observations on self assembly could ultimately be handled by a single predictive framework seems remote. That goal became less distant following extension of the ideas of Tanford[3, 8, 10] on dilute micellar systems to include larger assemblies like cylindrical micelles, vesicle and planar bilayers.

Those ideas[8, 10] linked together several notions: a hydrophobic free energy of association for surfactant tails, the principle of opposing surface forces at the surfactant-water, or oil-water interface, and geometric packing constraints, through the law of mass action. Out of that synthesis emerged a characterisation of self assembly in terms of a surfactant parameter that subsumed an older characterisation[11] involving hydrophilic-lipophilic balance. (The surfactant parameter, a measure of curvature at the interface, is set by geometrical packing and interfacial forces. It is v/al where v is the volume of the surfactant hydrocarbon region (which can include adsorbed oil), a is the area per surfactant head group and l the length of the tail region.) Of importance is the fact that critical micelle concentrations and aggregation numbers as a function of chain length, or salt concentration for ionic surfactants are provided in terms of this surfactant parameter. In principle it can be calculated.

The ideas were carried further to include microemulsions[8, 12], albeit limited to reverse swollen micelles, lamellar phases, or "normal" micelles swollen with oil. One role for cosurfactant was identified and partially quantified. These extensions took account of the vexed issue of vesicles, and again a whole set of previously disconnected observations began to fall into place systematically.

3.   RECONCILIATION WITH ION BINDING MODELS

Such ideas gave a global view of the self assembly process, but were qualitative only. Before these beginnings could have wider acceptance, there needed to be at least some further quantification of parameters. For the characterisation of self-assembly alluded to above appeared completely at odds with a much used phenomenological model for micelles called the ion binding model. This model - for ionic surfactants - described micelle formation by means of a pseudo-chemical equilibrium reaction between monomers (1) and counterions (2) to form an aggregate of N monomers, viz

$$X_N \rightleftarrows K X_1^N X_2^Q,$$

(where q = Q/N) is the fraction of "bound" counterions, and K is a constant. Experimentally q is constant, as a function of salt and cmcs also emerge correctly for this model given hydrophobic free energies of transfer. The conflict was resolved by invoking classical double layer theory to calculate surface free energies in the Tanford scheme[13]. The ion binding model then emerged as an asymptotic approximation which is a special case of the Tanford-like approach that holds only in the limit of q=1. It was this limit that had been most studied. It applies only to counterions like Li^+, Na^+, K^+, Cl^-, Br^-, I^-. By contrast for strongly hydrated counterions like OH^-, F^- and carboxylates, the ion binding model is meaningless.

The new theory did seem then to be on the right track. It was extended to highly concentrated dispersions of ionic surfactant, alcohol and water[14]. Numerical calculations based on Poisson-Boltzmann equation -albeit limited to spheres, cylinders, and planar bilayers- do give the ternary phase diagram correctly, except at low water content where hydration effects appear. The role of water structure is disguised at high water contents.

The fact that the Poisson-Boltzmann theory fails badly due to neglect of ion-ion correlations in many situations of interest[15, 16], and is a primitive model, i.e. ignores hydration effects, is irrelevant. So too is the restriction of the theory thus far to highly symmetric systems like spheres or cylinders. For dilute ionic systems and counterions like Na^+, K^+, Cl^+, Br^- it is known from direct force measurement[17] that the P-B theory provides a valid description of electrostatic interactions. From such studies it became clear that it is possible to extend the theory to include globular or other shaped micelles and it is evident that all the properties are subsumed into a (in principle) calculatable surfactant parameter.

4.   RENORMALISED VARIABLES : PHASE BEHAVIOUR OF NON-
       IONIC vs IONIC SURFACTANT

If this is accepted, the next stumbling block for a wider acceptance of such ideas comes from a consideration of the phase diagrams of non-ionic[18] and ionic surfactants. Plot temperature against concentration, and map out phase behaviour. Characteristically, non-ionics exhibit a cloud point with increased temperature. Except at high concentrations of some peculiar quaternary ammonium salts there is no evidence of cloud point phenomena in ionics. If the theory is to have any credibility at all, there must be universality in the phase diagrams. To see how this can come about consider first the non-ionics. As the temperature increases we have to explain how it is that at low surfactant concentration attractive forces between micelles can operate to give phase separation. The micelles are small and their separation is so large that no known forces exist attractive enough to account for the phenomena. However, if as temperature increases, the head group area decreases and/or chain length decreases the surfactant parameter (v/al) increases. The micelles then grow. If simultaneously the forces between such rod shaped micelles are attractive then phase separation will occur. [However weak, attractive forces between long rods are sufficient to do the job.] This indeed is exactly what happens. NMR studies show that at or near the cloud point head groups lose two molecules of hydration[19]. Also, chain length does decrease with increased temperature. Further direct measurements between monolayers of POE surfactants adsorbed onto hydrophobed mica surfaces show that at the cloud point the forces turn from repulsive to attractive[20].

We can then plot the phase diagram in terms of two different variables, the surfactant parameter replacing temperature, and interaction strength replacing concentration. We call them renormalised thermodynamic variables. If now we compare with ionics, we see that here a different situation arises. Ionic micelles decrease in size with temperature, so that by contrast v/al *decreases* with increasing temperature. To obtain universality in the thermodynamics we have to arrange that v/al *increases* with T exactly as for the non-ionics. This can be accomplished by ad- mixing double chained surfactants (v/al $\cong$ 1) with single chained (v/al $\cong$ 1/3), so that

$$(v/al)_{effective} = \frac{(X_D(1) + X_S(\frac{1}{3}))}{(X_D + X_S)} \quad ;$$

where X is the mole fraction of double (D) or single (S) chained surfactants. On the ordinate plot increasing $(v/al)_{effective}$ instead of temperature. On the abscissa plot again the fraction of water, or increasing added salt which weaken repulsive interactions. The experiment yields cloud point phenomena for the ionics!

The argument is not totally convincing, but strongly suggests that the interplay between curvature, as set by packing and intramolecular forces, and interactions between aggregates, rather than temperature and concentration, are indeed a more sensible way to think of microstructure and achieve universality in phase behaviour.

5.   INTERACTIONS

An enormous amount of work has been done on theories of interactions and direct measurements thereof. And the situation revealed by such measurements - between hydrophobic[21], or hydrophilic[22], surfaces with an intervening liquid has completely changed the face of colloid science meantime. A host of new forces - hydrophobic[21], hydration[23], ion specific[17, 24], due to ion-ion correlations[15, 16], due to surface-induced solvent structure, solvent-induced surface structure[22, 25] and surface dipole-dipole correlations[26] have emerged. While some progress has been made in quantifying these new and unexpected forces, a program that attempts to actually calculate such forces ab initio, and quantitatively is extremely difficult. We are beginning to understand how to exploit such surface forces to set surface interactions and curvature in those systems that self assemble[27].

At first sight awareness of such forces and their complexity, means that theoretical notions outlined so far appear to be so simple minded as to beggar the imagination of those schooled in more rigorous disciplines. This is not so great an inhibition as it seems.

6.   REVERSE MICELLES AND MICROEMULSIONS

For application and extension led to predictions for some ternary mixtures of ionic surfactant, water and oils that all fell into place[27]. Restriction to surfactants insoluble in both oil and water led immediately to further understanding. We know now how to tune up curvature - set on one side of the interface by oil, on the other by water and/or salt and how to modify that curvature with cosurfactant. Now bicontinuous phases of *constant* curvature (not just minimal curvature) can easily be made, and polymerised to make new materials.   Theoretical predictions were confirmed by scattering experiments[28], and even phase boundaries can be predicted[29] with the ubiquitous surfactant parameter which is, in these cases, directly measurable.

The nature of the oil, of the surfactant, its head group, of the counterion and of the cosurfactant have all been elucidated. All conspire to set microstructure, besides component volume fractions. For catalysis or particle synthesis within reverse micelles, awareness of these factors allows great flexibility. Further there is now an understanding that reactants themselves change microstructure. For example, methacrylate monomer added to a dispersion of spherical reverse micelles will change the microstructure to a bicontinuous phase. This is because the monomer affects water structure, head group interactions, and thereby curvature of the interface. On polymerisation that role for the monomer is removed. The system then reverts to disconnected water droplets surrounding polymer particles of prescribed size.

7.   FUTURE DIRECTIONS : HETEROGENEOUS MICROSTRUCTURE

So far theories of microstructure which seem to have been confirmed by experiment, are largely geometrical.   Further inroads into understanding microstructure are being made through the recognition and development of theories of fluctuation forces of bilayers[30].   Although in the main so far phenomenological we can expect a better link between these phenomenological approaches that will relate bending moduli to specific surface molecular forces. There are difficulties because theories of fluctuation forces in the harmonic approximation are applied to a situation which is intrinsically anharmonic, and therefore unacceptable.   But a larger problem than this remains.

In the context of self assembly the key question is that which plagues a nucleation theory of phase transitions: How can one write down a partition function (from which thermodynamics follows) that will average over size and *shape* of aggregates?   Geometric topology applied to elucidating minimal surfaces that approximate to the infinity of possible cubic phases is beginning to see good usage[31].   The broader area of elucidating the larger infinity of surfaces of constant curvature is just beginning to be recognised as an area of research of great elegance and difficulty.

Even were these problems to be solved there is a growing awareness of a still greater subtlety.   There is the reality of microstructure within microstructure[25].   To see this, consider a suspension of single-walled ionic vesicles.   Suppose that these vesicles are at a concentration of 10^{-5}M and have a radius of ~1000Å.   A moments thought will show that *inside* the vesicle charge neutrality demands a counterion concentration of ~10^{-2}M.   Any topologically closed bilayer will have the property that solution conditions can differ significantly between inside and outside.   An example is spontaneous vesicles of didodecyl dimethyammonium hydroxide which are in equilibrium with micelles in a kind of superaggregation phenomena[25].   (The analogue of the "reaction" N monomers $\rightarrow$ micelle is N micelles $\rightarrow$ vesicle.)   Because of the difference in solution conditions inside and outside, the "normal" multi-walled liposome collapses to a few- walled vesicle *inside* which resides the micelles[32].   That example is probably universal in biology.   It is likely, for example, that chloroplasts which were regarded as lamellar phases are actually few-walled containers inside which are cubic phases, deliberately designed by nature to be bicontinuous to allow the rapid ionic exchange required by the photosynthetic process.

Reverse micelles are but a tiny fragment of the total hierarchy of self assembled structures that we can see will in the future be accessible for templating of new materials. With progress in the study of reverse micelles, microemulsions, emulsions, and self assembly in general, physical chemistry is entering into a new arena, moving from a mastery of matter at a molecular level to the design and understanding of intermolecular systems and their interaction with light. These studies are bound to impact strongly on biochemistry and materials science.

Association Colloid Science used to be an arcane art and the bridge from physics and chemistry to biochemistry and biology but dimly perceived. This book is one of the first that takes us to a new level of sophistication and into that new arena.

REFERENCES

1. E.W. Montroll, On the Vienna School of Statistical Thought in Random Walks and their Applications in the Physical and Biological Sciences, ed. M.F. Schlesinger and B.J. West, AIP Conference Proceedings No. 109. Series editor H.C. Wolfe. (NBS/La Jolla Institute - 1982).

    The succeeding references (and the preceding) reflect (1) editorial deadlines which dictate that one quotes what is most easily accessible on one's desk, (2) real prejudice and bias. We have taken a deliberate and particular stance that reflects that bias. The balance will be redressed in abundance in succeeding papers of this book.

2. G.S. Hartley, *Trans. Faraday Soc.* 37, 130 (1941).

3. C. Tanford, The Hydrophobic Effect (John Wiley and Sons, New York, 1973).

4. L.E. Scriven, *Nature* (London) 263, 123 (1976).

5. V. Luzzati and F. Reiss-Husson, *Nature* 210, 1351 (1966).
    V. Luzzati and P.A. Spegt, *Nature* 215, 701 (1967).
    P. Ekwall, L. Mandell and K.J. Fontell, *J. Colloid Interface Science* 33, 215 (1970).
    K. Fontell, L. Mandell and P. Ekwall, *Acta Chem. Scand.* 22, 3209 (1968).
    K. Fontell, A. Ceglie, B. Lindman and B.W. Ninham, *Acta Chemica Scand.* A40, 247 (1986).

6. P. Kekicheff and B. Cabane, *J. Physique* 48, 1571 (1987).
    P. Kekicheff and B. Cabane, *Acta Crystallographica B* 44, 395 (1988).

7. M. Tenbyer and R. Strey, *J. Chem. Phys.* 87, 3195 (1987).

8.   D.J. Mitchell and B.W. Ninham, *J. Chem. Soc. Faraday Trans. 2* 77, 601 (1981).

9.   J.H. Fendler, *Chem. Rev.* 87, 877 (1987).
      X.K. Zhoa, S. Baral, R. Rolandi and J.H. Fendler, *J. Am. Chem. Soc.* 110, 1012 (1988).

10.   J.N. Israelachvili, D.J. Mitchell and B.W. Ninham, *J. Chem. Soc. Faraday Trans. 2* 72, 1525 (1976).
      H. Wennerstrom and B. Lindman, *Phys. Rep.* 52, 1 (1979).

11.   P.J. Becher, *J. Dispersion Sci. Technol.* 5, 81 (1984).

12.   K. Shmoda and S. Friberg, *Adv. Colloid Interface Sci.* 4, 281 (1975).

13.   D.F. Evans, D.J. Mitchell and B.W. Ninham, *J. Chem. Phys.* 88, 6344 (1984).

14.   G. Gunnason, B. Jonsson and H. Wennerstrom, *J. Phys. Chem.* 84, 3114 (1980).
      B. Jonsson and H. Wennerstrom, *J. Phys. Chem.* 91, 338 (1987).

15.   R. Kjellander and S. Marcelja, *J. Chem. Phys.* 82, 2122 (1985).
      R. Kjellander, *J. Chem. Phys.* 88, 7129 (1988).
      R. Kjellander and S. Marcelja, *J. Chem. Phys.* 88, 7138 (1988).
      Clay swlling & HNC:
      R. Kjellander, S. Marcelja and J.P. Quirk, *J. Colloid Interface Sci.* (in press).
      R. Kjellander, S. Marcelja, R.M. Pashley and J.P. Quirk, *J. Phys. Chem.* (in press).

16.   P. Attard, D.J. Mitchell and B.W.Ninham, *J. Chem. Phys.* 88, 4987 (1988).
      P. Attard, D.J. Mitchell and B.W. Ninham, *J. Chem. Phys.* (in press).

17.   R.M. Pashley, P.M. McGuiggan, B.W. Ninham, D.F. Evans and J. Brady, *J. Phys. Chem.* 90, 1637 (1980).
      R.M. Pashley and B.W. Ninham, *J. Phys. Chem.* 91, 2902 (1987).

18.   D.J. Mitchell, G. Tiddy, L. Waring, T. Boshtok and M.P. McDonald, *J. Chem. Soc. Faraday Trans. 1,* 79, 975 (1983).

19.   P.G. Nilsson and B. Lindman, *J. Phys. Chem.* 87, 4756 (1985).

20.   P.M. Claesson, R. Kjellander, P. Stenius and H.K. Christenson, *J. Chem. Soc. Faraday Trans. 1,* 87, 2735 (1986).

21.   J.N. Israelachvili and R.M. Pashley, *Nature* 300, 341 (1982).
      R.M. Pashley, P.M. McGuiggan, B.W. Ninham and D.F. Evans, *Science* 229, 1088 (1985).
      P.M. Claesson, C.E. Blom, P.C. Herder and B.W. Ninham, *J. Colloid Interface Science* 114, 234 (1986).

22.   H.K. Christenson, *J. Dispersion Sci. and Technology* 9, 171 (1988).
      J.N. Israelachvili and P.M. McGuiggan, *Science* 241, 795 (1988).

23.   J. Marra and J.N. Israelachvili, *Biochemistry* 24, 4608 (1985).

24.   J.L. Parker, H.K. Christenson and B.W. Ninham, *J. Phys. Chem.* 92, 4155 (1988).

25.   B.W. Ninham and D.F. Evans, *Faraday Disc. Chem. Soc.* 81, 1 (1986).

26.   P. Attard, D.J. Mitchell and B.W. Ninham, *Biophysics Journal* 53, 457 (1988).

27.   D.F. Evans, D.J. Mitchell and B.W. Ninham, *J. Phys. Chem.* 90, 2817 (1986).
      M. Kahlweit and R. Strey, *Angew Chem. Int. Ed. Engl.* 24, 654 (1985).

28.   I.S. Barnes, S.T. Hyde, B.W. Ninham, P-J. Derian, M. Drifford and T.N. Zemb, *J. Phys. Chem.* 92, 2286 (1988).

29.   S.T. Hyde, B.W. Ninham and T. Zemb, *J. Phys. Chem.* (1988) in press.

30.   W.Z. Helfrich, *Naturforsch* 28c, 693 (1973).
      D.A. Huse and S. Liebler, *J. Phys. France* 605 (1988).
      M. Winterhalter and W.Z. Helfrich, *J. Phys. Chem..* (in press).

31.   S. Anderson, S.T. Hyde, K. Larsson and S. Lidin, *Chem. Reviews* 88, 221 (1988).

32.   E. Radlinska, J.P. Dalbiez, B.W. Ninham and T.N. Zemb, *J. Phys. Chem.* (1988) in press.

STRUCTURE OF REVERSED MICELLES

D. LANGEVIN

1.   INTRODUCTION

Over the past 40 years, there have been many investigations of the formation of reversed micelles in non-polar solvents (1)-(5). Some of these studies address the problem of the determination of micellar size and shape. It was early recognized that the knowledge of the structure of the medium is an essential step for the understanding of its properties and for a better design of its applications. Nowadays, a number of powerful experimental techniques are available for structural investigations. The aim of this paper is to present these techniques and to illustrate their possibilities and limitations with examples of studies on reverse micelles.

Reverse micelles can be formed both in the presence and in the absence of solubilized water. However, if the medium is completely water free, there is no well defined critical micellar concentration, and the aggregates formed are very small and polydisperse (2). Let us take for instance the case of the extensively studied surfactant AOT (sodium di-2-ethylhexyl sulfosuccinate). Close packing of the surfactant polar heads leaves an empty volume in the center of the micellar core (radius 4.4 Å for icosaedral packing (6)) which can only be filled with hydration (or bound) water (Fig. 1). This explains why the presence of water is necessary to form a large surfactant aggregate. Even the non-ionic surfactants, which possess a long flexible polar part, do not form large monodisperse aggregates in the absence of water (7). In this paper we will focus on the properties of large and well defined micelles formed when water (or another polar solvent) is present. The reversed micelles considered here are then also microemulsions. We will not make differences between the terms swollen micelles and microemulsions, although certain authors use the first term for systems containing only bound water. Of course, the amount of bound water might be important for special properties or applications. For AOT, the maximum amount of bound water in the micelles correspond to a water-surfactant molar ratio $w = [H_2O] / [AOT]$ of about 10 (8). Above this amount, part of the water is "free". But there is no evidence of a particularly well defined transition at that point.

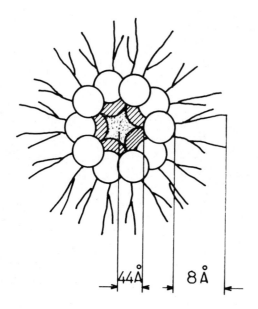

44Å   8Å

Fig. 1. Structure of an AOT reverse micelle

Because AOT is soluble in many non polar solvents over a wide range of concentrations, a large number of studies of the structure of reverse micelles have been made with this surfactant. They will be largely quoted here. We will also refer to studies of reversed micelles (or water in oil (w/o) microemulsions made with other surfactants : cationic, non ionic. We will include w/o microemulsions containing cosurfactants, which have also been widely studied. Some of these systems can exhibit a continuous structural inversion towards oil in water structures (o/w). This can happen for instance when either the oil/water ratio, or the salinity, or the temperature are varied. The inversion domain can be very broad and the structural modifications can begin in oil-rich systems (volume fraction of oil ~ 80 %). We will include in this paper a discussion about the corresponding structures which are no longer made of isolated micelles.

The paper is organized as it follows : first, we will introduce the general concepts necessary to describe the systems, and give a survey on the existing or conjectured structures. We will then present the experimental methods which fall into several groups : electron microscopy, scattering techniques (light, X-ray, neutron) and dynamic techniques (fluorescence, diffusion, centrifugation). We will illustrate the possibilities of each technique by existing experiments on reverse micelles and oil-rich microemulsions. A special emphasis will be put in the scattering techniques, to date the most powerful.

2. GENERAL CONCEPTS

2.1 Surfactant molecules packing

When dealing with large surfactant aggregates, surfactant molecular packing considerations become important (9). Geometrically, the surfactant molecule can be represented by a truncated cone which dimensions are determined by the respective ranges of hydrophilic and hydrophobic part interactions (Fig. 2). If the surfactant is ionic, the volume of the polar head is small.

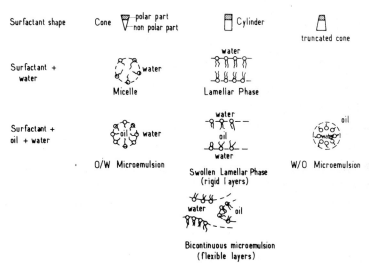

Fig. 2. Surfactant molecules types of packing for different values of the surfactant parameter

Then if v is the surfactant molecular volume, Σ the area per polar head and l the length of the hydrophobic part, the number $v/\Sigma l$, called the "surfactant parameter", gives a good idea of the shape of the aggregates which will form spontaneously :

$v/\Sigma l < 1/3$ spherical micelles in water
$1/3 < v/\Sigma l < 1/2$ rod-like micelles in water
$1/2 < v/\Sigma l < 2$ lamellar phases in water/or oil
$2 < v/\Sigma l < 3$ rod-like micelles in oil
$3 < v/\Sigma l$ spherical micelles in oil.

When both oil and water are present, o/w microemulsions will be formed when $v/\Sigma l < 1$, w/o microemulsions when $v/\Sigma l > 1$, lamellar phases when $v/\Sigma l \sim 1$ (10).

The following expressions for an hydrocarbon chain volume and extended length l_c are useful to calculate the surfactant parameter (11) :

$$v = 27.4 + 26.9\ n(\text{Å}^3)$$
$$l_c = 1.5 + 1.265\ n(\text{Å})$$

where n is the number of carbon atoms in the chain. The area per polar head is mainly determined by head interactions (9) and can be estimated as follows. The chemical potential of a surfactant molecule in the micelle is the sum of chain and head contributions. The head contribution is itself the sum of van der Waals attraction and electrostatic repulsion :

$$\mu = \gamma\Sigma + \frac{e^2 D}{2\varepsilon\Sigma} + g$$

where g is the chain contribution, γ an oil-water surface tension, e the charge per surfactant molecule, D the thickness of the double layer, and ε the local dielectric constant. By minimizing μ with respect to Σ, one obtains easily :

$$\Sigma = \sqrt{\frac{e^2 D}{2\varepsilon\gamma}}$$

with $D = 5$ Å, $\varepsilon \sim 40\varepsilon_0$, $\gamma \sim 50$ mN/m, one gets : $\Sigma \sim 60$ Å^2.

Let us estimate now $v/\Sigma l$ for AOT : $v \sim 600$ Å^3, $l \sim l_c \sim 8$ Å, $\Sigma \sim 60$ Å^2 gives $v/\Sigma l \geq 1$. This indicates that reverse micelles are favoured and that the water swelling ability is large as observed. However, when equal amounts of oil and water are mixed with a small amount of AOT, the micelles are formed in the oil phase only if a small amount of salt is added ($\geq$ 0.2 wt % of NaCl for heptane at 25°C) (12), i.e. when D is decreased by a small amount.

In order to be more predictive, some care has to be taken in the evaluation of the surfactant parameter. First, v represents not only the surfactant volume, but incorporates the amount of oil and water penetrating into the surfactant layer. It is known that short chains oils penetrate easily into surfactant layers : the spontaneous curvature of the layer will then be larger for reverse micelles in short chain alkane solvents than in longer chain ones. The limit of penetration seems to correspond to alkanes of the same chain length than the surfactant molecule (13). Similarly, when a surfactant like a short chain alcohol is incorporated to the micelles, v is increased. When the surfactant chains are loosely packed (area per chain larger than 20 Å^2), the chain length l is frequently less than l_c because of chain folding. The area per polar head Σ only is little affected by oil and/or alcohol penetration.

Finally, it must be pointed out that this simple picture neglects interaggregate interactions and applies only to dilute micellar dispersions. Structural evolution at large micellar concentration are dominated by interactions between the micelles.

2.2 Composition Constraints

We have seen that, due to packing considerations, the surfactant layer have a preferred spontaneous curvature C_0. The sign of C_0 is by convention positive for direct micelles, negative for reverse micelles. If $|C_0| \ll 1/l$, the "dry" micelle radius is much smaller than the optimum radius $R_0 = 1/|C_0|$. In the case of an inverse system, the micelle will be able to incorporate a large amount of water. When the radius reaches values close to R_0, the swelling process stops and the water is rejected in an excess phase. This is called the emulsification failure (14). It is due to the fact that the interfacial energy associated to the formation of a new phase is smaller than the bending energy of the surfactant layer. This last energy writes (15) :

$F_b = 2K(C - C_0)^2$ per unit area

where K is a bending elastic constant and $C = 1/R$ is the micelle curvature.

The systems where a w/o microemulsion coexists whith an excess water phase are called Winsor II systems, the Winsor I systems being those where an o/w microemulsion coexists whith an excess oil phase (3).

We have seen that the area per surfactant molecule Σ is reasonably constant. This allows to easily calculate the micellar radius when the composition of the system is known :

$$R = \frac{3\phi}{C_S\Sigma} \qquad [1]$$

ϕ is the micelle volume fraction, C_S the number of surfactant molecules per unit volume. The core radius is obtained by replacing ϕ by the water volume fraction ϕ_W :

$$R_W = \frac{3\phi_W}{C_S\Sigma} = \frac{3v_W}{\Sigma} w \qquad [2]$$

v_W being the molecular volume of water.

2.3 Intermicellar Interactions

Reverse micelles interact via van der Waals attractive forces and short range repulsive forces. Van der Waals attraction is frequently negligible and the overall interaction is of the hard sphere type (16). The hard sphere radius R_{HS} is equal or less than the micellar radius. R_{HS} is less than R when micelles interpenetration is allowed. This is accompanied by an additional attractive force which origin is still not well established : enhanced Van der Waals attraction (17), solvent induced attraction (18), ... The attractive part of the potential can become large enough, in some cases, to promote phase separation well before the emulsification failure. The coexistence curve is very assymmetrical, as in the case of polymer-solvent mixture : a micelle rich and a micelle poor phases are in equilibrium with each other. The limit for phase separation is a critical point, where the two micellar phases become identical.

The micellar attraction increases with droplet size. This explains why this type of critical point is found in the AOT-water-decane phase diagram but not in the AOT-water-octane one for instance (Fig. 3). Indeed R_O is larger in the decane system because oil penetration is smaller (see § 2.1).

Micellar attraction favours the formation of aggregates of micelles : dimers, trimers, ... If the micelle volume fraction is large enough, an aggregate of macroscopic dimensions can appear at the percolation threshold. This is accompanied by a steep increase of the electrical conductivity. This does not prove that the droplet aqueous cores become interconnected and that the droplet structure is disappearing. The same phenomenon will show up if electrical charges cross the interfacial layers of two neighbouring droplets by a hopping mechanism (19). Such a hopping mechanism is favoured by droplets interpenetration which reduces the hopping distance. The electrical percolation is indeed observed only in attractive droplets systems and is absent in hard-sphere like systems (20).

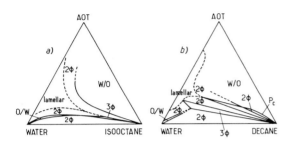

Fig. 3. AOT-water-oil phase diagrams: (a) isooctane   (b) decane

2.4 Case of Small Spontaneous Curvature

When C_O is close to zero, the micelles can in principle grow without limit. When the oil and water volume fractions are comparable, one expects to observe lamellar phases. When the water volume fraction is small, long rods aggregates might also be formed and lead to hexagonal phases. All these organized phases are observed in practice ; but other isotropic phases, which are not dispersions of isolated micelles, also exist. These isotropic phases are continuously connected to the micellar phases in the phase diagram (without first order transitions). These phases originate from the fact that thermal motion can destroy the long range orientational order of the rods or of the surface normals in the lamellar phase. The range of the order becomes finite and is characterized by the persistence length ξ_k of the rod or of the surface. In the last case (21) :

$$\xi_k = a \exp(2\pi K/kT)$$

where a is a molecular length, k the Boltzmann constant and T the absolute temperature. If K >> kT, ξ_k is large and macroscopic : the order is lamellar. If K ~ kT, ξ_k is microscopic : the system is disordered and sponge-like. But, like the lamellar structure, it is bicontinuous both in water and in oil. Geometrical arguments lead to a characteristic dispersion size :

$$\xi = 6\phi_0\phi_w/C_s\Sigma \qquad\qquad\qquad\qquad [3]$$

where ϕ_0 is the oil volume fraction. It is expected that $\xi \sim \xi_k$ (21).

Bicontinuous microemulsions can coexist with both oil and water excess phases. The corresponding systems are called Winsor III systems (3). The sequence of phase equilibria Winsor I $\rightarrow$ Winsor III $\rightarrow$ Winsor II can be promoted by varying salinity when the surfactant is ionic or temperature when the surfactant is non ionic : in this way, the spontaneous curvature of the surfactant layer C_0 is continuously varied from positive to negative values.

In the case of long flexible rods, the structure can be bicontinuous too, if the rods are interconnected. When the amount of water is increased, geometrical packing conditions lead to an evolution towards a disconnected droplets structure. The electrical conductivity drops then steeply, and the variation is entirely opposite to the case discussed in § 2.3. Structural studies of this type of system are presented in detail in chapter 1.

2.5 <u>Concentrated Systems</u>

We have seen that, when the oil and water volume fractions are similar, the structure can be lamellar or disordered bicontinuous. Other possibilities are cubic arrangements : bicontinuous cubic when $C_0 \sim 0$ or cubic crystals of spherical micelles if $|C_0| > 0$ (Fig. 4). Although the second possibility was frequently postulated for concentrated microemulsions (DBS (sodium dodecyl benzene sulfonate) - hexanol (22) and AOT (23)), there is now experimental evidence showing that the structure is rather glassy-like (24).

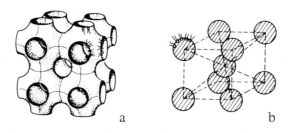

a   b

Fig . 4. Ordered cubic structures (a) bicontinuous . (b) micelles crystal

3. METHODS FOR STRUCTURAL DETERMINATION

3.1 Electron Microscopy

There have been many unsuccesful attempts to perform electron microscopy images of reverse micelles. The reason is that molecular exchanges are fast in these systems (see chapter 12) and that the cooling rates during the sample preparation are not fast enough to prevent structural evolution and/or phase separation. Observations of spherical micelles have frequently been reported even in systems which do not contain spheres at all.

In a recent study (25), likely to be free of artefact, a very fast cooling rate has been used to prepare freeze fracture replica of the samples. The systems are made with non-ionic surfactants where molecular exchanges are slow. Isolated droplets as well as droplets aggregates are clearly visible when ϕ_W is small (Fig. 5A). Bicontinuous structures are observed when $\phi_O \sim \phi_W$ (Fig. 5B). This technique will probably become widely used in the coming years.

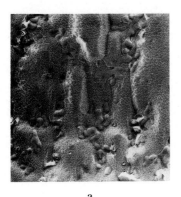

a   b

Fig. 5. Electron microscopy photographs. The shadow material specifically decorates the oil fracture face . (a) w/o microemulsion   (b) bicontinuous microemulsion .(after W.Jahn and R.Strey, ref 25)

3.2 Elastic Scattering : Light, X-Ray, Neutrons

Scattering is an indirect technique which gives the Fourier transform of the image of an object. It is not a priori complicated to inverse Fourier transform the data to get back to the object. Several difficulties are however present :
• The finite instrumental angular resolution and the instrumental noise limitate the accuracy on the scattering patterns.
• The explored Fourier space is limited : the maximum wave vector is $q_{max} = 2\pi/\lambda$ where λ is the wavelength of the scattered radiation and the minimum

wave vector $q_{min} = (2\pi/\lambda)\theta$ where θ is the minimum scattering angle of the apparatus used.

• Spatial correlations are averaged over the irradiated volume and the duration of the experiment. In micellar solutions, thermal motion produces large fluctuations of distances and orientations : the averaging reduces therefore drastically the amount of information contained in the scattering patterns.

There are many methods for X-ray and neutron scattering data treatment. The most commonly used is the one where an a priori model of the structure is adopted, its Fourier transform computed and compared with the data. When enough adjustable parameters are used, the agreement with the data can be satisfactory. But this is not a proof that the model structure is the only one that fits the data, and of course that the one used is the good one. Other more general but less straightforward methods can be used.

In most cases, the scattering pattern does not contain enough information to reconstruct a proper image of the structure. It can then be useful to label some parts of the objects to recognize them. This is called "contrast variation". It is relatively easy with neutron scattering where hydrogenated molecules can be replaced by deuterated molecules. Of course this operation should not be accompanied by structural changes. Although usually the changes are minor (26), it has been noted that, in some cases, the micellar size can vary to a large extent (27).

Several interesting ways to regain the information suppressed by orientational averaging have been explored recently. One can orient the micelles with a hydrodynamic shear flow (28). One can also investigate the structures of anisotropic phases of composition close to the micellar phases of interest.

Light scattering data contain much less information since $q_{max}R \ll 1$, R being the micellar radius. But the technique is more accessible, and provided some care is taken in the data interpretation, it can often be used with profit.

a- Scattering Theory (29)(30)(31)

The basic quantity measured in a scattering experiment is the differential cross section which is the ratio of the scattered intensity per unit solid angle to the incident intensity per unit surface :

$$\frac{d\sigma}{d\Omega} = \langle |\sum_{l=1}^{N} b_l \exp(i\vec{q}.\vec{r}_l)|^2 \rangle$$

b_l is the scattering length of the lth electron or nucleus in the sample, $\vec{r}_l$ its position vector, and $\vec{q}$ the scattering wave-vector.

$$q = \frac{2\pi}{\lambda} n \sin \frac{\theta}{2}$$

n is the refractive index in the sample and θ the scattering angle. The notation $I(q) = \frac{d\sigma}{d\Omega}$ is frequently used.

The differential cross-section contains an incoherent part which is q independent and is only appreciable for neutron scattering from hydrogen atoms. In the following, we will consider only the coherent scattering (b_l will be the scattering length for coherent scattering). Large angle X-rays or neutron scattering corresponds to distances $\vec{r}_l$ - $\vec{r}_{l'}$ of order 1 to 5 Å and to liquid diffuse rings. Small angle scattering : $q \sim 0.006$ - 0.6 Å^{-1} corresponds to distances 10 - 1000 Å which yield the interesting information about the micellar structure : SAXS (small angle X-ray scattering), SANS (small angle neutron scattering). In this range, the molecules can be treated as a continuous medium and the scattering is described by the density of scattering length

$$\rho(\vec{r}) = \frac{1}{V} \sum_l b_l \, \delta(\vec{r} - \vec{r}_l)$$

V being the irradiated volume.
The scattered amplitude is the Fourier transform of ρ :

$$A(\vec{q}) = \int_V \rho(\vec{r}) \exp(i\vec{q}.\vec{r}) \, d^3r \qquad [4]$$

For particles dispersed in a uniform solvent, the scattering density of the solvent ρ_s can be substracted (a spatially uniform medium does not produce scattering outside the beam). For a particle i :

$$A_i(\vec{q}) = \int_{V_i} (\rho_i(\vec{r}) - \rho_s)\exp(i\vec{q}.\vec{r}) \, d^3r$$

If we separate in the scattering cross section the terms coming from the same particle and those coming from different particles whose center of mass are located respectively at $\vec{r}_i$ and $\vec{r}_j$, we get :

$$I(\vec{q}) = \langle|A(\vec{q})|^2\rangle = \langle\sum_i |A_i(\vec{q})|^2\rangle + \langle\sum_i \sum_{j\neq i} A_i(\vec{q})A_j^*(\vec{q}) \sum_{l\in i} \sum_{l'\in j} expi\vec{q}(\vec{r}_l - \vec{r}_{l'})\rangle$$

Let us assume that the particles are identical and spherical. The first term is just the square of the single particle amplitude times the particles number N. The second contains the interparticles interactions. It is related to the pair correlation function

$$g(r) = \frac{V}{N^2}\langle\sum_{l\in i}\sum_{i\neq l'\in j} \delta(\vec{r}-\vec{r}_l + \vec{r}_{l'})\rangle$$

which Fourier transform is the interparticle term of the structure factor :

$$S(q) = 1 + \frac{N}{V} \int_V g(r)\ exp(i\vec{q}.\vec{r})d^3r$$

Finally $I(q) = N\,|A(q)|^2 S(q)$ [5]

If the particles are spherical but polydisperse, rigorous expressions can only be given for hard spheres (32). In the other cases, approximations must be made. One can neglect the corrrelations between the size and the positions of the particles : interactions can still be described by a single structure factor and :

$$I(q) = N\,\{|\langle A(q)\rangle|^2\,S(q) + \langle|A(q)|^2\rangle - |\langle A(q)\rangle|^2\}$$ [6]

This expression can also be applied to non spherical particles if orientational correlations are neglected. The averages refer then to orientational averages. The above "decoupling" approximations fail when the solution is not dilute. Indeed micelles of different size interact via different potentials. A practical consequence is that the crude approximation :

$$I(q) = N <|A(q)|^2> S(q) \qquad\qquad [7]$$

is better in this case. S(q) is the effective one-component structure factor. The effective one component radius for S(q) is $R^3 = \dfrac{1}{N} \sum\limits_i R_i^3$

b- Interaction potential

These equations show clearly that the scattering is affected both by intraparticle structure and by interparticle interactions. The determination of particle size cannot be done if particle interactions are unknown. Two methods are currently used for this purpose.

b.1 Calculation of g(r) from the interparticle potential W (30) ; fortunately g(r) is not very sensitive to the shape of the potential. A commonly used potential is the sum of a repulsive hard sphere potential and an attractive square well. W contains three adjustable parameters : hard sphere radius R_{HS}, interaction strength ε and range δ

$$W = \infty \qquad\qquad \text{for} \qquad r < 2R_{HS}$$

$$W = -\varepsilon kT \qquad\qquad \text{for} \qquad 2R_{HS} < r < 2R_{HS} + \delta$$

$$W = 0 \qquad\qquad \text{for} \qquad r > 2R_{HS} + \delta$$

For AOT microemulsions [33], R_{HS} is comparable to the micellar size. δ is of the order of 3 Å, coherent with the fact that the attraction is only important when two micelles are interpenetrating : the difference between the two AOT chains lengths is about 1.5 Å. ε is proportional to R, as predicted by theory (17). The Yukawa potential, which allows to obtain S(q) in analytical form, has also been used :

$$W = \infty \qquad\qquad\qquad \text{for} \qquad r < 2R_{HS}$$

$$W = -2\varepsilon kTR_{HS}\frac{e^{-\gamma r/2R_{HS}}}{r} \qquad \text{for} \qquad r > 2R_{HS}$$

For AOT microemulsions γ is temperature independent ($\gamma \sim 1.4$) but ε increases with temperature as a critical point is approached (34).

    b.2 Dilution : At large interparticle separation $S(q) \simeq 1$; this method is easy with ternary systems where dilution is achieved simply by adding oil. It is more complicated in systems with alcohol because the cosurfactant is not only present in the surfactant film, but also in the oil and water phases. Dilution procedures have been worked out to solve the problem (35)(36)(37). It must be reminded that dilution is inoperant if the micellar attraction is too large (close to critical points) (38), and of course if the structure is bicontinuous.

    In the case of light scattering, there is usually no angular variation of the scattered intensity. Indeed the wave vectors q are small (10^{-5} - 10^{-3} Å^{-1}) and the particles are seen as a continuous medium : $qR \ll 1$. The only information extracted is

$$I = N|A (q \rightarrow 0)|^2 S(q \rightarrow 0) \qquad\qquad [8]$$

$A (q \rightarrow 0) = (\bar{\rho} - \rho_s)v \qquad \vec{\rho}$ being the average scattering density in the particle

$$S(q \rightarrow 0) = \frac{N}{V} kT \chi = kT(\frac{\partial \pi}{\partial \phi})^{-1} \qquad \phi = \frac{N}{V} v$$

where v is the particle volume and χ the isothermal osmotic compressibility : $\chi = (1/V)(\partial V/\partial \pi)$, π being the osmotic pressure. It is hopeless to try to calculate χ without information on $S(q \neq 0)$. Dilution is therefore the only applicable method. When dealing with a number of different dilutions containing different volume fraction ϕ of the same micelles, information about the interaction potential can be obtained. Assuming again that the potential is the sum of a hard sphere potential plus a small attractive term, one can write (16) :

$$\pi = \pi_{HS} + \frac{kT}{v} \frac{\alpha}{2} \phi^2$$

π_{HS} being the osmotic pressure of a hard sphere system, well approximated by the Carnahan-Starling formula (16). The parameter α is related to the interaction potential W(r) through :

$$\alpha = \frac{4\pi}{kT} \frac{1}{v} \int_{2R_{HS}}^{\infty} W(r)r^2 dr$$

This expression usually fits well the data (fig 6) (16)(39)(40), excepted close to critical points where the attraction is large and cannot be described by the above perturbation expansion. It has also been used to describe the extrapolation to q=0 of SANS data (41).

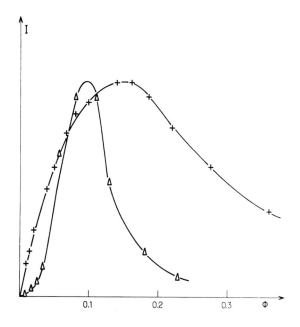

Fig. 6. Light scattering spectra for two w/o microemulsions; + hard sphere droplets, Δ attractive droplets

b.3 Analysis of the high q scattering. The structure factor $S(q)$ exhibits generally several oscillations and tends assymptotically towards 1 at high q. It has been shown for hard spheres that $S(q) \sim 1$ for $qR \geq 2$ if the volume fraction ϕ is less than 10 % (42). The high q part of the scattering pattern is therefore only related to $A(q)$ and it is used frequently to avoid dilution or $S(q)$ computations.

c- Particle size. Once the contribution of particles interactions has been removed, particle size can be obtained in many ways for X-ray and neutron scattering experiments. For light scattering experiments, the only applicable method is c.2.

c.1 - Fitting with $A(q) = 3A(0)(\sin qR - qR\cos qR)/(qR)^3$ for a sphere, or other expressions if the particle shape is simple (shells, ellipsoids, cylinders,...).

c.2 - Measuring the absolute value of $d\sigma/d\Omega$. This requires a calibration usually done with water for X-rays and neutrons and benzene with light. This gives the absolute value of A(0)

$$A (q\rightarrow 0) = v (\bar{\rho} - \rho_s)$$

For light, $\bar{\rho} - \rho_s$ is proportional to the refractive index increment of the solution and must be measured independently. It can be measured or calculated with neutrons. For X-ray it can only be calculated : it is proportional to the number of electrons per unit volume. This method gives the particle volume v .

c.3 - The procedure can be refined by using contrast variation. It gives then interesting information about polydispersity. For monodisperse objects, A is zero for $\bar{\rho} = \rho_s$. For polydisperse objects, A can only be zero if the objects have the same composition. Otherwise, no solvent can cancel the sample scattering. The height of the minimum of |A| versus ρ_s measures the dispersion in the composition of the particles.

This method has shown that the polydispersity of reverse micelles is very small : AOT with light scattering (43), SDS (sodium dodecyl sulfate) pentanol systems with neutron scattering (44). In the first case, $\bar{\rho}$ varies with particle size (because the [H_2O]/[AOT] ratio w varies) and the scattered intensity drops to almost zero around w ~ 30. This is because at this particular composition, the average refractive index of the micelle is equal to the refractive index of an alkane solvent.

The composition of the particle can be determined from the matching point. This is not an obvious question because one does not always know which component belongs to the particle and which one to the solvent.

However the solvation cannot be extracted in this way : only the particle volume for which $\bar{\rho} \neq \rho_s$ contributes to the scattering.

The first systematic SANS study by this method on AOT reverse micelles yielded the average volumes scattering densities and radius for water cores and surfactant layers (45). Addition of salt was shown to produce a growth of the aqueous cores, i.e. a Σ decrease as expected (46) (§ 2.1).

c.4 - Radius of gyration. Its general definition makes use of the "Patterson Function"

$$P(r) = \int I(q) \exp(i\vec{q}.\vec{r}) \frac{d^3q}{(2\pi)^3}$$

$$R_g^2 = \frac{\int r^2 P(r) 4\pi r^2 dr}{\int P(r) 4\pi r^2 dr}$$

R_g is frequently measured using the Guinier law, valid at small q :

$$I(q) \simeq I(0) \left(1 - \frac{q^2 R_g^2}{3}\right)$$ [9]

for instance, for spheres, $R_g^2 = \frac{3R^2}{5}$.

c.5 - High q determinations. These determinations are interesting when $S(q)$ is unknown, because in this range $S(q) \approx 1$. The first zero of $A(q)$ corresponds to $qR \sim 4.5$ for spheres. The variations of $A(q)$ around this value are sharp, and even a small polydispersity transforms the zero into a minimum. The position of the minimum being not strongly affected by the interactions, $I(q)$ also exhibits a minimum for the same q value. This is a very rough but rapid measurement of R. It is of course only justified if there are good reasons to think that the particles are spherical.

Integrals of the intensity times large powers of q are also not very sensitive to $S(q)$. For instance

$$P(r\to O) = \int_0^\infty \frac{q^2}{2\pi^2} I(q) dq = (\bar{\rho} - \rho_s)^2 v$$

Therefore $v = \frac{I(q\to 0)}{P(r\to 0)}$ [10]

This method does not require calibration.

In the very high q range ($qR \gg 1$), small, flat-like portions of the particles are probed. Scattering follows Porod's law :

$$q^4 I(q\to\infty) \to 2\pi(\bar{\rho} - \rho_s)^2 NS$$ [11]

where S is the surface area of the particle. This relation is valid even if there are no individual particles but sharp interfaces separating two media of different scattering densities. NS is then replaced by the total interfacial area. The above equation has been used for bicontinuous microemulsions to calculate the area per surfactant molecule (Fig. 7).

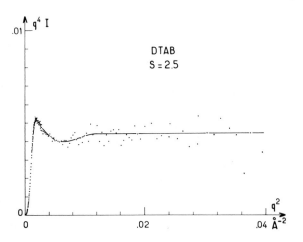

Fig. 7.   Porod's plot for a bicontinuous microemulsion X ray scattering spectrum.

d- Droplets aggregation and percolation

The study of aggregation is very difficult with scattering techniques : a dispersion of hard spheres and a network of connected spheres give very similar scattering curves (Fig. 8). This is probably why no structural differences were observed between a KO-hexanol system that is hard-sphere like (16) and a KO-pentanol system that shows electrical percolation and which structure might ressemble Fig. 8-b (47).

Studies of SDS-pentanol and KO-pentanol and hexanol systems have shown that monodisperse ellipsoids better fit the scattering curves than polydisperse spheres (41)(47). Since the Σ values are then too large, it was proposed that the apparent anisotropy of the micelles is in fact due to the presence of aggregates of micelles. In dilute systems, less than 10 % of dimerized droplets would lead to the same scattering spectra (41).

In other systems, the fit with polydisperse spheres can be satisfactory. But the polydispersity is high : 20 - 40 %, much larger than the one deduced from contrast variation, sedimentation or quasielastic light scattering experiments (§ 3.4). It was suggested that the polydispersity is mainly apparent and due to the presence of aggregates. The aggregates being transient, some surfactant reservoir could be present to allow Σ to remain constant. The presence of dry micelles was reported in the AOT-water-heptane system by SAXS and attributed to this effect (48)

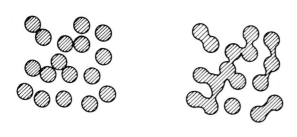

a   b

Fig. 8. Schematic representation of a hard sphere dispersion (a) and of a network of connected spheres (b)

e- Critical behaviour

When attractive interactions are important, the small angle X-ray and neutron scattering curves are strongly modified (Fig. 9). $S(q{\to}0)$ increases and the scattered intensity becomes large. Close to a critical point $S(q)$ can be approximated by a modified Ornstein-Zernike formula

$$S(q) = 1 + \frac{NkT}{V} \frac{\chi}{1+q^2\xi_c^2} \qquad [12]$$

ξ_c being the correlation length. Critical point theories predict that $\xi_c = \xi_0 \, \varepsilon^{-\nu}$ and $\chi = \chi_0 \varepsilon^{-\nu}$ where ε is the distance to the critical point $((T-T_c)/T_c$ in a temperature scan), ξ_0 and χ_0 scale factors and ν and γ critical exponents $\gamma \approx 2$ ν. This expression has been used together with the expression of $<|A(q)|^2>$ for polydisperse spheres to fit the data close to T_c in the AOT/water/decane system

(49). The fit is excellent and gives $\xi_0 = 11 \pm 2$ Å ; $v = 0.72 \pm 0.04$; $\gamma \sim 2v$; R = 44.6 Å. R does not vary appreciably with temperature. The phenomenon ressembles to a liquid-gas critical point, in which the micelles would play the role of molecules. This is consistent with the large value of the scale factor ξ_0.

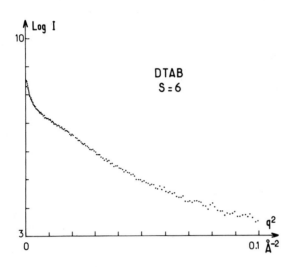

Fig. 9. X ray scattering spectrum for a microemulsion close to a critical point.

An alternative data analysis can be made with a computation of S(q) using the interparticle potential (§3.b). A description of the fluctuating critical domains has also been made in terms of fractal aggregate structure in the AOT-dodecane-water system (50). Light is better adapted to analyse critical behaviour which, as said above, only affects the small q region. When ξ_c is large enough (≥ 100 Å), angular variations of the scattered light intensity are observable . A number of light scattering experiments have been made close to critical points with reverse micelles : AOT-water-decane (51)(52), SDS-alcohol systems (53)(54). In some cases, the critical exponents are the classical Ising exponents for simple fluids (52)(53), in other cases they are different and can evolve continuously with sample composition (54). The origin of the discrepancies is still not understood.

f- Concentrated systems ; measurement of interfacial film curvature. In concentrated systems, strongly pronounced interaction peaks become visible in the X-ray and neutron scattering intensity curves. The problem is : are droplets still present or is the structure evolving for instance towards a bicontinuous structure ? Fig. 10 shows a spectrum corresponding to a bicontinuous microemulsion.

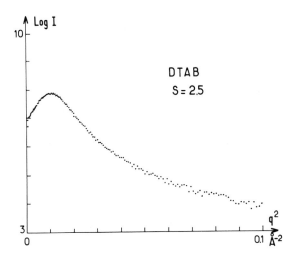

Fig. 10. X ray scattering spectrum for a bicontinuous microemulsion (same than in fig. 7)

This kind of curves can be fitted with the theoretical expression for polydisperse interacting spheres (55). But in reality there are no spheres present, as demonstrated by independent measurements with other techniques. I(q) can also be fitted by a lamellar structure factor broadened by a Debye-Waller factor representing a random dispersion of small lamellar domains (56). It can be fitted as well by a simple empirical expression proposed recently (57) :

$$I(q) = \frac{1}{1 + c_1 q^2 + c_2 q^4}$$

One must therefore be extremely careful in interpreting the data. We will describe below two methods used to check for consistency.

f.1- One method is to study samples of different composition. For instance, if the oil-water ratio is one and if the micelles are locally ordered in a body centered cubic lattice, the wave vector of the peak should satisfy :

$$q_{max} \sim \frac{\phi_s}{(1-\phi_s)^{2/3}}$$

ϕ_s being the surfactant volume fraction (23). This relation is fulfilled in AOT-water-decane systems.

If the local order is rather lamellar like, at constant w one should have

$$q_{max} \sim \frac{(1 - \phi_o)}{\phi_s}$$ instead of $q_m \sim \phi_s^{1/3}$ in the preceeding case. The linear variation of q_m versus the oil volume fraction ϕ_o has been observed in the AOT-octane-water system (58). It is of course surprising to observe such very different local structure in two very similar systems ; it is also difficult to understand why the system with the highest spontaneous film curvature leads to lamellar order, while spherical droplets remain present in the other system. Further work is still needed to understand this point.

In alcohol containing systems, droplets appear to remain present in some cases (22)(47). In other cases, the structure becomes bicontinuous. One then has, if ϕ_s is small :

$$q_m = \frac{\pi}{\xi} = \frac{\pi c_s \Sigma}{6 \phi_o \phi_w}$$

ξ being the characteristic dispersion size (eq [3]. The variation of q_m with ϕ_o and ϕ_w is very different than for droplets (oil droplets $q_m \sim \phi_o^{-1}$ water droplets $q_m \sim \phi_w^{-1}$) and lamellae (q_m = Cste).The experiments are in agreement with a parabolic variation of q_m^{-1} versus ϕ_o (or ϕ_w) (59)(60)(61). The Σ value deduced from the q_m value is in agreement with the determination using Porod's law.

f.2- Contrast variation method. We have seen that the spectra can be interpreted as due to scattering from polydisperse spheres. When spheres are present, the spectra relative to deuterated water and oil are different. If both oil and water are deuterated, the spectra are characteristic of spherical shells (surfactant film scattering). In bicontinuous microemulsions for which $\phi_o = \phi_w$, the spectra for deuterated water and oil are identical. The film spectra do not show peaks as shells would do (55).

The following detailed interpretation of the data allows to deduce the surfactant film curvature C. For a system made of three homogeneous volume, oil (o), water (w) and surfactant film (f), one finds by separating the contributions of each volume in I(q) :

$$I(q) = (\rho_w - \rho_o)^2 \, S_{ww}(q) + 2(\rho_w - \rho_o)(\rho_f - \rho_o) \, S_{wf}(q) + (\rho_f - \rho_o)^2 \, S_{ff}(q)$$

[14]

The structure factors S_{ij} correspond to the self and cross correlations of species o, w and f

$$S_{ij}(q) = < \int_{r \in i \ r' \in j} e^{i\vec{q}(\vec{r}-\vec{r}')} \, d^3r \, d^3r' >$$

$$= \int < \delta\phi_i(0)\delta\phi_j(\vec{r}) > e^{i\vec{q}\cdot\vec{r}} d^3r$$

$\delta\phi_i(\vec{r})$ being the ith-component volume fraction fluctuation. For water in oil spheres, if $d \ll R$:

$$\frac{S_{wf}(q \to 0)}{S_{ww}(q \to 0)} = \frac{3d}{R}$$

d being the film thickness. For oil in water spheres, $S_{wf}(0)/S_{ww}(0) = -3d/R$. If the film has a zero spontaneous curvature $S_{wf}(0) = 0$: this is because a fluctuation $\delta\phi_w$ does not change ϕ_s contrary to the sphere case where a fluctuation of $\delta\phi_w$ produces a variation of radius and hence an increase of ϕ_s if $C > 0$, a decrease if $C < 0$ (62). This method has been used in SDS-butanol bicontinuous microemulsion where it was found that, when $\phi_o = \phi_w$, $C = 0$ (63), and in AOT microemulsions where $C^{-1} = 76$ Å for $\phi_o = \phi_w$ (64).

3 - TIME RESOLVED FLUORESCENCE PROBING

This method measures the mean micellar aggregation number and is not sensitive to intermicellar interactions (65). The method is described in detail in chapters 5, 8, and 10. Being dynamic in nature, it leads also to interesting information about micellar exchanges. Many experiments on reverse micelles have already been performed with success (66)-(72). Let us mention that a recent extensive study of AOT reverse micelles has recently been performed investigating the role of solvent, salt and temperature (73). An alternative method based on fluorescence polarization has also been used on AOT reverse micelles (74).

Because it is very sensitive to size variations (N ~R^3), small increase in droplets radii have been evidenced close to the critical point in the decane-AOT-water-system. The method has also been used to check the validity of the dilution procedures, and to make a correlation between micellar interactions and molecular exchanges (69). As noticed earlier, exchanges are more important if the alcohol chain length is shorter (66). By changing the alcohol chain length, one can go from a hard sphere system to an attractive system exhibiting percolation phenomena (20). Molecular exchanges become very important close to a critical point (69). Up to now, the relation between exchanges and interactions remain unexplained.

4 - MEASUREMENTS OF MICELLAR TRANSPORT PROPERTIES : DIFFUSION AND SEDIMENTATION COEFFICIENTS

These measurements have to be used with extreme caution. Indeed the transport coefficients are affected by interparticle interactions in a way which is not fully understood presently. Dilution is therefore necessary, and only the limit of infinite dilution can be exploited. Except for neutron spin echo techniques, there is no spatial resolution at the particle scale : a structural model must be used, without internal check for consistency. One interest of these measurements is to give an accurate estimation of the particles polydispersity (44). The polydispersity measured in that way is frequently less than the one extracted from SANS and SAXS spectra : less than 10 % compared to 20 % in the SDS-butanol system (53)(55). As explained in § 3.2.d, the discrepancy is likely to arise from the presence of particles aggregates.

a- Translational diffusion and sedimentation. The diffusion and sedimentation coefficients of a single spherical particle are related to its hydrodynamic radius via the Stokes-Einstein relation :

$$D_0 = \frac{kT}{6\pi\eta_0 R_H} \qquad s_0 = \frac{M_{app}}{6\pi\eta_0 R_H} \qquad [16]$$

η_0 being the solvent viscosity and M_{app} the apparent particle mass. R_H represents the radius of the sphere which carries eventually bound solvent during the hydrodynamic motion. The comparison between R and R_H is therefore interesting because it contains information about solvation.

In practice, one never studies a single droplet, but many of them that interact via the interaction forces already discussed in §3.2.b and undirectly via hydrodynamic interactions. At this stage, one can distinguish between two types of diffusion coefficients.

a.1- Tracer self-diffusion

When a tracer is attached to the micelles, its diffusion coefficient is independent of direct interactions :

$$D_S = \frac{kT}{f} \tag{17}$$

f is a friction coefficient which is time dependent. The characteristic time is the one needed for a droplet to change its environment, i.e. to diffuse along one interparticle distance.

Short time self diffusion coefficients are measured with quasi-elastic neutron scattering (spin echo techniques) (75)(76). Long time self diffusion coefficients are measured with forced Rayleigh scattering (77), fringe pattern photobleaching (78) and spin echo NMR techniques (79). Although the D_S versus ϕ variation can be predicted theoretically, the experimental data are far from being understood. The role of transient droplets aggregates has been shown to be important (78).

a.2- Mutual diffusion

The mutual diffusion coefficient D_C is associated with the relaxation of particles concentration fluctuations. It depends on direct interactions and on hydrodynamic interactions as well

$$D_C = \frac{\partial \pi / \partial \phi}{f'} \tag{18}$$

f' is a friction factor which contains different hydrodynamic interactions than f. D_C is also time dependent in the same way than D_S. Quasi-elastic light scattering QELS probes the long time mutual diffusion coefficient. Its concentration dependence has been extensively studied in a number of reverse micellar system (39)(40)(80)(81)(82). For small ϕ :
$D_C = D_0(1 + \beta\phi)$
β is related to the interaction potential W Interacting spheres theories have been successfully applied to AOT systems (82). In SDS-alcohol systems, the agreement is not as good (39)(80).

a.3- Sedimentation coefficient

The friction coefficient f' also appears in sedimentation coefficient which can be measured with ultracentrifugation techniques : $s = M/f'$. By combining this expression with eqs 8 and 18, one finds that $I(q = 0)D_C/s$ should be constant. Again this is not verified in alcohol containing systems (39)(80).

The discrepancies have been attributed to molecular exchanges. The extrapolation to $\phi = 0$ of all the measured transport coefficients should give the particle hydrodynamic radius. All these extrapolations give indeed the same radius. For reverse micelles $R_H \geq R$. The difference between R and R_H is thought to be due to solvent penetration. As expected from the theory (17), it is larger for attractive droplets. For hard sphere droplets $R_H \sim R \sim R_{HS}$.

a.4- Behavior at larger concentration

Close to a critical point, D_C becomes q dependent, and goes to zero. This is described by the Kawasaki theory and allows to determine the correlation length ξ_C

$$D(q=0) = \frac{kT}{6\pi\eta\xi_C}$$
$$D(q) = D(q=0) \, K(q\xi_C)$$

where η is the microemulsion viscosity and K a universal fonction. The ξ_C values found in this way are in agreement with the ones deduced from static light scattering experiments (53).

The behaviour of diffusion coefficients at large volume fractions is still mostly not understood. In bicontinuous microemulsions for instance, D_C is about 10 times smaller than $kT/6\pi\eta\xi$, ξ being the dispersion size.

In AOT systems the correlation function measured in QELS experiments becomes strongly non exponential. This has been attributed to the vicinity of a glass transition around $\phi \sim 0.62$ (24).

b- Rotational diffusion

The rotational motion of the structural elements can be studied with transient electric birefringence. For a spherical particle, the rotation time is

$$\tau = \frac{8\pi\eta_0 R^3}{6kT}$$

A spherical particle is not oriented in an electric field and gives no birefringence response. But the birefringence measured in reverse micellar systems is frequently large (83)(84)(85). The response time τ is in the range 1-10 μs, for systems with radii R $\sim$50 Å. Estimations of τ with the above formula lead to a much smaller value $\tau \sim$30 ns . The observations are thus clearly in favour of the presence of an appreciable fraction of large anisotropic particle aggregates.

TABLE
Structural data relative to AOT w/o microemulsions

	n-heptane					n-decane				
W	N	R_W (Å)	R_H (Å)	Σ (Å²)	ref	N	R_W (Å)	R_H (Å)	Σ (Å²)	ref
0						22	9		51	(6)
8	115	19		38	(46)	95	18		41	(71)
10			35		(12)					
11.4	108	21		50	(71)					
13						245	28		41	(71)
15	180	27		50	(71)					
	160	26		52	(46)					
18	255	32		51	(71)	405	37		43	(71)
20		40			(48)					
21.2	314	36		53	(71)	424	40		48	(71)
23.6	350	39		55	(71)					
24			56		(12)					
25			58		(12)		31			(70)
26.3	405	42		56	(71)	590	48		49	(71)
30	520	48		56	(71)	830	56		48	(71)
33							57			(49)
							55			(70)
41			89		(12)		67			(49)(70)
								78		(80)
46			93		(12)					
49			89		(12)		78			(49)
							102			(70)

The amount of aggregates has been found larger in attractive systems. The aggregates lifetime τ_a and the fraction P of collisions leading to dimer formation can also be extracted from the measurements : for BHDC (benzyl hexadecyl dimethyl ammonium chloride) systems, τ_a is in the range of 10 μs and increases when attraction increases ; P is about 5 % (85).

4. CONCLUSION

We have presented the main experimental procedures used for reverse micelles structural determination. We have illustrated the problems encountered in these determinations with many examples. Some of the problems are specific to alcohol containing systems in which molecular exchanges are very important. Ternary systems like the widely studied AOT systems have a behavior close to the more conventional colloidal systems. We have reported in the table some structural data relative to this particular surfactant.

Structural determination is specially difficult in the domain where the micellar concentration is high, and where in some cases structural changes leading to the disappearance of micelles can occur. The problems are not fully solved, but the research in the field is presently very active.

REFERENCES
1   P. Stenius in P.L. Luisi and B.E. Straub eds. "Reverse Micelles", Plenum, New York ,1984.
2   C.R. Singleterry, J. Amer. Oil. Chem. Soc., 32 (1955) 446.
     R.C. Little and C.R. Singleterry, J. Phys. Chem., 68 (1964) 3453.
3   P.A. Winsor, "Solvent Properties of Amphiphilic Compounds", Butterworths, London,1954.
     P.A. Winsor, Chem. Rev. 68 (1968) 1.
4   P. Ekwall, L. Mandell and K.J. Fontell, J. Coll. Int. Sci. (1970) 215.
     P. Ekwall, Adv. Liq. Cryst., 1 (1975) 1.
5   J.H. Fendler, Acc. Chem. Res., 9 (1976) 153.
     J.H. Fendler, E.J. Fendler, "Catalysis in Micellar and Macromolecular Systems", Academic Press, New York, 1975.
6   M. Kotlarchyk, J.S. Huang and S.H. Chen, J. Phys. Chem., 89 (1985) 4382.
7   P. Jones, E. Wyn-Jones and G.I.T. Tiddy, J. Chem. Soc. Faraday Trans I, 83 (1987) 2735.
8   H.F. Eicke and P. Kvita in the book ref 1.
9   J.N. Israelachvili, D.J. Mitchell and B.W. Ninham, J. Chem. Soc. Far. Trans II, 72 (1976) 1525.
10   D.J. Mitchell and B.W. Ninham, J. Chem. Soc. Faraday Trans II, 77 (1981) 601.

11   C. Tanford, J. Phys. Chem., 76 (1972) 3020.
12   R. Aveyard, B.P. Binks, S. Clark and J. Mead, J. Chem. Soc. Faraday
     Trans I, 82 (1986) 125.
13   D.W.R. Gruen and D.A. Haydon, Pure and Appl. Chem., 52 (1980) 1229.
14   S.A. Safran and L.A. Turkevich, Phys. Rev. Lett., 50 (1983) 1930.
15   W. Helfrich, Z. Natursforsch., 28 (1973) 693.
16   A.A. Caljé, W.G.M. Agterof and A. Vrij in K.L. Mittal eds. 'Micellization,
     Solubilization and Microemulsions" vol. 2, Plenum, New York 1977.
17   B. Lemaire, P. Bothorel and D. Roux, J. Phys. Chem., 87 (1983) 1023.
18   P.A. Pincus and S.A. Safran, J. Chem. Phys., 86 (1987) 1644.
19   S.A. Safran, I. Webman and G.S. Grest, Phys. Rev. A 32 (1985) 506.
20   A.M. Cazabat, D. Chatenay, D. Langevin and J. Meunier, Faraday Discuss.
     Chem. Soc., 76 (1983) 291.
21   P.G. de Gennes and C. Taupin, J. Phys. Chem., 86 (1982) 2294.
22   D.J. Cebula, R. H. Ottewill, J. Ralston and R. Pusey, J. Chem. Soc. Faraday
     rans I, 77 (1981) 2585.
23   M. Kotlarchyk, S.H. Chen, J.S. Huang and M.W. Kim, Phys. Rev. Lett., 53
     (1984) 941.
24   S.H. Chen and J.S. Huang, Phys. Rev. Lett., 55 (1985) 1888.
25   W. Jahn and R. Strey in J. Meunier, D; Langevin and N. Boccara eds   "
     Physics of Amphiphilic Layers", Springer Berlin 1987 p. 353; J. Phys.
     Chem. 92 (1988) 2294.
26   N.J. Chang and E.W. Kaler, J. Phys. Chem. 89 (1985) 2996.
27   S. Candau, E. Hirch and R. Zana, J. Coll. Int. Sci., 88 (1982) 428
     R. Zana, C. Picot and B. Duplessix, J. Coll. Int. Sci. , 93 (1983) 43.
28   J.B. Hayter and J. Penfold, J. Phys. Chem., 88 (1984) 4589.
29   O. Glatter and O. Kratky "Small Angle X Rays Scattering", Academic
     Press 1982.
30   J. Hayter in V. Degiorgio and H. Corti eds. "Physics of Amphiphiles :
     Micelles, Vesicles and Microemulsions", North Holland, Amsterdam 1985.
31   B. Cabane in R. Zana ed. "Surfactant Solutions : New Methods of
     Investigation, M. Dekker, 1987.
32   A. Vrij, J. Chem. Phys., 71 (1979) 3267
     P. Van Beurten and A. Vrij, J. Chem. Phys., 74 (1981) 2744
33   J.S. Huang, S.A. Safran, M.W. Kim, G.S. Grest, M. Kotlarchyk and N.
     Quirke, Phys. Rev. Lett., 53 (1984) 592.
34   S.H. Chen, Physica, 137B (1986) 183.
35   J.H. Schulman, W. Stoeckenius and L. Prince, J. Phys. Chem., 63 (1959)
     1677.
36   A. Graciaa, J. Lachaise, A. Martinez, M. Bourrel and C. Chambu, C.R.
     Acad. Sci. ser. B, 282 (1976) 547.
37   J. Biais, P. Bothorel, B. Clin and P. Lalanne, J. Coll. Int. Sci., 80 (1981)
     136.
38   A.M. Cazabat, J. Phys. Lett., 44 (1983) L-593.
39   A.M. Cazabat and D. Langevin, J. Chem. Phys., 74 (1981) 3848.

42

40   S. Brunetti, D. Roux, A.M. Bellocq, G. Fourche and P. Bothorel, J. Phys.
     Chem., 87 (1983) 1028.
     T. Dichristina, D. Roux, A.M. Bellocq and P. Bothorel, J. Phys. Chem., 89
     (1985) 1433.
41   R. Ober and C. Taupin, J. Phys. Chem., 84 (1980) 2418.
42   N.W. Ashcroft and J. Lekner, Phys. Rev., 83 (1966) 145.
43   M. Zulauf and H.F. Eicke, J. Phys. Chem., 83 (1979) 480.
44   M. Dvolaitzky, M. Guyot, M. Lagues, J.P. Le Pesant, R. Ober, C. Sauterey
     and C. Taupin, J. Chem. Phys., 69 (1978) 3279.
45   C. Cabos and P. Delord, J. Appl. Cryst., 12 (1979) 502.
46   C. Cabos and P. Delord, J. Phys. Lett., 41 (1980) L)455.
47   E. Caponetti, L.J. Magid, J.B. Hayter and J.S. Johnson Jr., Langmuir, 2
     (1986) 722.
48   A.N. North, J.C. Dore, J.A. Mc Donald, B.H. Robinson, R.H. Heenan,
     A.M. Howe, Coll. and Surf., 19 (1986) 21.
49   M. Kotlarchyk, S.M. Chen, J.S. Huang, J. Phys. Chem., 86 (1982) 3275 ;
     Phys. Rev. A, 28 (1983) 508.
50   A.N. North, J.C. Dore, A. Katsikides, J.A. Mc Donald and B.H. Robinson,
     Chem. Phys. Lett., 132 (1986) 541.
51   J.S. Huang and M.W. Kim, Phys. Rev. Lett., 47 (1981) 1462.
52   P. Honorat, D. Roux and A.M. Bellocq, J. Phys. Lett., 45 (1984) L-961.
53   A.M. Cazabat, D. Langevin, J. Meunier and A. Pouchelon, J. Phys. Lett.,
     43 (1982) L-89, Adv. Coll. Int. Sci., 16 (1982) 175.
54   G. Fourche, A.M. Bellocq and S. Brunetti, J. Coll. Int. Sci., 89 (1982) 427.
     A.M. Bellocq, P. Honorat and D. Roux, J. Phys. (Paris) 46 (1985) 743.
55   A. de Geyer and J. Tabony, Chem. Phys. Lett., 113 (1985) 83 ; 124 (1986)
     357.
56   C.G. Vonk, J.F. Bellman and E.W. Kaler, J. Chem. Phys., submitted.
57   M. Teubner and R. Strey, J. Chem. Phys., 87 (1987) 3195.
58   C. Cabos, P. Delord and J. Marignan, Phys. Rev. A, to appear.
59   L. Auvray, J.P. Cotton, R. Ober and C. Taupin, J. Phys., 45 (1984) 913.
60   F. Lichterfeld, T. Schmeling and R. Strey, J. Phys. Chem., 90 (1986)
     5762.
61   O. Abillon, B.P. Binks, C. Otero, D. Langevin and R. Ober, J. Phys.
     Chem.92 (1988) 4411.
62   B. Widom, J. Chem. Phys., 81 (1984) 1030.
63   L. Auvray, J.P. Cotton, R. Ober and C. Taupin, J. Phys. Chem., 88 (1984)
     4586.
64   J.S. Huang and M. Kotlarchyk, Phys. Rev. Lett., 57 (1986) 2587.
65   R. Zana in the book of ref. 31
66   S.S. Atik and J.K. Thomas, J. Phys. Chem., 85 (1981) 3291.
67   S.S. Atik and J.K. Thomas, J. Amer. Chem. Soc., 103 (1981) 3543.
68   F. Gelade and F.C. de Schryver, J. Photochem., 18 (1982) 223.
69   P. Lianos, R. Zana, J. Lang and A.M. Cazabat in K.L. Mittal and
     P.Bothorel eds. "Surfactant in Solution", vol. 6, Plenum New York, 1986.
70   A.M. Ganz and B.E. Boeger, J. Coll. Int. Sci., 109 (1986) 504.

71   J.M. Furois, P. Brochette and M.P. Pileni, J. Coll. Int. Sci., 97 (1984) 552.
      P. Brochette, T. Zemb, P. Mathis and M.P. Pileni , J. Phys. Chem. 91
      (1987) 1444.
72   M.P.Pileni, P. Brochette, B. Hickel and B. Lerebours, J. Coll. Int. Sci., 98
      (1984) 549. C. Petit, P. Brochette and M.P. Pileni, J. Phys. Chem., 90
      (1986) 6517.
73   J. Lang, A. Jada and A. Malliaris, J. Phys. Chem., in press.
74   E. Keh and B; Valeur, J. Coll. int. Sci., 79 (1981) 465.
75   J.B. Hayter, R.H. Ottewill and P.N. Pusey in V. Degiorgio and M. Corti
      eds : "Physics of Amphiphiles", North Holland, Amsterdam 1985.
76   J.S. Huang, S.T. Milner, B. Farago and D. Richter, Phys. Rev. Lett., 59
      (1987) 2600.
77   W. Dozier, M.W. Kim and P.M. Chaikin, J. Coll. Int. Sci., 115 (1987)
      545.
78   D. Chatenay, W. Urbach, A.M. Cazabat and D. Langevin, Phys. Rev. Lett.,
      54 (1985) 2253.
79   P. Guering and B. Lindman, Langmuir, 1 (1985) 464.
80   W.M. Brouwer, E.A. Nieuvenhuis and M. Kops-Werkhoven, J. Coll. Int.
      Sci., 92 (1983) 57.
81   J.D. Nicholson and J.H.R. Clarke in K.L. Mittal and B. Lindman eds :
      "Surfactant in Solution", vol. 3, Plenum, New York 1984.
82   M.W. Kim, W. Dozier and R. Klein, J. Chem. Phys., 84 (1986) 5919.
83   H.F. Eicke and Z. Markovic, J. Coll. Int. Sci., 79 (1981) 151 ; 85 (1982)
      198.
84   P. Guering and A.M. Cazabat, J. Phys. Lett., 44 (1983) L-601.
85   P. Guering, A.M. Cazabat and M. Paillette, Europhys. Lett., 2 (1986) 953.

STRUCTURAL CHANGES OF REVERSE MICELLES AND MICROEMULSIONS BY ADDING SOLUTES OR PROTEINS

M.P.PILENI

1. INTRODUCTION

Several models [1-12] have been proposed to describe the solubilization of protein in reverse micelles. The models proposed take into account the size of the protein. In the case in which the protein size is smaller than the water pool two differents models are proposed: some groups claim that the presence of the protein gives rise to larger surfactant aggreggates [1,2,6,9,10,12] than those formed in the absence of protein. Other groups suggest that the protein [3,4] has no effect whatsover on the size of the aggregate . In the case in which the protein size is larger than the water pool, very few models have been proposed [5]. Most of the groups using probes in reverse micelles estimate that these probes do not affect the structure of the droplet. From spectroscopic data, the average location of the probe has been estimated. Only one model [7] uses probes instead of protein to determine the structural changes in reverse micelles.

In the present paper we will try to show that most of the experiments presented, for protein having a size comparable to or smaller than the water pool content, are not in contradiction with each other. The structural changes can be examined at low (w below 20) and at high water content (w above 20).

2. RESULTS AND DISCUSSIONS

2.1 Change of the water pool radius with the water content:

Assuming the droplets are spherical, the water pool radius, r_w, is given by :
$$r_w = 3V/4S$$

in which V and S are respectively the total polar volume and the total surface available. Using the surface per head polar group of AOT equal to 60 A^2 [13] and the water molecule volume equal to 30 A^2, the relation between the water pool radius and the water content, $w = \{H_2O\}/\{AOT\}$, is:

$$r_w = 1.5w$$

This relation has been checked by small angle X ray scattering and by neutron scattering, (see Fig. 1).

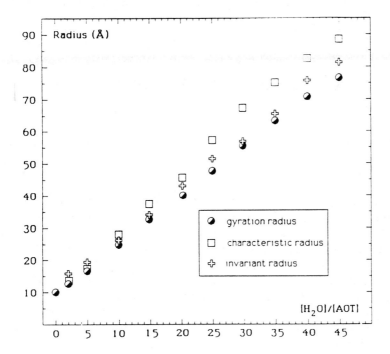

Fig. 1. Variation of the water pool radius with the water content tested by small angle x ray scattering

2.2 Small structural changes of the reverse micellar systems by adding probes or small proteins:

Small structural changes usually occur when the size of the protein is closed to that of the water pool of the droplet whereas significant changes are obtained when the radius of the protein is larger than that of the water pool. In this section, the data obtained at high water content (above 20) and at low water content (below 20) will be reviewed.

2.2. High water content

2.2.1 Geometrical model tested by small angle X ray scattering.[7]. Because of the a geometrical relationship between the water pool radius and the water content, it is possible to look at the change of the water pool radius brought about by adding solutes. It can be expected that the water pool radius changes with the location of the solute in the droplet:

-The average localization of guest molecules soluble only in hydrocarbons (see Fig. 2A), will remain in the outer bulk phase and cannot, therefore, perturb the micellar structure; physical and chemical measurements on the solubilization are then expected to be similar to those obtained in homogeneous solutions. From structural measurements no changes are expected in case A. Similar data are expected for probes located at the external interface, (see Fig. 2B).

-For molecules located in the water pool, (see Fig. 2C and 2D), the addition of a certain amount of solute to the water pool causes an increase in the droplet volume, dV, which is due to the molecular volume of the added solute, while the total interfacial area, S, remains constant if it is assumed that the total surfactant is bound. The water pool, is thereby determined by the ratio of the total polar volume V+dV to the total interfacial area, S. The variation in the micellar mass, concomitant with the solubilization of the hydrophilic component, is equivalent to the perturbation observed on addition of the same volume of water. If dV is significant with respect to V, the addition of a solute to the water pool causes an increase in the characteristic radius.

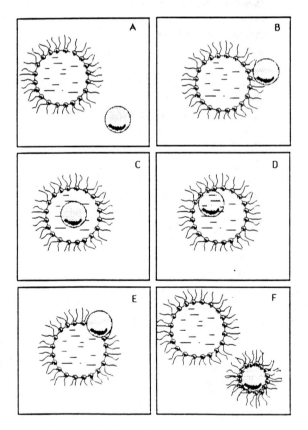

Fig. 2. Model of different location of probes or proteins in reverse micelles [7]

-Molecules located at the interface and acting as a surfactant, (see Fig. 2E), cause an increase in the interfacial area, dS, for an unchanged polar volume which induces a decrease in the water pool radius.

-Molecules which are able to create their own aggregates as has been previously proposed by Menger and Yamada[14] induce the formation of small aggregates around the solute,(see Fig. 2F), In such a case,nearly two micellar populations are expected.

This geometrical model has been tested by SAXS, at w up to 20: On adding ether, mainly solubilized in the bulk non polar phase, no changes in the size of the water pool are observed (case A). On adding either methylviologen or chymotrypsin or ribonuclease at pH 4 and at pH 11, an increase in the water pool radius is obtained. From this result it can be deduced that the viologen and chymotrypsin and ribonuclease are located in the water pool. However the average location could be either at the internal interface or inside the water pool. Using the model described in the chapter entitled "Hydrated electrons in reverse micelles", it is observed that the quenching rate constant of hydrated electrons by methylviologen, or ribonuclease at pH 4 depends on the surface whereas that of chymotrypsin or ribonuclease at pH 11 depends on the volume of the micellar droplet. From the data obtained by SAXS and by pulse radiolysis[16], it can be deduced that methylviologen and ribonuclease at pH 4 are mainly located at the internal interface of the droplet whereas chymotrypsin and ribonuclease at pH 11 are located in the bulk water pool. These last data are in good agreement with those already published using chymotrypsin or ribonuclease [15]. Addition of copper lauryl sulfate or cytochrome c to the micellar solution induces a decrease in the size of the droplet which indicates, according to the proposed model, that they are both located at the interface acting as a surfactant. These data are confirmed by the model proposed using hydrated electrons [16] as a probe and copper lauryl sulfate or cytochrome c as a quencher. In the case of cytochrome c more quantitative data have been obtained [11]: From a plot of $I.q^4$ against q for micellar solutions containing various cytochrome c concentrations, (see Fig. 3a), it can be seen that the minimum is shifted to larger q values and the Porod limit increases when the cytochrome c concentration increases. These experiments have been performed at various water pool contents. The shift of the minimum indicates a decrease in the water pool radius and the increase of the Porod limit an increase in the total surface available. These data indicate that the size of the droplet decreases and the total surface available increases which are consistent with the geometrical model. Fig.3b shows that the decrease in the water pool radius is greater at high water content and become unobservable at low water content.

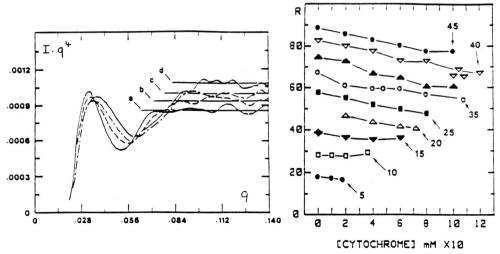

Fig. 3.   a- scattering measurement of cytochrome c in reverse micelles at differents cytochrome concentrations

b- variation of the gyration radius of the reverse micelles at various cytochrome c concentrations and at various water content

2.2. 2   Quasi elastic light scattering[8], QELS. The insertion of myelin basic protein into reverse micelles has been studied by QELS. The light scattering measurements give an average value of the hydrodynamic radius of the filled and empty
micelles. The hydrodynamic radius of filled and empty micelles respectively can
only obtained by calculation. The hydrodynamic radius of the micelles increases
with increasing their occupancy. The hydrodynamical r adii of empty and filled
droplets are respectively equal to $r_w + 1$ and $r_f + 1$. Where 1 is the length of the surfactant and equal to 12A for AOT and r_w and r_f the water pool radii of empty and
filled droplets. By increasing the occupancy of protein, at w= 24.6, the hydrodynamic radii do not change indicating that, in contradiction to what is obtained at low
water content, the hydrodynamic radii of filled and empty micelles are similar.

2.2.3   Fluorescence recovery after fringe photobleaching[12], FRAPP. A
fundamental difference between light scattering and FRAPP experiments is the fact
that QELS measurements lead to an average value of the hydrodynamic radius of
filled and empty micelles while FRAPP allows the size determination of the filled
micelles. The principle of this technique is the following: the fluorescence of the
dye molecule bound to a protein can be destroyed by intense laser illumination. In
the first step, the sample is illuminated with a high intensity fringe pattern produced
by two crossed laser beams. This creates a non uniform dye concentration distribution, since some of the dye molecules located in the bright fringes are irreversibly

destroyed. Then the relaxation of the concentration profile is monitored with the same fringe pattern but at a much weaker intensity in order to prevent a further bleaching of the dye. The system is monitored by measuring the fluorescence intensity as a function of time. From the diffusion coefficient obtained from the characteristic time of the fluorescence recovery curve, the hydrodynamic radius can be calculated. The hydrodynamic radius does not vary after incorporation of protein. So in this case, the filled and empty micelles have the same hydrodynamic radius.

2.3 At low water content

The changes of the micellar structure by adding protein at low water content have been determined by several techniques :

2.3.1 <u>Ultracentrifugation method</u>[3, 4, 9].The method was first developped by Martinek and coworkers [3, 4] and then used under other experimental conditions by P.L.Luisi and coworkers[9]. The method is based on the analytical centrifugation method. It utilizes two dyes which are monitored simultaneously with an UV scanner device. The first dye is completely water soluble whereas the second dye sticks exclusively to the micellar interface acting as a cosurfactant. By monitoring the first dye which probes only the water pool of the reverse micelles, the partition of water in micelles containing protein and in empty micelles is determined. By measuring the absorption of sedimenting filled and empty micelles, the quantity of water associated with filled and empty micelles can be deduced. At the same time, in a second cell, by monitoring the absorption of the dye located exclusively at the interface, the amount of AOT engaged in empty and filled micellar solutions can be evaluated. Thus, the distribution of water and of the surfactant in filled and empty micelles can be deduced. Either phenolate or picric is used as the acid for the water soluble dye and either potassium dichromate or 2 naphtoic acid as the interfacial dye. The main assumption of this method is, in the ultracentrifuge runs, that the concentration of the water soluble dye is directly proportional to the concentration of the water in the droplet and the concentration of the dye sticking to the interface is proportional to the surfactant concentration in the droplet. These droplets can be empty or full of protein. In others words, the measurements should quantitatively reflect the partition of surfactant between the filled and the empty micelles. Indeed this assumption would not be true in the case of specific interaction between dye and protein. When the protein is solubilized in reverse micelles a second boundary appears on the sedimentogram (see Fig. 4). It is noteworthy that the slow sedimenting fraction of micelles is characterized by the sedimentation coefficient equal to that of empty micelles. The quickly sedimenting heavy fraction consists then of filled micelles.

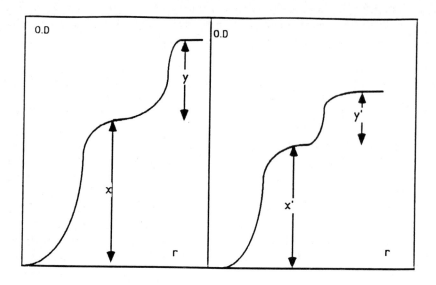

Fig. 4. The typical sedimenting boundaries in reverse micelles[3, 4, 9]

Using the water soluble dye:

$\{H_2O\}_e = x/x+y. \{H_2O\}_t$

$\{H_2O\}_f = y/x+y. \{H_2O\}_t$

Using the cosurfactant dye:

$\{AOT\}_e = x'/x'+y' \{AOT\}_t$

$\{AOT\}_f = y'/x'+y' \{AOT\}_t$

where $\{H_2O\}_e$, $\{H_2O\}_f$, $\{H_2O\}_t$, $\{AOT\}_e$, $\{AOT\}_f$, $\{AOT\}_t$ are respectively filled, empty and total water concentrations and filled, empty and total AOT concentrations.

The picture which emerges from the data is as follows : for solubilization in reverse micelles at w below to 15, a protein needs more water than that originally present in the water pools. As a consequence, the reverse micelles which are left empty will have a water to surfactant ratio lower than before. This is probably due to the fact that reverse micelles existing at low w values are too small to entrap a protein molecule and its hydration shell without altering their own size and shape. Pier Luisi and coworkers[9] found a change in the size using myelin basic protein; lysozyme and chymotrypsin for w in between 5.6 to 13.7. Martinek and coworkers [3,4] using chymotrypsin found a change in the size below 12 and no change up to 15.

This method has its limitations. It does not work well when the degree of occupancy is small or for a large w value when the difference in water or surfactant content between filled and empty micelles is too small

2.3.2 <u>Small angle neutron scattering, S.A.N.S</u>[10]. Chen and coworkers[10] developed a simple model for the filled and empty micelles. The model is similar to that proposed previously[1] assuming that two micellar species may be present in solution simultaneously. The empty micelles are assumed to be spherical consisting of a polar aqueous core surrounded by a surfactant head shell. The other species consists of spherical reverse micelles containing spherical protein located concentrically within the water pool of the micelle which in turn is surrounded by the surfactant shell. The protein chosen is cytochrome c. Assuming the protein is spherical, the gyration radius is found equal to 28A. Multiple occupancy of these micelles is assumed not to be permitted. Thus the insertion of protein in the reverse micelle polar core causes a significant reapportioning of the surfactants and water between the filled and unfilled micelles. Chen and coworkers consider the empty and the filled micelles separately. For these two classes of micellar solution, the micellar concentrations, the water and surfactant aggregation numbers are estimated. The geometrical constraints on the volumes and surfactant areas of the two micellar classes are taken into account. Then the amount of AOT engaged in empty and filled micellar solutions is evaluated which, together with material balance constraints, is used to interpret the results of SANS measurements. By optimizing the fit of the model to the data, subject to the constraints previously presented, the aggregation numbers of the surfactant and of water are determined. From these available values, all other parameters may be calculated. It is deduced that, at low water content (below 20), the cytochrome c is placed in the polar core of the micelle, a significant redistribution of the surfactant species and water occurs between the filled and empty micelles which leads to an increase in the relative size of the droplet containing cytochrome c .

The two models proposed by Chen and Pileni[7,10,11] using cytochrome c in reverse micelles are not in contradiction. The first one is applicable only at w above 20 whereas the data presented in the second one are given only for w below 20.

2.3.3 <u>Quasi elastic light scattering, QELS and FRAPP</u>[8,12]. Experiments similar to those presented previously have been performed using myelin basic protein at w= 5.6. The hydrodynamic radius of the filled droplet is significantly larger than the radius of the empty one. It is found that about three empty droplets (containing only water molecules) are required to build up a droplet of sufficient size to accomodate a single protein molecule. Using the FRAPP technique , at low water content, the radius of filled micelles increases indicating a constant change in the structural properties of the system at the molecular level.

3. STRUCTURAL CHANGES USING A LARGE PROTEIN SUCH AS RHO-DOPSIN [5]:

The rhodospin is solubilized in the inverse lipid micellar solution. By small angle X-ray scattering, the gyration radius is determined in the absence and in the presence of protein. In the absence of protein from a straight line of the Guinier plot, the gyration radius of the inverse micelles is found equal to 22A. In the presence of rhodopsin, the Guinier plots show two distinct components: a component with a very large gyration radius for which the lower limit is about 160A and a second component parrallel to that obtained previously in the absence of protein with a gyration radius equal to 22A. The simplest plausible model that can be envisaged and one that explains both the chemical and the known functional properties of the rhodopsin is the following: a model based on the existence of reverse lipid micelles surrounding the protein polar moities while the protein hydrophobic domains act as cross-links between the reverse micelles presented.

4. CONCLUSION

From the data published in the last few years it seems reasonable to conclude that the addition of probes or of small proteins to reverse micelles generally does not strongly perturb the structure of the reverse micelles. However it has been shown that at low water content (below w=20), the filled micelles are larger than the empty micelles. This usually occurs when the size of the protein including its hydration shell is smaller than that of the water pool of the droplet. At relatively high water content ($20 < w < 60$), the simple shell and core model assuming single occupancy of the reverse micelles can be considered. However such small proteins can drastically change the structure of the reverse micelles by changing the usual experimental conditions [17].In the case of big proteins the structure is drastically changed[5].

REFERENCES

1   F.J.Bonner, R.Wolf and P.L.Luisi, J.Solid Phase Biochem., 5, 255, (1980).

2   C.Grandi, R.E.Smith and P.P.Luisi, The J.Biol.Chem., 256, 837, (1981).

3   A.V.Levashov, Y.L.Khmelnitsky, N.L.Klyachko, V.Y.Cheernyak and K. Martinek, Analytic Biochem., 118, 42, (1981).

4   A.V.Levashov, Y.L.Khmelnitsky, N.L.Klyachko, V.Y.Cheernyak and K. Martinek, J.Colloid Int. Sc., 88, 444, (1982).

5   V.R.Ramakrishnan, A.Darszon and M.Montal, The J.Biol. Chem., 258, 4857, (1983).

6   B.H.Robinson, C.Toprakcioglu, J.C.Dore and R.D.Tack, J.Chem.Faraday Trans. I 80, 413, (1984).

7   M.P.Pileni, T.Zemb and C.Petit Chem.Phys.Letters 118, 414, (1985).

8   D.Chatenay, W.Urbach, A.M.Casabat and M.Waks, Biophys.J., 48, 893, (1985).

9   G.G.Zamperi, H.Jackle and P.L.Luisi J.Phys.Chem., 90, 1849, (1986).

10   E.Sheu, K.E.Goklen, T.A.Hatton and S.H.Chen , Biotechnology progress
2,   175, (1986).

11   P.Brochette, C.Petit and M.P.Pileni J.Phys.Chem., 92, 3505(1988).

12   D.Chatenay, U.Wladimir, C.Nicot, M.Vacher and M.Waks, to be published

13   60A per head polar group

14   F. Menger and Yamada, J.Am.Chem.Soc., 101, 6731, (1979).

15   R.Wolf and P.L.Luisi, Biochem.Biophys.Res. Commun., 89, 209, (1979).

16   C.Petit, P.Brochette and M.P.Pileni, J.Phys.Chem., 90, 6517, (1986).

17   J.P.Huruguens and M.P.Pileni, unpublished data.

WHAT CAN BE EXPECTED FROM
NMR IN REVERSED MICELLES ?

A. LLOR and T. ZEMB

ABSTRACT:
We give a review of NMR studies on reversed micellar systems since 1970. We emphasise general principles through examples which have led to relevant physico-chemical results in the area. NMR techniques or theories are not detailed in order to focus primarily on the information obtained on the micelles.

1. INTRODUCTION:

We will consider here that an inverted micellar system is a thermodynamically stable solution of aggregates of surfactants + ions + polar substances (like water) into an hydrophobic solvent. This definition may not be consistent with others in this book, but allows one to choose between more or less restrictive concepts depending on the meaning of the word "aggregate". In most of the following considerations, "aggregates" will be rather well defined objects of finite size and often of spheroidal shape (micelles). The micelles may be diluted within some limits, and are primarily defined by their aggregation number N (average number of surfactant molecules per micelle) and the molecular ratios of the various compounds (w will always define the water/surfactant ratio). Now, in many phase diagrams of these systems, there are continuous paths connecting solutions of truly independent objects with microemulsions, where aggregates coalesce to form a more or less continuous structure. In this case we shall prefer the term "microemulsion" to "micellar system".

The litterature covered by the present review was published since 1970 (1-38). The recently published reviews in the field (42-44) are devoted to micelles and microemulsions in general (direct or reversed). Ref. (42) and (43) focus on the latest works, while (44) is rather exhaustive.

Before any detailed discussion of the publications, we shall briefly recall the application of NMR concepts to micellar systems. We shall give some examples from the literature to illustrate the physico-chemical results NMR can provide, and give some present trends in the field.

Like any other spectroscopy, NMR can be used within two different approaches: the phenomenological or the interpretative strategies. The former consists in recording extensively the spectroscopic parameters of interest, and look for correlations between some appropriate reference state and the studied system. In the latter, the spectroscopic parameters are given an interpretation within the framework of a general theory which describes the physical system (in the case of electronic spectroscopy for instance, this would be the role of ab initio calculations of the electronic excited states). This is of course a very schematic description since a wide continuous spectrum exists between the extreme cases, depending on the number, kind and level of the assumptions introduced in the interpretations. At first glance, an interpretative result seems more interesting but, as we will examplify later, it may have low operative value (that is low predicting power in general). Since detailed decriptions are available elsewhere (45,46), we shall briefly summarise the NMR spectroscopic parameters and the mainframe within which their interpretation must be done.

2. A BRIEF SURVEY OF NMR CONCEPTS:

NMR deals with five main interactions:
- Chemical shift, which reflects the "shielding effect" of the external magnetic field at the nucleus due to the surrounding electrons.
- "J" coupling, or indirect coupling between nuclei, proceeds through the contact couplings of the nuclei with the molecule's electrons.
- Quadrupolar coupling is proportional to the electric field gradient induced by the surrounding charges (in the case of nuclei of spins >1/2 whose distributions of charges are not spherical).
- Dipolar coupling through space between nuclei, or a nucleus and a lone electron's moment, is similar to the interaction between magnetic needles.
- Last, the Zeeman interaction, which is usually not of spectroscopic relevance since it is only related to the magnetic field intensity. However it provides the first and most striking feature of NMR: the intensities of the observed signals are always proportional to the numbers of nuclei in the sample, allowing thus stoechiometric and imaging measurements. Since all of the other interactions are observed as perturbations over the Zeeman, this property is always preserved.

The intrinsic interactions are of very local character: except for the dipolar coupling, all of them are due to the interaction with the closely surrounding electrons (barely more than three covalently bonded atoms away). The dipolar interaction, although not restricted by chemical links, has a $1/r^3$ geometrical dependence which in most cases reduces its range to a few neighbouring layers of nuclei.

Except for the dipolar coupling, the interactions' magnitudes are of difficult and tedious interpretation (it involves the description of the electronic structure of the molecule). But they are almost exclusively determined by the molecular structure, so they can be considered as constant fingerprints which are tabulated for a wide range of molecules. Important exceptions for micellar systems are the chemical shifts of hydrogen bonded protons or of phosphorus in substituted PO_4 groups, which depend of many parameters like temperature, pH, ionic strength, etc. In contrast to all other interactions, dipolar couplings are simply and directly related to the distances between the nuclei through universal constants.

Compared to other standard spectroscopies, all the intrinsic interactions are small (up to about 1MHz), and in any case they must be averaged over all the molecular motions which are faster. When dealing with an isotropic media, like a micellar solution, these motions average out all of the anisotropic interactions, leaving only the isotropic parts of the chemical shift and J-coupling interactions (this is the standard case of high resolution NMR in liquids). We must however keep in mind that in closely related phases like liquid cristals or some highly viscous microemulsions, the anisotropic motions do not average out all the interactions (then the remaining anisotropy is a source of information on angular order of the molecules in the system).

The influence of the averaged interactions is reduced to relaxation, a second order effect which reflects the molecular motions. As a rule of thumb, the relaxation rates can be written (45,46):

$$1/T \approx H^2 . \tau / (1+\omega^2\tau^2) \qquad [1]$$

where H is the interaction intensity, ω the Zeeman frequency, and τ an effective molecular correlation time. Because, in general, the local interactions have short range but also strong dependance on orientation (anisotropy), $1/\tau$ is almost always a molecular reorientation or exchange rate. Since H is usually known, the measurement of a relaxation time gives information on the molecular motions. However restrictive assumptions on the molecular motions are needed to derive equation [1], some of them rather approximative, and in a micellar system many different movements may superimpose (like surfactant motion inside the aggregate motion). Careful analysis is then needed to avoid misleading interpretations, and a detailed dependence of relaxation over the frequency ω may happen to be necessary. Geometrical parameters are also often needed but may be deduced from other information sources (such as X-ray or neutron scattering, or even plain sterical considerations). When H is not previously known, it is very seldom possible to determine it from the relaxation experiments.

3. NMR AND INVERSED MICELLAR SYSTEMS:

For NMR studies, the most relevant features of micellar systems are:
- microheterogeneity: the different kinds of molecules are confined into specific regions (namely water, oil and interfacial film). In each of the regions, the proximity of the interface produces various deviations of the NMR parameters from their bulk values. To describe these effects, the simplest model is to introduce

fictitious subregions for the molecules far or close to the interface, where the parameters are assumed to be constant. A very well known example is the "pseudo-phase" model (44).

- dynamic structure: all the molecules are in a liquid-like phase. This ensures that all the NMR interactions are averaged over all the possible positions and sites arising from the Brownian motions. In exceptional cases the averaging rate may be lower than a caracteristic NMR interaction, providing direct spectroscopic parameters for the various sites in the system.

- macroscopic isotropy: this means that we have a liquid-like case where the only non-vanishing interactions are chemical shifts and scalar couplings (excepted for the special cases just mentioned above, where powder patterns may be observed).

- microscopic dynamical anisotropy: the surfactant film is locally anisotropic and it influences all the neighbouring molecules. All the induced effects are averaged by the film motions and the system becomes isotropic as we mentioned. However the local anisotropy influences the averaging process of the nuclear interactions and it must be included in any proper description of relaxation rates.

The "pseudo-phase" like models are the most extensively used interpretation methods and may be applied to almost any NMR parameter: chemical shifts, J-couplings, line widths, relaxation rates, etc. Q being one of these quantities, its value in the simplest case of a two site system, is given by:

$$Q = x.Q_1 + (1-x).Q_2 = Q_2 + (Q_1-Q_2).x \qquad [2]$$

where x is the relative amount of site 1, and Q_1, Q_2 are the values of Q in the pure sites 1 and 2. Provided that Q_1 and Q_2 are measurable, this leads to a direct relationship between Q and x. One of the earliest applications of the method is the determination of CMCs. The two sites are the "free" and "aggregated" surfactants (Fig. 1). However, the interpretation may be seriously hampered when Q_1 or Q_2 are not independent of x.

In any case, the regions of the sample to be discriminated must display a gradient between them of some stable NMR quantity, and their relative amounts must be alterable by changing some parameter of the system (like concentrations). The gradient may be intrinsic to the system or created by some artificial mean. A very common method in this last case consists in adding a paramagnetic species which induces relaxation or chemical shift in its close environment. Another important example is the the case of aromatic probes whose ring currents may shift resonances of surrounding molecules. A probe molecule, observable by NMR and known for its selective affinities (hydrophobic, hydrophilic, amphiphilic, etc), may also provide an indirect labelling of the regions.

A very interesting extension of the pseudo-phase model appears whenever the dynamical process, which averages the NMR parameter gradient in the sample, is comparable to the involved NMR time scale. Then, it is possible to give exchange rates between the labelled pseudo-regions of the sample. This is done creating an off-equilibrium distribution of the labelling quantity, and monitoring the subsequent decay. Exchange processes have been successfully measured in related systems (like vesicles), but they are usually much too fast in micellar systems for any local labelling quantity (like chemical shift or relaxation). An exception is provided by the measurement of surfactant exchange rates in DLPC/water/benzene reversed micelles labelled with paramagnetic ions (27).

Although technically different but conceptually related, is another application of utmost importance: the pulsed field gradient method. The labelling distribution is provided by an external macroscopic field gradient (which translates on the spectrum as an effective chemical shift gradient). Thus, the exchange rates between the various regions of the sample give the macroscopic diffusion coefficients of the monitored chemical species.

In a more involved approach, any NMR parameter derived from the pseudo-phase methods (as Q_1 or Q_2 in equation [2]) may be related to detailed local descriptions if the interactions are known. As we already mentioned in the previous section, only relaxation rates due to dipolar and quadrupolar interactions may be in general quantitatively explained. Two situations are found whether the relaxation rate refers to a region close to or far from the interfacial film. In the former case the local molecular reorientations are isotropic on the NMR time scales and average out the interactions, whereas in the later the averaging is incomplete and many slower motions of the surfactant film or aggregate are also involved in the relaxation process. The correlation times for small molecules or segmental motions are seldom of interest since they differ very slightly from their bulk sample values. Of more valuable information are the slow motion correlation times (of the molecules or of the aggregates) and their local order parameters. However, many assumptions, and some times unknown parameters, are needed to extract them from raw data. In exploring the cascade of correlation times for instance, it is of utmost importance to collect data at various NMR frequencies (three and more).

Also related to relaxation is the recently introduced technique of "nuclear spin quenching" (39) where the modulation of an NMR interaction (in this case the scalar coupling of ^{17}O with protons) is due to the exchange process of a quencher between the micelles (H_3O^+ in this example). Exchange rates as well as droplet sizes (related to quencher distribution) can be obtained.

4. PHYSICO-CHEMICAL RESULTS DEDUCED FROM NMR:

Let us summarise the informations NMR can provide:
- All standard NMR spectroscopic informations commonly used in chemistry (detection of a given amount of a given molecule in the sample);

- Compartment ratios (for instance free/aggregated molecules), partition coefficients, and proximity effects between molecules (with paramagnetic or ring current probes);
- Slow exchange rates between different compartments;
- Macroscopic diffusion constants;
- Local or semi-local reorientation correlation times and orientation order parameters.

All of these concepts, except the first one which is not specific to the micellar systems, are illustrated by a selection of published examples reproduced in Fig. 1 to 9. As can be seen, the informations NMR can provide are loosely related to the general microstructure of the system (although very sensitive to anisotropy effects), but are much more detailed on dynamical effects. The description of the dynamical processes can take place and have a meaning only within a geometrical framework. As a general rule, previous to any NMR study, a reasonable background of informations should be available from other techniques of higher sensitivity to structure (such as X-ray or neutron scattering).

Most of the mentioned techniques have been extensively applied to direct micelles to explore such different aspects as self-diffusions, reorientations, segmental motions, order parameters, ion bindings, hydration of polar heads, solubilisations, etc (44). Comparatively, much fewer works have been published on inverted micelles, or even microemulsions. This is probably due to the following reasons:
- Ionic concentrations in water cores of inverted micelles are much higher than in direct systems. The well known structuring effects of water may thus be subject to considerable variations with the molar ratio w, outruling simple pseudo-phase like interpretations in many cases. In microemulsions, the interpretation of many relaxation data is difficult because of scarce and poor structural informations to describe the interfacial film.
- Many properties of the surfactant films have been much more easily and extensively studied on liquid crystalline phases, since, for many NMR properties, they behave very similarly to those of micellar systems (47).
- Last, inverted micellar systems are less common than direct ones, and the surfactants they involve most often have double branched chains and special polar heads. This makes systematic studies of chain length or counter ion variations difficult.
Thus, most of the publications we present are devoted to AOT, double chained phosphates and phosphatidylcholines.

4.1 Aggregation numbers and exchange rates:

Since the initial association processes in inversed micelles are not yet understood, and because of the low corresponding critical micelle concentrations (if any), aggregation studies from NMR are rather scarce.

To our knowledge, direct determination of a CMC has only been reported in the diethyhexylphosphate/water/benzene system (at w=3.5) (Fig. 1) (26). This was possible since chemical shifts of ^{31}P in phosphate groups are very sensitive to hydration effects. Otherwise, aggregate concentrations have been probed through the chemical shift of methanol solubilized in the dodecyl ammonium propionate / carbon tetrachloride system (3).

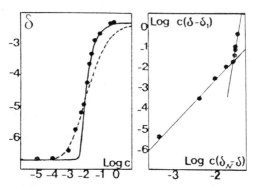

Fig. 1. CMC determination in the system diethyhexylphosphate/water/benzene (at w=3.5) using a pseudo-phase model for the water proton chemical shift (26). In this system, the aggregation up to N=11.3 proceeds throught a pre-aggregation step to N=2 as shown by the dotted simulated line. A standard two-sites aggregation model would give the solid line. N values can be determined by the detailed profile of the plot, taking into account the mass law which describes the aggregation process. The determination of N displays very poor precision at high values (above 40) where very steep profiles are measured.

A measurement of exchange rates of surfactant between micelles is available on a dilinoleyl- phosphatidylcholine system (DLPC) (Fig. 2) (27). Here again ^{31}P NMR of the phosphate group was used.

Most striking is the recent observation of water exchange between micelles using the "nuclear spin quenching" technique (39). In this experiment the [17]O lineshape is monitored as a function of the added H_3O^+ concentration (Fig. 3). As is well known from bulk water studies (48,49), the line width is pH dependent since the scalar couplings with the protons in a molecule are modulated by the H_3O^+ and OH^- catalysed hydrogen exchange with the surrounding molecules. In the case of a inverted micellar solution, the line width behaviour also depends on the exchange process of the H_3O^+ quenchers between the aggregates. An important feature of this experiment compared to other probe exchange techniques (for instance photosensitive), is that the added chemical species induces much smaller perturbations in the system, and that its exchange rate is identical to that of water. Applied to the AOT / water / iso-octane at w=30, the method gives coalescence rates around $8 \times 10^6 M^{-1} s^{-1}$ (in agreement with previous data from other techniques) and a droplet radius of about 80Å (at variance with many previous studies which indicate a 50Å radius).

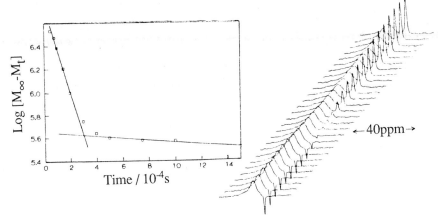

Fig. 2. Magnetisation transfer experiment on ^{31}P in the dilinoleyl-phosphatidylcholine (DLPC) / water / benzene system with Pr^{3+} ions added as paramagnetic probe (at w=18, [DLPC]=3.13x10^{-5}M and [Pr^{3+}]/[DLPC]=1/101) (27). The chemical shift induced by Pr^{3+} in the aqueous core allows the use of a selective excitation sequence to create an off-equilibrium magnetisation gradient between micelles containing and not containing the probe. The recovery behaviour of the magnetisation gives an exchange rate of the surfactant molecules between the aggregates (k=0.8x10^{6}M^{-1}s^{-1}).

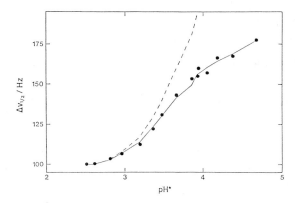

Fig. 3 ^{17}O line width in the AOT / water / iso-octane system at w=30 as a function of pH (39). The solid line fits the experimental data to a model of water protons exchange which depends of the fusion-fission kinetic of the aggregates.

4.2 Hydration and related studies:

Although hydration, ionic or polar head effects are easily observed on many nuclear species, the interpretations are difficult and somewhat disapointing. Since the interactions (mainly quadrupolar and dipolar) as well as the correlation times are hydration dependent, the relaxation rates and line widths are usually explained on a qualitative basis. Pseudo-phase like considerations seldom apply properly.

Counterion binding and hydration as a function of the water content have been investigated using Na NMR in sodium octanoate / pentanol (4) or hexanol (22) /water, AOT / water / heptane (5), diethyhexylphosphate / water / benzene (26), and using Br NMR in cetyltrimethylammonium bromide / n-hexanol /water (2). The polar heads behaviours have been explored in specific cases suitable for NMR studies: H and ^{13}C chemical shifts in polyoxyethylene dodecyl ether / water / benzene (14), J-couplings for polar head conformations in AOT / water / i-octane, chloroform, methanol and cyclohexane (15, 24), ^{14}N relaxation in cetyltrimethylammonium bromide or sulfate / water / hexanol (33).

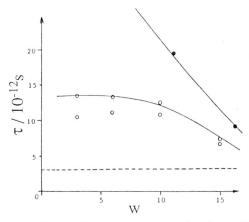

Fig. 4. The correlation time of water molecules in the AOT/water/cyclohexane system as a function of w=[H$_2$O]/[AOT] (34). These data were obtained from proton relaxation rates 1/T$_1$ with help of an isotopic substitution method which separates contributions from different interactions. The values found are close to the bulk values in concentrated ionic solutions (the dotted line gives the value in pure water). The line in the upper right corner was drawn from (5). The difference is due to interactions modulated by a collective movement in the 0.5 to 2ns range, also found by other methods or in similar systems (28). This motion is almost independent of w and not yet clearly explained.

The structure and state of water in reversed micelles and particularly at low water content has received considerable attention.The proton chemical shifts have been explored extensively in AOT / water / n-heptane (12), methanol, chloroform, isoctane and cyclohexane (15, 24, 31). However, relaxation data are necessary to give a more quantitative picture. In an earliest work on AOT / water / heptane (5), the relaxation rate of water protons was explained with a single dipolar intermolecular interaction which led to the misleading conclusion of a strongly structurated media with long correlation times in the low w micelles. These conclusions were revisited (Fig. 4) (34) in AOT / water / cyclohexane, and it is now established that the water behaviour in small reversed micelles is close to that of the corresponding bulk ionic solution. In a recent work (Fig. 5) (40), the longitudinal relaxations of ^2H and ^{17}O at higher water contents (w= 15 to 50) were accounted for in a two site model. The reorientation correlation time in the interfacial layer was found to be about 3 to 5 times longer than in bulk water, and almost independent of w provided that w>15. This value roughly corresponds to the hydration layer of the interfacial film. [2]H, [17]O and proton relaxation studies of water were also caried out in diethylhexylphosphate / water / cyclohexane (26) where much stronger bonding to the phosphate head groups takes place.

The effects of added salts of Na, K, Mg and Ca on proton and ^{31}P in egg phosphatidylcholine / benzene / water (35) have been measured. An extensive exploration of the water and chain proton relaxation rates throughout the phase diagram of 1/3 sodium dodecyl sulfate / 2/3 butanol / water / toluene displayed some interesting qualitative features (6,7).

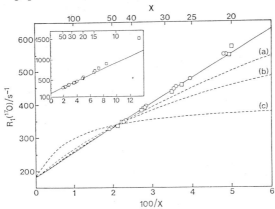

Fig. 5. The spin-lattice relaxation time of water [17]O in the AOT / water / iso-octane system as a function of w (40). The 1/w behaviour is consistent with a pseudo-phase model (with "free" and "bonded" molecules). The onset of linearity (shown in insert) gives an estimate of the hydration layer thickness.

4.3 Solubilisation:

Ring currents have seldom been used in reverse systems, but their utlity as probe localisation method has been proved by studies in direct micelles (44). By monitoring chemical shifts of the surfactant polar heads, aromatic solubilisates were localized in the vicinity of interfacial films: benzene in sodium oleate / butanol or decanol / water (1), and benzene and nitrobenzene in AOT / water / isooctane (21).

A solubilised phosphate buffer provides a convenient probe of the pH in the core of the micelles by observation of the ^{31}P chemical shift in sodium octanoate / hexanol / water (11), in AOT / water / octane (20), or the ^{31}P relaxation in egg phosphatidylcholine / benzene / water (19). In the sodium octanoate system, the ^{31}P shift has been calibrated first in phosphate buffers and the conclusion is that the internal pH of these reverse micelles is about 8 (11). In AOT micelles, the pure water core has been replaced by a phosphate buffer. Above w=7, the mobility of

Na+ counterions is independant of the water content and the apparent pH deduced from the ^{31}P shift of the water core of the reverse micelle is about 7, slightly decreasing when the water content increases: 7.5 at w=7 and 7.2 at w= 50.

The interactions with the surfactant films of amino-acids or peptides solubilized in egg phosphatidylcholine / benzene / water (18) and AOT / water / isooctane (32) have also been explored by proton chemical shifts and relaxations.

4.4 Macroscopic self diffusion:

Self diffusion studies are not very common since special spectrometer equipment is needed (50). However their contribution is of high relevance particularly in the microemulsion phases where direct, inverted or bicontinuous structures may be difficult to distinguish by other techniques (except for the trivial case of macroscopic connectivity of the water pools which can be detected by conductivity studies).

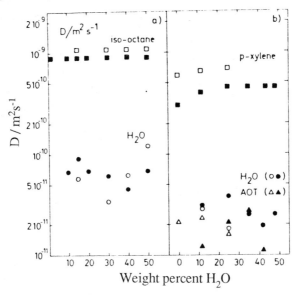

Fig. 6. Pulsed field gradient measurements of self diffusion coefficients in AOT/water/iso-octane and p-xylene systems (23). The diffusion constants of water and AOT are equal within experimental error and much lower than those of the solvent, showing common macroscopic transport through the diffusion of the aggregates. In bicontinuous phases, the water diffusion constant is considerably higher due to percolation paths in the structure. The diffusion constants can be related to the size of the aggregates using brownian and hydrodynamic motion theories.

The AOT / water / p-xylene or i-octane was studied at 25°C (Fig. 6) (23), and revisited later as a function of temperature (30). The self-diffusion of water closely follows that of the surfactant confirming the existence of well defined aggregates and of slow water exchange rates between them. This is not the case in some cetyltrimethylammonium bromide / hexanol / water systems where drastic changes in water self diffusion appear upon dilution of the microemulsion (33). These two studies show that different surfactants exhibit different connectivity thresholds when they turn into bicontinous structures: up to now, the driving forces of these behaviours have not been identified, although they are certainly related to the interfacial film stiffness.

4.5 Segmental motions, conformations and order parameters:

The segmental motions of the surfactant chains have been rather extensively studied.The prime method is the interpretation of ^{13}C relaxation rates either in the diamagnetic case as in AOT / chloroform or benzene (8,9), AOT / water / iso-octane (10), AOT / water / benzene, carbon tetrachloride or cyclohexane (16, 25), AOT / water / p-xylene (Fig. 8) (28), sodium p-octylbenzenesulfonate / butanol / water / toluene (29), or with help of paramagnetic ion probes like Gd^{3+} in AOT / water / benzene (37), or Mn^{2+} in sodium diethylhexylphosphate / water / cyclohexane (38).

The segmental motions are described by local reorientation times for each carbon, which, as expected, are shorter at the chain tails and longer near the polar heads. For a given surfactant, their orders of magnitude are very similar in most systems, ranging, in the case of AOT, from 10ps at the chain tail, up to 200ps near the polar head.

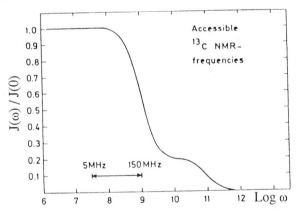

Fig. 7. An example of normalised spectral density (which describes the frequency dependence of the relaxation times) shows the influence of two correlation times at 10ps and 1ns for a local order parameter of 0.2 (28). This is the "two-step" model, generally used in micellar systems.

Except in strongly hindered situations (like water free AOT micelles) these correlation times have a poor informative value. However, it has been recognised from the earliest studies that the segmental motions display a local anisotropy and do not average out the ^{13}C-H dipolar interactions. Then, the relaxation process is also sensitive to slower collective motions like surfactant reorientation in the film or aggregate rotation. In earliest studies this fact was included using an effective correlation time (a common procedure in isotropic fluids), but the interpretation was revised later (28, 29, 38) with help of the "two step" model: local order parameters (Fig. 7, 8) describe the anisotropy of the segmental motions which share a common slow motion correlation time; the frequency dependence of the relaxation rates allows the simultaneous determination of all the parameters. The slow motion correlation times thus obtained can seldom be associated with aggregate rotation which are longer by one or two orders of magnitude (except for small micelles (28, 38)). In fact the model should include the influence of aggregate rotation in a three step model; this was done for the sodium deoxycholate / water system (36).

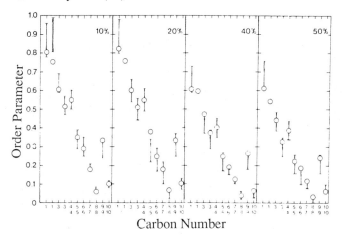

Fig. 8. Order parameter profiles along the AOT chains in the AOT/water/p-xylene (at w=3.8, 8.45, 22.5 and 34) (28). A two step spectral density profile (Fig. 7) was used to fit the relaxation rates assuming a common slow correlation time at each w value (ranging from 0.9 to 2ns). The corresponding motion is much slower than the aggregate overall reorientation, and not clearly identified.

In a recent work (41), interesting informations on the slow motions of the interface were obtained through a careful analysis of the ^2H and ^{17}O transverse relaxation rates of water in the AOT / water / iso-octane system (at w=15 to 50). Compared measurements between the two nuclei, at controlled pH values, and variable frequency data gave an unambiguous evidence for a slow motion controlled quadrupolar relaxation effect. This reflects an incomplete averaging of the quadrupolar interaction over fast motions due to a local anisotropy of the system. Detailed calculations showed that the effect arises from a shape anisotropy of the aggregates (either frozen or fluctuating around a mean spherical shape). Estimates could be extracted for the axial ratios in the cases of oblate or prolate aggregates (Fig. 9), and found to be about 2 to 3. This kind of information could a priori be obtained from order parameter studies of the surfactant molecules, but as explained above, a complex cascade of slow motions makes interpretations difficult. In the case of small molecules like water all the motions are fast and average the anisotropy over the aggregate.

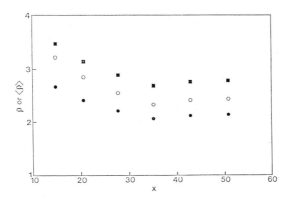

Fig. 9 Axial ratios of the aggregates in the AOT / water / iso-octane system as a function of w and for various shape and dynamical assumptions (oblate, prolate or fluctuating). These parameters were obtained through analysis of the ^2H and ^{17}O transverse relaxation rates of water (41).

6. CONCLUSION:

The relevance of the results NMR has provided or might provide in the future, must be evaluated bearing in mind the important problems arising in the study of inversed micellar systems. In our opinion, three types of NMR studies must be highlighted, mainly for the unability of other techniques to get similar informations.

The "pulsed-field-gradient spin-echo" technique, which allows simple and straighforward interpretations of possible connectivities of water or oil "phases", will probably become a standard structural investigation method in the future.

Up to now, the effect of a solute on the micellar structure is not well understood, although a large number of dispersed observations has been made. General trends and origin of the driving forces are unknown, and no theory is able to predict even on a relative basis what the solubility of a given compound should be in a reversed micelle. In view of this, chemical shifts induced by ring currents of aromatic probes are very useful tools for solute localisation studies.

Quantitative results relying on often complex interpretations of relaxation data have added little knowledge to the physics of the interfacial film in reversed micelles. However, extensive and careful analysis are in some cases tractable, and then yield geometrical (anisotropy) and dynamical informations (slow motions) on the interface layer. This is examplified in a recent work (Fig. 9) (41), were departure from sphericity of the droplet was clearly identified. This kind of study may lead to some understanding of one of the crucial characteristics of interfacial films: stiffness.

REFERENCES:
1   G. Gillberg, H. Lehtinen and S. Friberg, J. Colloid Interface Sci., 33 (1970) 40-53.
2   G. Lindblom, B. Lindman and L. Mandell, J. Colloid Interface Sci., 34 (1970) 262-271.
3   A. Kitahara, K. Konno and M. Fujiwara, J. Colloid Interface Sci., 57 (1976) 391-392.
4   J.B. Rosenholm and B. Lindman, J. Colloid Interface Sci., 57 (1976) 362-378.
5   M. Wong, J.K. Thomas and T. Nowak, J. Am. Chem. Soc., 99 (1977) 4730-4736.
6   J. Biais, B. Clin, P. Lalanne and B. Lemanceau, J. Chim. Phys. Phys.-Chim. Biol., 74 (1977) 1197-1202.
7   P. Lalanne, J. Biais, B. Clin, A.M. Bellocq and B. Lemanceau, J. Chim. Phys. Phys.-Chim. Biol., 75 (1978) 236-240.
8   M. Ueno, H Kishimoto and Y. Kyogoku, Chem. Lett., (1977) 599-602
9   M. Ueno, H Kishimoto and Y. Kyogoku, J. Colloid Interface Sci., 63 (1978) 113-119.
10 H.F. Eicke and P.E. Zinsli, J. Colloid Interface Sci., 65 (1978) 131-140.
11 H. Fujii, T. Kawai and H. Nishikawa, Bull. Chem. Soc. Jpn., 52 (79) 2051-2055.
12 J. Rouviere, J.-M. Couret, M. Lindheimer, J.-L. Dejardin and R. Marrony, J. Chim. Phys., 76 (1979) 289-296.
13 H. Kondo, T. Hamada, S. Yamamoto and J. Sunamoto, Chem. Lett., (1980) 809-812.
14 H. Christenson, S.E. Friberg and D.W. Larsen, J. Phys. Chem., 84 (1980) 3633-3638.
15 A.N. Maitra and H.F. Eicke, J. Phys. Chem., 85 (1981) 2687-2691.
16 C.A. Martin and L.J. Magid, J. Phys. Chem., 85 (1981) 3938-3944.
17 P. Stilbs, Chem. Scr. 19 (1982) 93-94.
18 C.A. Boicelli, F. Conti, M. Giomini and A.M. Giuliani, Spectrochim. Acta, 38A (1982) 299-300.
19 C.A. Boicelli, F. Conti, M. Giomini and A.M. Giuliani, Chem. Phys. Lett., 89 (1982) 490-496.
20 H. Fujii, T. Kawai, H. Nishikawa and G. Ebert, Colloid Polym. Sci., 260 (1982) 697-701.
21 A. Maitra, G. Vasta and H.-F. Eicke, J. Colloid Interface Sci., 93 (1983) 383-391.
22 H. Fujii, T. Kawai, H. Nishikawa and G. Ebert, Colloid Polym. Sci., 261 (1983) 340-345.
23 P. Stilbs and B. Lindman, J. Colloid Interface Sci., 99 (1984) 290-293.
24 A. Maitra, J. Phys. Chem., 88 (1984) 5122-5125.
25 L. J. Magid and C.A. Martin, in: P.L. Luisi and B.E. Straub (Eds), Reverse Micelles, (Proc. Eur. Sci. Found. Workshop), 4th, Plenum, New-York, 1984, pp181-193.
26 A. Faure, A.M. Tistchenko, T. Zemb and C. Chachaty, J. Phys. Chem., 89 (1985) 3373-3378.
27 L. Barclay, C. Ross, J.B. Balcom and B.J. Forrest, J. Am. Chem. Soc., 108 (1986) 761-766.
28 J. Carnali, B. Lindman, O. Soederman and H. Walderhaug, Langmuir, 2 (1986) 51-56.
29 O. Soederman and H. Walderhaug, Langmuir, 2 (1986) 57-63.
30 S. Geiger and H.-F. Eicke, J. Colloid Interface Sci., 110 (1986) 181-187.
31 A. De Marco, E. Menegatti and P.L. Luisi, J. Biochem. Biophys. Methods, 12 (1986) 325-333.
32 A. De Marco, L. Zetta, E. Menegatti and P.L. Luisi, J. Biochem. Biophys. Methods, 12 (1986) 335-347.

33   T. Waernheim and U. Henriksson, Acta Chem. Scan., A40 (1986) 292-298.

34   A. Llor and P. Rigny, J. Am. Chem. Soc. 108 (1986) 7533-7541.

35   C.A. Boicelli, M. Giomini and A.M. Giuliani, in: E. Drioli and M. Nakagaki, Membr. Membr. Processes, (Proc. Eur.-Jpn. Congr. Membr. Membr. Processes), Plenum, New-York, 1986 pp361-370.

36   L.E. Kay and J.H. Prestegard, J. Am. Chem. Soc., 109 (1987) 3829-3835.

37   H. Okabayashi, K. Taga, K. Tsukamoto, K. Matsushita, O. Kamo and K. Yoshikawa, Colloids Surf., 24 (1987) 337-352.

38   A. Faure, T. Alhnas, A.M. Tistchenko and C. Chachaty, J. Phys. Chem., 91 (1987) 1827-1834.

39   G. Carlström and B. Halle, Mol. Phys., 64 (1988) 659.

40   G. Carlström and B. Halle, Langmuir, in press.

41   G. Carlström and B. Halle, submitted for publication.

42   K. Vos, C. Laane and A.J.W.G. Visser, Photochem. Photobiol., 45 (1987) 863-878.

43   B. Lindman, Z. Phys. Chem. (Munich), 153 (1987) 15-26.

44   C. Chachaty, Prog. N.M.R. Spectroscopy, 19 (1987) 183-222.

45   A. Abragam, The Principles of Nuclear Magnetism, Oxford University Press, London 1961.

46   J.K.M. Sanders and B.K. Hunter, Modern NMR Spectroscopy, Oxford University Press, London 1987.

47   J. Seelig, in: P.L. Luisi and B.E. Straub (Eds), Reverse Micelles, (Proc. Eur. Sci. Found. Workshop), 4th, Plenum, New-York, 1984, pp209-220.

48   S. Meiboom, J. Chem. Phys., 34 (1961) 375.

49   S.W. Rabideau H.G. Hecht, J. Chem. Phys., 47 (1967) 544.

50   P. Stilbs, Progress NMR Spectroscopy 19 (1987) 1.

FLUORESCENCE QUENCHING IN MICELLAR SYSTEMS

M. VAN DER AUWERAER AND F.C. DE SCHRYVER

K.U. Leuven Chemistry Department
Celestijnenlaan 200 F
3030 Leuven   Belgium

1. Introduction.

Micelles and reversed micelles are entities whose dimensions are intermediate between molecules and macroscopic objects. In photochemical and photophysical experiments the concentration of the probe and the quencher will be such that the average number of quenchers and probes present in an individual micelle will be close to one. As a consequence the actual values of this occupation number, which must always be an integer number can exhibit large relative fluctuations between different micelles. When the quenching of the fluorescence of a probe solubilized in a micelle is investigated (1), the quenching process can be considered as a diffusion of the quencher to the micelle followed by the diffusion of the quencher on or in the micelle to form an encounter complex (2). Fluorescence quenching in micellar systems can be influenced by several distributions: the distribution of micellar sizes (3), the distribution of probe and quencher between the micellar phase and the solvent (4), the distribution of the quencher and the probe over the volume of the micelle or the reversed micelle (5,6), the distribution of the quencher over the micelles (7,4). When the distribution of the quenchers in micellar systems is discussed usually the latter distribution is meant. These distributions influence the fluorescence quenching in micellar and reversed micellar systems identically and lead to kinetic expressions that are formally the same (8) for both systems. Due to the more abundant information related to the fluorescence quenching in aqueous micellar systems this contribution will start from the treatment of the different aspects determining the fluorescence quenching in aqueous micelles.

2. A general model for fluorescence quenching

In the present contribution it is our aim to discuss the luminescence decay of probes solubilized in micellar systems. When a micelle containing one excited probe and n quenchers M_n^* is considered following processes (4,7) can alter the concentration of M_n^*.

$$M_n^* \xrightarrow{\quad k_0 \quad} M_n: \qquad \text{monomolecular decay of the probe}$$

$$M_n^* \xrightarrow{\quad nk_q \quad} M_n: \qquad \text{quenching of the excited probe}$$

$$M_n^* \xrightarrow{\quad nk^- \quad} M_{n-1}^* + Q_{aq}: \qquad \text{exit of a quencher from } M_n$$

$$M_n^* + Q_{aq} \xrightarrow{\;[Q_{aq}]k^+\;} M_{n+1}^*: \qquad \text{entrance of a quencher}$$

$$M_n^* + M_{j-1} \xrightarrow{\;n[M_{j-1}]k_{ex,q}\;} M_{n-1}^* + M_j: \qquad \text{exchange of a quencher}$$

$$M_n^* + M_{j+1} \xrightarrow{\;(j+1)[M_{j+1}]k_{ex,q}\;} M_{n+1}^* + M_j: \qquad \text{exchange of a quencher}$$

$$M_n^* + M_j \xrightarrow{\;[M_j]k_{ex,p}\;} M_j^* + M_n: \qquad \text{exchange of a probe}$$

$$M_n^* + M_j \xrightarrow{\;[M_j]k_{fus}\;} M_i^* + M_k \;,\; j+n=i+k: \text{fusion-fission}$$

This general scheme is based on the assumptions that the exit rate of the quencher and the rate of the intramicellar quenching process are proportional to the number of quenchers present in the micelles and that the entrance rate does not depend upon the number of quenchers already present in a micelle. The first and the third assumptions will be discussed in section two while the second assumption will be discussed in the section three.

3. Distribution of the probes or the quenchers over the micelles

The most simple model (7) for the distribution of the quenchers (or the probes) between the micelles is based upon two assumptions:
1) The rate with which a quencher (or a probe) leaves the micelle is not influenced by the interaction between the quenchers (or probes) and is therefore proportional to the number of quenchers in the micelles. This rate, v_n^-, is given by equation 1 for a micelle containing n quenchers
2) The rate with which a quencher (or a probe) enters a micelle from the aqueous phase does not depend upon the number of quenchers in the micelle; this rate, v_n^+, is given by equation 2 for a micelle containing n quenchers.

$$v_n^- = nk^-[M_n] \tag{1}$$

$$v_n^+ = [Q_{aq}][M_n]k^+ \tag{2}$$

$[M_n]$: the concentration of micelles containing n quenchers

$[Q_{aq}]$: the concentration of the quencher in the aqueous phase

k^-: the exit rate constant of quencher bound to a micelle

k^+: the entrance rate constant of a quencher into a micelle

Application of equation 1 and 2 to the occupation number of the micelle (for either the quencher or the probe) leads to a Poisson distribution for the probability P_n to find n quenchers or probes in the micelle (equation 3)

$$P_n = \exp(-\mu)\mu^n/n! \qquad\qquad [3]$$

μ: the average number of quenchers or probes per micelle.

When the exchange of solubilized molecules occurs by a collision mechanism (9,10) rather than by an exit-entrance mechanism an identical distribution is obtained when the two basic assumptions (the rate with which a quencher goes from one micelle to another is proportional to the number of solubilized molecules in the former and independent upon the number of solubilized molecules in the latter) are respected (11).

When fluorescence quenching in micellar systems is investigated two different types of quenching can be distinguished. In the first case (12) quenching occurs by a molecule that is chemically different from the probe. In the second case the quencher is identical to the probe and quenching occurs by excimer (13) formation. In the first case the average numbers of probes per micelle (μ_{probe}) is kept much smaller than one in order to make the probability to have more than one probe per micelle negligible (e.g. when μ_{probe}, the average number of probes per micelle. equals respectively 0.1 or 0.01 this probability equals respectively $4.7 \ 10^{-3}$ and $4.9 \ 10^{-5}$). In the second case μ_{probe} can become larger than one. In most luminescence techniques the intensity of the exciting radiation is kept low enough to make the probability that a probe is excited much smaller than one. However when the quenching of an excited state is studied by a transient absorption technique where excitation occurs by a pulsed laser, characterized by high peak powers, one has to verify if this condition (1) is still fulfilled.

The assumptions leading to equation 1-3 are valid as far as the quenchers do not change appreciable the properties of the micelle. This will be the case for micelles where the number of quenchers is considerably smaller than the aggregation number. However when the number of quenchers bound to a micelle starts to increase one can expect those quenchers to change the properties of the micelle, including the partition coefficient of the quencher between the solvent and the micellar phase. Dorrance and Hunter(14) took into account a de-creasing affinity for the quencher when the number of quenchers per micelle increases by setting a maximum m to the number of quenchers that can be solubilized in a micelle. In this model P_n is given by equation 4.

$$P_n = \frac{(\mu/m)^n \ (1-\mu/m)^{m-n} m!}{n! \ (m-n)!} \qquad [4]$$

The processes leading to equation 4 can be particularly important for reversed micelles at low R (R: the ratio of the number of water molecules to the number of surfactant molecules in an reversed micelle) values where important deformations of the aggregate by the probe or quencher are expected (15).

Rothenberger et al.(16) suggested that in case the distribution of the quencher is between the micellar phase and a phase of submicelles instead of between the micellar phase and the aqueous phase the exit rate does not depend upon the number of quenchers per micelle. This would lead to a distribution given by equation 5.

$$P_n = \frac{\mu^n}{(1+\mu)^{1+n}} \qquad [5]$$

This assumption can however be criticized, even if the rate with which a submicelle leaves a micelle does not depend upon the number of quenchers per micelle the probability that this submicelle contains a quencher will depend upon the number of quenchers per micelle. If this probability would be proportional to the number of quenchers per micelle the distribution would again be given by equation 5.

A combination of the assumption of Dorrance and Hunter and that of Rothenberger et al. would lead to an even more complex distribution function where P_n can no longer be given by a closed analytical expression (16b).

4.   Distribution of the quencher (or probe) between the solvent and the
      micelle

When the distribution of a probe or quencher between the solvent and the micellar phase is considered a distinction must be made for aqueous micelles between neutral (organic) molecules and charged species. For the neutral species a detailed model was developed by Almgren and Grieser[5a]. As these molecules are not subjected to long range (electrostatic) interactions a simple two-phase model can be used, in which they are either bound to the micelle or completely free in the solvent. Binding to the micelle occurs by hydrophobic interactions. This is demonstrated by the fact that the binding constant increases exponentially with the area of the hydrophobic part of the molecule(17). The electrochemical potential of the neutral species in the solvent, outside the micelle does not depend upon the distance from the micelle. Using the cell-model (18) for the micellar solution this means that r_b, the distance from the center of the micelle - where the mean first passage time for a quencher to become bound again to the micelle, equals the time to migrate to the cell boundary (r_c) - is very close to the radius of the micelle (r_b). The influence of the micellar concentration on the exchange rate of quenchers or probes during the excited state decay time indicated that the exchange occurs by the exit of a quencher to the solvent and the independent entrance of the quencher or probe into another micelle.

For ionic species the situation is strongly different. Here the ionic quenchers move around in the field of the electrical potential due to the charge of the micelle, the counterions and eventual coïons (free surfactant or added salt). The distance dependence of this field can be obtained by solving a non-linear Poisson-Boltzman equation (18,19). Under all experimental circumstances divalent counterions will be strongly bound to the micelle, the divalent counterions should be considered as quenchers that are totally micellized. The exchange of divalent ions during the excited state decay time should rather be considered as the migration of the quencher from a micelle to the micelle in a neighboring cell. The radius r_b (cfr supra) will be very close to r_c (the radius of the cell). At no moment during the exchange process can the quencher be considered as diffusing freely. When a quencher or probe leaves the influence of one micelle it has a much larger probability to become bound to the parent or a neighboring micelle than to leave the cell of either the parent and the neighboring micelle. The rate with which the counterions become free from the electric field of the micelle is given by equation 6

$$v_n^{-\prime} = nk^{-\prime}[M_n] \qquad\qquad [6a]$$

where $k^{-\prime}$ is the rate with which a bound counterion becomes free from the electrostatic interaction of the micelle. $k^{-\prime}$ will be larger than k^{-} as a fraction of the quenchers, that become free from the long range interaction of one micelle bind, again to the parent micelle before leaving the cell. If a fraction γ of the ions that become free from the electrostatic attraction of the parent micelle is captured by the Coulomb field of a neighboring micelle the rate with which the ions are captured by the neighboring micelle is given by equation 6b where N_{coor} represents the coordination number of the packing of the cells.

$$n\gamma k^{-\prime}[M_n]/N_{coor} \qquad\qquad [6b]$$

For divalent ions γ will be very close to one.

As each cell is surrounded by N_{coor} cells the total rate with which quenchers coming from other micelles bind to a micelle is given by equation 7

$$v^+ = \mu\gamma k^{-1} \qquad\qquad [7]$$

For monovalent quenchers the situation is intermediate between that of neutral molecules and multivalent ions (20). Formally (21) the fluorescence quenching can be described by the same equations as those used for the neutral molecules. However the different rate constants will depend on the concentration of surfactant as this parameter influences the potential drop over the Gouy Chapman layer surrounding each micelle.

Reversed micelles (22) on the other hand are neutral entities present in a dielectric medium. Since probes and quenchers are frequently ionic (8a,23) or polar uncharged (8b,24) species the same formalism as used for the uncharged probes and quenchers in aqueous micelles can be applied. Below the percolation threshold (25) ionic species must be considered as nearly completely micellized. They will only be able to exchange during micellar collisions (26,27), which leads to a situation intermediate between the exchange and a fusion-fission process (28). In principle they should also be able to leave the micelle as ion pairs (29) (with eventually a surfactant as counterion) or as part of a larger aggregate (30) or submicelle.

5.   Immobile probe and quenchers

5.1   Monodisperse micelles

5.1.1. The probe and the quencher are totally micellized

When the probe and quencher do not exchange between the micelles, the time dependence of the concentration of the excited state in a micelle containing n quenchers is given by equation 8

$$\frac{d[M_n^*]}{dt} = -(k_0 + nk_q)[M_n^*] \qquad\qquad [8]$$

This leads to the expression of equation 9 for the fluorescence decay of an excited probe in a micelle with n quenchers

$$[M_n^*](t) = [M_n^*](0)\exp[-(k_0 + nk_q)t \qquad\qquad [9]$$

Assuming a Poisson distribution of quenchers the total fluorescence decay of the solution is then given by equation 10.

$$I(t) = \Sigma[M_n^*](t)$$
$$= [M^*](0) \exp[-k_0 t - \mu(1-\exp-k_q t)] \qquad [10]$$

Equation 10 is equivalent to following expression.

$$I(t) = A_1 \exp\{-A_2 t - S_3[Q][(1-\exp(-A_4 t)]\} \qquad [11]$$

As S_3 equals $1/[M]$ when the quencher is totally micellized a simultaneous analysis (31) of several fluorescence decays obtained at different quencher concentrations allows a fast and accurate determination of $1/[M]$ and of the aggregation number.

The assumptions leading to equation 10 and 11 were proved to be valid for the fluorescence quenching of naphthylacetic acid by nitrate and iodide ions (8b) or of acridinium (28) by Co^{2+} in reversed micelles of AOT (sodium dihexyl sulfosuccinate) in heptane at low R-values (R is the ratio of the number of ware molecules to the number of surfactant molecules). Under these conditions an aggregation number of 53 was obtained R = 6

A situation where the quencher and the probe are considered immobile while the quencher is not totally micellized is not realistic as this would always lead to values of μk^- which are of the order of k_q or k_0.

5.1.2. The quenching process is reversible

If the quenching process is reversible, as is e.g. the case for excimer and exciplex (32) formation the kinetic scheme given in section 2 has to be extended by following steps

$$M_n^* + n\,Q \xrightarrow{nk_{fo}} (M_nQ)^* : \qquad \text{exciplex (excimer) formation}$$

$$(M_nQ)^* \xrightarrow{k_{di}} M_n^* + n\,Q: \qquad \text{exciplex (excimer) dissociation}$$

$$(M_nQ)^* \xrightarrow{k_1} M_n + n\,Q: \qquad \text{exciplex (excimer) decay}$$

The fluorescence of an excited probe in a micelle containing n quenchers is given by equation 12

$$[M_n^*] = \frac{[M_n^*](0)\{(k_1+k_{di}-\lambda_{1n})\exp(-\lambda_{1n}t)-(k_1+k_{di}-\lambda_{1n})\exp(-\lambda_{2n}t)\}}{\lambda_{2n}-\lambda_{1n}} \quad [12]$$

where λ_{1n} and λ_{2n} are the roots of

$$\lambda_n^2-[k_0+k_1+nk_{fo}+k_{di}]\lambda_n+k_0(k_1+k_{di})+nk_{fo}k_1= 0 \quad [13]$$

Eventual ionic dissociation of the exciplex (13, 34c) would lead to an increase of k_1. The direct irreversible formation of a solvated ionpair (k_{et}) would introduce an additional irreversible quenching process (2b). This process (4,13,34) would occur with a rate nk_{et}. The equations 13-16 remain valid when k_0 is replaced by k_0+nk_{et}. The total fluorescence intensity at a time t is given by $\Sigma[M_n^*]$. If the probe and the quencher are different molecules $[M_n^*](0)$ equals $[M^*](0)P_n$ and the fluorescence decay of the excited probe is given by equation 14

$$I(t) = [M^*](0)\Sigma P_n\frac{\{(k_1+k_{di}-\lambda_{1n})\exp(-\lambda_{1n}t)-(k_1+k_{di}-\lambda_{1n})\exp(-\lambda_{2n}t)\}}{\lambda_{2n}-\lambda_{1n}} \quad [14]$$

When intramicellar excimer formation is considered the probe and the quencher are identical molecules, in that case equation 14 has to be replaced by equation 15:

$$I(t) = [M^*](0)\Sigma nP_n\frac{\{(k_1+k_{di}-\lambda_{1n})\exp(-\lambda_{1n}t)-(k_1+k_{di}-\lambda_{1n})\exp(-\lambda_{2n}t)\}}{\mu(\lambda_{2n}-\lambda_{1n})} \quad [15]$$

The fluorescence decay of the excimer is under those conditions given by:

$$I(t) = [M^*](0)\Sigma n(n-1)P_n\frac{\{(k_{fo}(\exp(-\lambda_{1n}t)-\exp(-\lambda_{2n}t)\}}{(\lambda_{2n}-\lambda_{1n})} \quad [16a]$$

If the decay of an intramicellar exciplex is considered equation 16a has to be replaced by equation 16b

$$I(t) = [M^*](0)\Sigma P_n\frac{\{(k_{fo}(\exp(-\lambda_{1n}t)-\exp(-\lambda_{2n}t)\}}{\mu(\lambda_{2n}-\lambda_{1n})} \quad [16b]$$

The fluorescence decays obtained for intramicellar exciplex (12c,33) and excimer (13) formation in aqueous micelles could be analyzed in the framework of respectively equations 14 and 15. Although intramicellar excimer (35) and exciplex (24) formation was also observed in reversed micelles the experimental conditions only allowed analysis of the fluorescence decays of the locally excited state as a one exponential decay. This was due on the one hand to the fact that for the systems investigated k_{di} was always considerably larger than k_1 and on the other hand to fast exchange of the quencher between the micellar aggregates and the solvent.

5.2 The influence of micelle polydispersity

The decay laws derived in the preceding section are useful to determine micellar properties as long as the micelles are relatively monodisperse. Considering reversed micelles this will be the case for anionic surfactants (35). The application of equations 10 and 11 in polydisperse systems can however give rise to less reliable micellar parameters (3).

In polydisperse systems the fluorescence decay can be considered as the superposition of fluorescence decays $I_s(t)$ of micelles with different aggregation numbers s. As in each ensemble of micelles with a given aggregation number a Poisson distribution is obtained for the quenchers the fluorescence decay of micelles with aggregation number s is given by equation 17 for a completely micellized quencher:

$$I_s(t) = [M_s^*](0)\exp\{-k_0t-[Q]s[1-\exp(-k_q(s)t)]/([surf]-cmc)\} \qquad [17]$$

Replacing $[Q]/([surf]-cmc)$ by $S_3^{\frac{1}{3}}$ equation 21 can be simplified to equation 18

$$I_s(t) = [M_s^*](0)\exp\{-k_0t-S_3^{\frac{1}{3}}s[1-\exp(-k_q(s))t)]\} \qquad [18]$$

According to Almgren et al. two situations must be considered, a "static" picture where the residence time of probe and quencher in the micelle is much shorter than the lifetime of the micelle and a "dynamic picture" where the lifetime of the micelle is much shorter than the residence time of the probe and the quencher. Although polydispersity has up to now not been considered for reversed micelles it could become important for cationic reversed micelles where the aggregation number depends on the surfactant concentration.

5.2.1 The "static" case

In this case the probability to excite a probe in a micelle with aggregation number s is proportional to the actual aggregation number, s, of the micelle. This leads to equation 19 for the fluorescence decay $I(t)$

$$I(t) = [M^*](0)\Sigma\ sP_{sw}I_s(t) \tag{19}$$

where $[M^*](0)$ equals the total concentration of excited probes at $t = 0$; P_{sw} equals $P_s/\Sigma sP_s$ and P_s equals the probability that the aggregation number of a micelle equals s.

Combining equation 18 and 19 yields equation 20

$$I(t) = [M^*](0)\ \Sigma sP_{sw}\exp\{-k_0t-S\tfrac{1}{2}s[1-\exp(-k_q(s)t)]\} \tag{20}$$

Fluorescence decays assuming $k_q(s)$ to be proportional to the inverse of s, ($k_q(s)= k_q'/s$), and assuming a gaussian or an exponential distribution for P_s were obtained by Almgren. For the intercept of $\ln(I(t))$ at long times ($\ln(I(0)-\ln(I(\infty))$) and for $dI(t)/dt$ at time zero analytical expressions could be obtained. The inverse proportionality of k_q and s made it impossible to describe the fluorescence decays by an analytical expression. However assuming that $k_q(s)$ is inversely proportional to s, the initial decay rate constant is still given by $k_0 + k_q'[[Q_m]/[S_m]]$. This indicates that the expression for the fluorescence decay will approach equation 10 with μ and k_q replaced by respectively $<s>_Q$ and $k_q'[Q_m]/(<s>_Q[S_m])$.

$$<s>_Q= \ln\{I(0)/I_\infty(0)\} = \ln\Sigma sP_{sw}-\ln\Sigma s\exp\{[Q_m]/[S_m]s\}P_{sw} \tag{21}$$

$[Q_m]$: Concentration of the micellized quencher
$[S_m]$: Concentration of the micellized surfactant

For low ratio's of the concentration of the quencher and the surfactant, $<s>_Q$ becomes equal to the weight average aggregation number.

For a gaussian distribution with width σ ,$<s>_Q$ is given by equation 22

$$<s>_Q = <s>_N + \sigma^2(1/<s>_N - [Q_m]/\ 2[S_m])$$

$$= <s>_W - \sigma^2[Q_m]/2[S_m] \tag{22}$$

This means that for low values of $[Q_m]$ over $[S_m]$ the aggregation number approaches $<s>_W$ and becomes smaller when $[Q_m]$ over $[S_m]$ increases.

For an exponential distribution of the aggregation number $\langle s \rangle_Q$ is given by equation 23

$$\langle s \rangle_Q = s_0 + [S_m]\{2 \ln(1+[Q_m]\sigma/[S_m]) - \ln(s_0+\sigma+[Q_m]s_0\sigma/[S_m]) - \ln(s_0+\sigma)\}[S_m]$$

$$[23]$$

where s_0 is the the minimal value of the aggregation number.

For both cases the initial decay rate constant is given by $k_0 + k_q'[Q_m]/[S_m]$ and is thus independent of σ and the shape of the distribution.

5.2.2. The "dynamic" case

In this case the Poisson distribution of quenchers is not only valid for each ensemble of micelles with a given aggregation number but also for the total distribution of quenchers the average number of quenchers per micelle follows a Poisson distribution determined by $[Q_m]\langle s \rangle/[S_m]$ and hence $\langle s \rangle_Q$ equals $\langle s \rangle_N$. Contrary to the static limit this equality holds for all values of $[Q_m]/[S_m]$. The fluorescence decay will however not longer be given by Tachiya's expression (1b,4b) due to the dependence k_q on s and equation 11 has to be replaced by

$$I(t) = \Sigma s P_{sw} \exp\{-k_0 t - [Q_m]\langle s \rangle/[S_m][1-\exp(-k_q t/s)]\}/\Sigma s P_{sw}$$ [24]

6. Mobile probe or quencher
6.1 The quencher is partially micellized
6.1.1. Assuming a Poisson distribution of quenchers

For a neutral quencher that is only partially micellized (4) the entrance rate of quenchers from the solvent can be of the order of magnitude of k_0. In that case equation 8 has to be extended by equation 1 and 2.

$$\frac{d[M_n^*]}{dt} = -(k_0 + nk_q + nk^-)[M_n^*] + k^+[M_{n-1}^*][Q_{aq}]$$ [25]

Since for each value of n an equation of the type of equation 25 can be obtained the equations giving $I_n(t)$ are the solutions of an infinite system of linear differential equations with a generating function given by equation 26.

$$G(s,t) = \Sigma I_n(t)s^n$$ [26]

In this case I(t) is given by G(1,t)

$$I(t) = \exp\{-(k_0 + S_2[Q])t - S_3[Q][1-\exp(-(P_4)t)]\} \qquad [27]$$

$$\text{with } S_2 = \frac{k_q k^+}{(k^- + k_q)(1+K[M])} \qquad [28]$$

$$S_3 = \frac{k_q^2}{(k^- + k_q)^2(1/K + [M])} \qquad [29]$$

$$P_4 = k^- + k_q \qquad [30]$$

From the ratio of S_3/S_2 and P_4, k^- and k_q can be obtained. This allows to obtain "corrected" values of S_2 and S_3: S_{2c} and S_{3c}.

$$\text{with } S_{c2} = \frac{k^+}{(1 + K[M])} = \frac{S_2(k^- + k_q)}{k_q} \qquad [31]$$

$$S_{3c} = \frac{1}{(1/K + [M])} = \frac{S_3(k^- + k_q)^2}{k_q^2} \qquad [32]$$

A plot of S_{3c}^{-1} versus the difference between the surfactant concentration and the cmc yields a linear relationship with a slope equal to $1/N_{AGG}$ and an intercept equal to $1/K$. A plot of S_{2c}^{-1} versus the difference between the surfactant concentration and the cmc yields a linear relationship with a slope of $1/k^-$ and an intercept of $1/k^+$. If this kinetic model is valid the values of k^-, k^+ and K must be in agreement with $K = k^+/k^-$.

For neutral quenchers the kinetic model given in this section can be applied for different probe-quencher combinations in micellar micellar systems. It allows a simultaneous determination of k_q, k^-, k^+ and the aggregation number (36). Also the quenching of pyrenesulfonic acid (PSA) (36) by nitromethane and diiodomethane in reversed micellar systems should in principle be described by equation 27-29. However due to the fact that in the latter two systems k^- was considerably larger than k_q the observed fluorescence decay was, for the range of quencher concentrations investigated mainly exponential.

6.1.2. Assuming a binomial distribution of quenchers

For a binomial distribution of the quenchers over the micelles the fluorescence decay is given by equation 33:

$$[M^*](t)=[M^*](0)\exp(-k_0t)\left\{\frac{(1-s_2)(m-\mu+\mu s_1)}{m(s_1-s_2)}\exp\left[-\frac{\mu(1-s_1)}{m-\mu}k^-t\right]\right.$$

$$\left.-\frac{(1-s_1)(m-\mu+\mu s_2)}{m(s_1-s_2)}\exp\left[-\frac{\mu(1-s_1)}{m-\mu}k^-t\right]\right\}^m \qquad [33]$$

where s_1 and s_2 are the roots of equation 34

$$s^2k^-+[(m-\mu)(k^-+k_q)-\mu k^-]s-(m-\mu)k^-= 0 \qquad [34]$$

6.2. Exchange of the quencher
6.2.1. Assuming a Poisson distribution of quenchers

In addition to the exchange of the quencher between the micelle and the solvent quenchers can be exchanged (9,37,38) between two micelles. If one assumes that the rates with which this process occurs is proportional to the number of quenchers present in the micelle from which the quencher leaves and independent of the number of quenchers in the micelle in which the quencher enters the rate with which a quencher leaves a micelle containing n quencher for a micelle containing $j-1$ quenchers is given by $nk_{ex,q}[M_n^*][M_{j-1}]$

The total rate with which quenchers leave a micelle containing n quenchers is, assuming a Poisson distribution of quenchers , given by equation 35

$$v_{ex}^- = \sum_j nk_{ex,q}[M_n^*][M_j] \qquad [35]$$

$$= k_{ex,q}[M]\Sigma n[M_n^*]$$

Equation 35 is equivalent to equation 1 with k^- replaced by $k_{ex,q}[M_j]$.

The rate with which a quencher enters a micelle containing n quenchers from a micelle containing j+1 quenchers is than given by $(j+1)[M_n^*][M_{j+1}]$

The total rate with which a quencher enters a micelle containing n quenchers is then given by equation 36 or 37

$$v_{ex}^+ = k_{ex,q} \Sigma(j+1)[M_n^*][M_{j+1}]$$

$$= k_{ex,q}[M_n^*][Q]K/(1+K[M]) \tag{36}$$

or

$$v_{ex}^+ = k_{ex,q}[M_n^*][Q_m] \tag{37}$$

with $[Q_m]$ the concentration of the quenchers bound to a micelle. Equation 36 and 37 are equivalent to equation 2 with $k^+[Q_{aq}][M]$ replaced by $k_{ex,q}[Q_m]$. Therefore a solution of the same type as equation 27 can be expected

$$I(t) = \exp\{-(k_0 + S_2[Q])t - S_3[Q][1-\exp(-(P_4)t)]\} \tag{38}$$

with
$$S_2 = \frac{\beta k_{ex,q} k_q}{(k_{ex,q}[M] + k_q)} \tag{39}$$

$$S_3 = \frac{\beta k_q^2}{(k_{ex,q}[M] + k_q)^2} \tag{40}$$

$$P_4 = k_{ex,q}[M] + k_q \tag{41}$$

In equations 38-41 β corresponds to the fraction of quenchers bound to the micelle. For ionic quenchers solubilized in reversed micelles β will be very close to one. Originally this exchange process was interpreted as a collision between two micelles followed by an exchange of a quencher. Although such an interpretation remains valid for reversed micelles one should be careful in applying it to the quenching by counterions of aqueous micelles (18).

Equation 28 and 39 can actually be replaced by the common expression given in equation 42:

$$S_2 = \beta k_t^- k_q/(k_t^- + k_q)[M] \tag{42}$$

with
$$k_t^- = k^- + k_{ex,q}[M] \tag{43}$$

In the same way equation 29 and 40 can be replaced by equation 44

$$S_3 = \frac{\beta k_q^2}{(k_t^- + k_q)^2 [M]} \qquad [44]$$

For reversed micelles and hydrophobic quenchers solubilized in aqueous micelles, where the binding of the quenchers requires an adsorption of the quenchers on the micelles, a distinction between k^- and $k_{ex,q}$ can easily be made . This is however no longer the case for counterions that are subjected to long range electrostatic interactions with the micelle.

When k_t^- becomes large compared to k_q $S_3[Q]$ approaches zero (39) and $S_2[Q]$ will be given by equation 45.

$$S_2[Q] = \beta k_q k_t^- [Q] / \{[M](k_q + k_t^-)\} \qquad [45]$$

As β is given by $K[M]/(1+K[M])$ equation 45 can be written as

$$S_2[Q] = K k_q k_t^- [Q] / \{(1+K[M])(k_t^- + k_q)\} t$$

$$= k_q k_t^- [Q] / \{(1/K+[M])(k_t^- + k_q)\} \qquad [46]$$

The fluorescence decay is in this case given by equation 47

$$I(t) = P_1 \exp\{ -P_2 - k_q k_t^- [Q] t / [(1/K+[M])(k_t^- + k_q)]\} \qquad [47]$$

If k_t^- is much larger than k_q or if $k_{ex,q}[M]$ is much larger than k^-, a plot of the decay rate constant versus the quencher concentration will yield a linear relationship with a slope given by $k_q/(1/K+[M])$. Plotting the inverse of this slope versus the micellar concentration yields again a linear relationship with a slope equal to $1/k_q$ and an intercept equal to $1/k_q K$.

This monoexponential decay was originally observed (39,40) for quenchers that bind weakly to the micelle (e.g. Hg_2Cl_2) or inefficient quenchers (Cs^+). It can also be observed for the quenching by coions. Initially, small values of k_q and the limitation of the experimental conditions to low concentrations of the quenchers lead to the opinion that fluorescence quenching in reversed micelles always had to be described in the framework of equation 47. The use of more efficient quenchers and the extension to larger quencher concentrations indicated that at low and moderately values of R the nonexponential decay (equation 11 or 38) was more appropriate However at sufficiently large values of R the fluorescence decay could under all experimental conditions be described by equation 47.

Experimentally it can sometimes be difficult to obtain reliable decay parameters to an observed fluorescence decay when the ratio k_q/k_t^-, although small, is still too large to allow the use of equation 47. In this case the use of global analysis, which proved to give reliable decay parameters even for very small values of $S_3[Q]$, can be expected to be a valuable tool to obtain the different decay parameters (31).

6.2.2 Assuming a binomial distribution of quenchers

In this case the rate with which a quencher migrates from a micelle containing an excited probe and n quenchers to a micelle containing j quenchers is given by equation 4 8.

$$v_{ex,q}^- = k_{ex,q} n(1-j/m)[M_n^*][M_j] \tag{48}$$

The rate with which a micelle containing an excited probe and n quenchers obtains a quencher from a micelle containing j quenchers is given by equation 49

$$v_{ex,q}^+ = k_{ex,q} j(1-n/m)[M_n^*][M_j] \tag{49}$$

The fluorescence decay of the ensemble of micelles is than given by equation 50

$$[M_m^*](t)=[\overset{*}{M}_m](0)\exp(-k_0 t)\left\{ \frac{(1-s_2)(m-\mu+\mu s_1)}{m(s_1-s_2)}\exp\left[-\frac{\mu(1-s_1)}{m-\mu}k_{ex,q}[M]t\right]\right.$$

$$\left.-\frac{(1-s_1)(m-\mu+\mu s_2)}{m(s_1-s_2)}\exp\left[-\frac{\mu(1-s_1)}{m-\mu}k_{ex,q}[M]t\right]\right\}^m \tag{50}$$

where s_1 and s_2 are the roots of equation 51

$$s^2 k_{ex,q}[M]+\{(m-2\mu)k_{ex,q}[M]+mk_q\}s-(m-\mu)k_{ex,q}[M]= 0 \tag{51}$$

According to Iwamura et al. (41) equation 50 and 51 should be used rather than equation 38 to describe the fluorescence decay of pyrene quenched by metal ions in aqueous micelles. Although this limiting value of the number of solubilizates per micelle is a realistic (15) boundary condition in reversed micelles at low values of R no attempts have been made up to now to analyze the fluorescence quenching in reversed micelles in this framework.

6.2.3. Assuming a reversible quenching process

If the quenching is reversible as can be the case for exciplex or excimer formation and if k_t^- is considerably larger than k_{fo} the rate of complex $((QM)^*)$ formation will mainly be determined by the rate with which quenchers bind to the micelle

$$\frac{d[M^*]}{dt} = -\{k_0+k_{fo}k_t^-[Q]/[(1/K+[M])(k_t^-+k_q)]\}[M^*]+k_{di}[(MQ)^*] \qquad [52]$$

$$\frac{d[(MQ)^*]}{dt} = k_{fo}k_t^-[Q]/[(1/K+[M])(k_t^-+k_q)][M^*]-(k_{di}+k_1)[(MQ)^*] \qquad [53]$$

Under those conditions $I(t)$ is given by equation 54.

$$I(t) = I(0)\frac{(k_1+k_{di}-\lambda_1)\exp(-\lambda_1 t)-(k_1+k_{di}-\lambda_1)\exp(-\lambda_2 t)}{\lambda_2-\lambda_1} \qquad [54]$$

$$\lambda^2-[k_0+k_1+k_{fo,eff}+k_{di}]\lambda+k_0(k_1+k_{di})+k_{fo,eff}k_1 = 0 \qquad [55]$$

where $k_{fo,eff}$ equals $k_{fo}k_t^-[Q]/[(1/K+[M])(k_t^-+k_q)]$.

The fluorescence decay of the exciplex, $I_E(t)$, is under those conditions given by

$$I_E(t) \approx \frac{k_{fo,eff}[\exp(-\lambda_1 t)-\exp(-\lambda_2 t)]}{\lambda_2-\lambda_1} \qquad [56]$$

The fluorescence decays obtained for intramicellar exciplex formation between ω-naphthyl alkanoic acids and triethylamine or m-DCB (meta-dicyanobenzene) solubilized in AOT reversed micelles can be analyzed (24,42) according to equation 54-56.

6.3. Exchange of the probe

When a quencher is exchanged the number of quenchers in micelle containing an excited probe and n quenchers can only change to n-1 or n+1. When however the excited probe is exchanged (43) between a micelle containing n quenchers and a micelle containing m quenchers the number of quenchers that can quench the probe change from n to m. This rate will be proportional with the micellar concentration and with the probability that the micelle with which the probe is exchanged contains m quenchers.

The change of the number of micelles containing an excited probe and n quenchers, due to this process, is given by equation 57 and 58.

$$v_{ex,p}^- = k_{ex,p} [M_n^*]\sum_j[M_j]$$

$$= k_{ex,p} [M_n^*][M] \qquad [57]$$

$$v_{ex,p}^+ = k_{ex,p}[M_n]\sum_j[M_j^*]$$

$$= k_{ex,p}[M]P_n\sum_j[M_j^*] \qquad [58]$$

Equation 57 and 58 differ in a qualitative way from equation 1 and 2. This has as a consequence that it is no longer possible to obtain an analytical solution for $G(1,t)$. Equation 25 has to be replaced by equation 59

$$\frac{d[M_n^*]}{dt} = -(k_0+nk_q+k_{ex,p}[M])[M_n^*]+e^{-\mu}\mu^n[M]k_{ex,p}\sum[M_j^*]/n! \qquad [59]$$

Contrary to what happens when the quenchers are exchanged between the micelles, the quenchers are no longer Poisson distributed over the micelles at all times after excitation. However as long as one can assume that all times after excitation the standard deviation of this distribution ($\sigma^2(\mu(t))$) is proportional to the average number of quenchers per micelle $\mu(t)$, the fluorescence decay can be approximated by an expression similar to equation 11. The meaning of A_2, S_3 and A_4 has however been changed.

$$S_2 = k_0 + \langle x\rangle_s k_q \qquad [60]$$

$$S_3 = \mu(1-\langle x\rangle_s/\mu)^2 \qquad [61]$$

$$A_4 = k_q/(1-\langle x\rangle_s/\mu) \qquad [62]$$

$\langle x\rangle_s$ is the average number of quenchers per micelle in the exponential,part of the decay. When the quenchers are exchange this number is given by $\mu k_q/(k_q+k^-)$. For the exchange of the probe a analytical expression for $<x>_s$ as a function of μ, k_q and $k_{ex,p}$ does not exist. Contrary to what is observed when the quencher migrates or is exchanged the ratio of $<x>_s/\mu$ decreases when μ becomes larger. This effect will become more important for larger values of $k_{ex,p}/k_{ex,q}$. For values of $k_{ex,p}/k_{ex,q}$.close to one $<x>_s$ decreases about 20 % when μ increases from zero to five. For the quenching of large hydrophobic molecules (e.g. pyrene or methyl-pyrene) by ions or small and less hydrophobic organic molecules $k_{ex,p}/k_{ex,q}$ will always be smaller than one one and the values obtained for the intramicellar rate constant and the aggregation number will not deviate strongly from the correct values. However in reversed micelles the exchange of the hydrophobic probe will often be faster than that of the often hydrophilic quencher leading to important differences between $k_q^2/(k_t + k^-)^2$ and $<x>_s/\mu$ at high quencher concentrations. When the fluorescence quenching of an aromatic molecule, eventually carrying anionic or cationic substituents, is investigated in reversed micelles it often occurs that the probe is bound to the interface while the ionic quenchers are present in the water pool of the micelle. Under those conditions exchange of the less hydrophobic probe can occur faster than that of the quenchers. To avoid distortions of the fluorescence decay from equation 38 due to the qualitative difference between the exchange of an excited probe and that of a quencher it is necessary to keep the average number of quenchers per micelle below one (43,44). Whether a combined exchange of probe and quencher prevails in reversed micelles over a fusion-fission process will be discussed in the next section.

6.4. Fusion-fission processes

When "sticky" collision between micelles are possible the number of quenchers in a micelle containing an excited probe can also change by a fusion fission process. When a collision between a micelle containing an excited probe and n quenchers and a micelle containing an excited probe and j quenchers is considered, the excited probe can be after the collision in a micelle containing any number of quenchers between zero and j+n.

If it can be assumed that the actual distribution of the quenchers over the micelles, which for t>0 is no longer a Poisson distribution, follows a distribution characterized by a standard deviation ($\sigma^2\mu(t)$) proportional to the average number of quenchers per micelle $\mu(t)$ at all times after excitation, the fluorescence decay of the ensemble of micelles can be given by an expression identical to equation 11. The different decay parameters are given by equation 60-62. However contrary to the previous case $<x>_s$ can be approximated by equation 63

$$\langle x \rangle_s = k_{fus}\{1-exp(-\mu/2)\}/k_q \qquad\qquad [63]$$

In agreement with the previous model $\langle x \rangle_s/\mu$ decreases when μ becomes larger. For the same value of k_{fus}/k_q the decrease of $\langle x \rangle_s/\mu$ is however slower than when only the probe migrates. While the exchange of probes or quenchers is likely to occur at micellar collisions for probes and quenchers present at the interface of reversed micelles the fusion-fission mechanism is according to Eicke (10b) and Zana (23a) predominant for ionic probes and quenchers solubilized in the water core of the reversed micelle.

6.5. The probe is partially micellized

Two possibilities can be distinguished.:

6.5.1. The quencher is only present in the solvent

For the concentrations of the excited probe in the micellar phase, $[P_m^*]$ and in the solvent $[P_{aq}^*]$ the following expressions can be written:

$$\frac{d[P_m^*]}{dt} = -(k^-+k_{0m})[P_m^*]+k^+[M])[P_{aq}^*] \qquad\qquad [64]$$

$$\frac{d[P_{aq}^*]}{dt} = k^-[P_m^*]-(k^+[M]+k_{0aq}+k_q[Q])[P_{aq}^*] \qquad\qquad [65]$$

with boundary conditions

$$[P_m^*](0) = [P^*](0)\frac{K[M]}{1+K[M]} \qquad\qquad [66]$$

$$[P_{aq}^*](0) = [P^*](0)\frac{1}{1+K[M]} \qquad\qquad [67]$$

$[P_{aq}^*]$: concentration of the excited probe in the solvent.
$[P_m^*]$: concentration of the excited probe in the micellar phase.
$[P^*] = [P_{aq}^*]+[P_m^*]$: total concentration of the excited probe
k_{0m}: decay time of the probe in the absence of quenchers in the micellar phase.
k_{0aq}: decay time of the probe in the absence of quenchers in the solvent.

The solution of equation 64 and 65 leads to a two exponential decay of the excited probe. Assuming an identical molar extinction coefficient and fluorescence rate constant for $[P_m^*]$ and $[P_{aq}^*]$ the fluorescence decay of the excited probe is given by equation 68

$$I(t) \approx [P^*] = [P^*](0)\{(C-\frac{B-A}{2})\exp[(\frac{B-A}{2})t]-(C-\frac{B+A}{2})\exp[(\frac{B+A}{2})t]\} \qquad [68]$$

A,B and C are given by

$$A = \{(k_{0aq}+k_q[Q]+k^+[M]-k_{0m}-k^-)^2-k^+k^-[M]\}^{1/2} \qquad [69]$$

$$B = -(k_q[Q]+k^+[M]+k_{0aq}+k_{0m}+k^-) \qquad [70]$$

$$C = -\{k_q[Q]+k^+[M]+k^-+(k_{0aq}+k_{0m})/2\} \qquad [71]$$

At low micellar concentrations and high quencher concentrations $k^+[M]$ is much smaller than $k_q[Q]$. Under those conditions k^- becomes the rate determining step of the quenching and the fluorescence decay is given by equation 72

$$I(t) \approx [P^*](0)\exp[-(k_{0m}+k^-)t]\{1+\frac{k^-}{k_{0aq}+k_q[Q]}[1-\exp(-(k^-+k_q[Q])t)]\} \qquad [72]$$

At even higher quencher concentrations $k_q[Q]$ becomes considerably larger than k_{0aq} and the fluorescence decays exponentially with a decay time equal to $(k_{0m}+k^-)^{-1}$.

These equations can also be applied to the decay of excited triplet states quenched by quenchers in the solvent (e.g. the system anthracene/cationic surfactant/ Cu^{++} or hydroxyphenylacetic acid/AOT/Tb^{3+}) (4a,5a,10b,45). While the situation described in this section can be encountered in aqueous micelles , especially when triplet quenching by coions is considered, no examples are known up to now for reversed micelles.

6.5.2. The quencher is only present in the micellar phase.

This situation resembles the situation where the probe is exchanged between two micelles. The fluorescence decay of the excited probe is governed by the following equations

$$\frac{d[P^*_{m,n}]}{dt} = -(k^-+k_{0m}+nk_q)[P^*_{m,n}]+P_nk^+[M][P^*_{aq}]$$ [73]

$$\frac{d[P^*_{aq}]}{dt} = k^-[P^*_m]-(k^+[M]+k_{0aq})[P^*_{aq}]$$ [74]

P_n: the probability that a probe contains n quenchers
$[P^*_{m,n}]$: the concentration of micelles with an excited probe and n quenchers
$[P^*_m]$: the total concentration of micelles with an excited probe
The boundary conditions are given by equation 66 and 67.

Summing over all values of n equation 73 becomes:

$$\frac{d[P^*_m]}{dt} = -(k_{0m}+k^-)[P^*_m]-k_q\Sigma n[P^*_{m,n}]+k^+[M][P^*_{aq}]$$ [75]

Multiplying equation 73 by n and summing over all values of n yields equation 76

$$\frac{d\Sigma n[P^*_m]}{dt} = -(k_{0m}+k^-)\Sigma n[P^*_m]-k_q\Sigma n^2[P^*_{m,n}]+k^+\mu[M][P^*_{aq}]$$

[76]

Now $\langle x \rangle$ and $\langle x^2 \rangle$ can be defined as:

$$\langle x \rangle = \Sigma_n n[P^*_{m,n}]/[P^*_m]$$ [77]

$$\langle x^2 \rangle = \Sigma_n n^2[P^*_{m,n}]/[P^*_m]$$ [78]

$\langle x \rangle$ and $\langle x^2 \rangle$ are both a function of time; $\langle x \rangle$ changes from μ at time zero to a stationary value $\langle x \rangle_s$ at t = ∞.
Using equation 77 and 78 equation 75 and 76 can be written as:

$$\frac{d[P^*_m]}{dt} = -(k_{0m}+k^-)[P^*_m]-\langle x \rangle k_q[P^*_m]+k^+[M][P^*_{aq}]$$ [79]

$$\frac{d\Sigma n[P^*_m]}{dt} = -(k_{0m}+k^-)\langle x \rangle[P^*_m]-k_q\langle x^2 \rangle[P^*_m]+k^+\mu[M][P^*_{aq}]$$

[80]

Furthermore:

$$\frac{d\Sigma n[P_m^*]}{dt} = [P_m^*]\frac{d\langle x\rangle}{dt} + \langle x\rangle\frac{d[P_m^*]}{dt}$$ [81]

Using equation 79 and 80 in equation 81 leads to

$$\frac{d\langle x\rangle}{dt} = -(\langle x^2\rangle - \langle x\rangle^2)k_q + k^+[M](\mu - \langle x\rangle)[P_{aq}^*]/[P_m^*]$$ [82]

If one assumes that (as in the more classical systems) $\langle x^2\rangle - \langle x\rangle^2$, the standard deviation on the number of quenchers per micelle is at all times proportional to $\langle x\rangle$, $\langle x^2\rangle - \langle x\rangle$ can be replaced by $\alpha\langle x\rangle + (1-\alpha)\mu$. Note that if the probe is totally micellized and if only exchange of the quencher occurs, α equals one. If this approximation is used equation 82 becomes:

$$d\langle x\rangle/dt = -\alpha\langle x\rangle(k_q + k^+[M][P_{aq}^*]/[P_m^*]) - \mu\{(1-\alpha)k_q - k^+[M][P_{aq}^*]/[P_m^*]\}$$ [83]

Now equation 74, 75 and 83 determine the temporal behavior of $[P_{aq}^*]$, $[P_m^*]$ and $\langle x\rangle$. By solving this system of differential equations numerically the fluorescence decay can be obtained. If one assumes that k^- and $k^+[M]$ are sufficiently larger than the difference between k_{0aq} and k_{0m} and if the excited probe is mainly present in the micellar phase the fluorescence intensity will decay exponentially after some time with a decay rate constant given by equation 84

$$k_{0m} + \langle x\rangle_s k_q = \frac{k_{0m}K[M] + k_{0aq}}{1 + K[M]} - \frac{k^+[M]}{k_{0aq} + k^+[M]} + \frac{k_qK[Q]}{1 + K[M]}$$ [84]

where K is the distribution constant of the probe between the aqueous and the micellar phase. The situation described in this section could occur when the fluorescence quenching of aromatic molecules by ions or strongly polar molecules is considered in reversed micelles (33c,45,46).

7. The intramicellar quenching process (5,6)

When Fick's equation is solved for the boundary conditions (immobile probe in the center of the micelle and freely diffusing quencher) corresponding to a diffusion controlled intramicellar quenching process in a micelle containing one probe and one quencher, a solution in the form of a power series as given by equation 85 is obtained

$$g(t) = \sum_i \Gamma_i \exp[-\alpha_i Dt/A] \qquad [85]$$

where α_i and Γ_i are constants depending upon the encounter distance and the micellar length or radius. ($\sum \Gamma_i = 1$), D is the mutual diffusion coefficient of probe and quencher and A the area of the micellar surface

At long times (t> $A/\alpha_2 D$) equation 85 can be approximated by equation 86

$$g(t) = \Gamma_1 \exp[-\alpha_1 Dt/A] \qquad [86]$$

As g(t) gives the probability that a probe, excited at t = 0, has not been desactivated after a time t, in a micelle containing one quencher this probability amounts $g(t)^N$ in a micelle containing an excited probe and N quenchers. This means that one of the basic assumption leading to the derivation of the equations describing the fluorescence decay in a micellar solution is always valid, even if the simplification of replacing equation 85 by equation 86 may not be made. When equation 85 may not be replaced by equation 86 all the expressions given for a Poisson distribution of quenchers over the micelles remain valid as long as $\exp(-k_q t)$ is replaced by the correct expression for g(t).

In most kinetic models the intramicellar quenching process is described by a first order kinetics (exponential decay or equation 86). When the quenching is assumed to be a diffusion controlled process the intramicellar quenching rate constant can be estimated using the mean reaction time approximation for several micellar shapes and several locations of the probe and the quencher. The situation described by equation 85 is rather unlikely in aqueous micelles (5a). It can however be applied to the quenching of a negatively charged probe, e.g. pyrenetetrasulfonic acid (PTSA), by cationic quenchers in reversed anionic micelles (44) if one assumes that the quenching occurs by migration of the counterions rather than by that of the probe. In this case however also the influence of the electric potential present in the reversed micelle (8c,47) on the diffusion process has to taken into account. The expression given by equation 85 would also be valid for the quenching of an immobile cationic probe by an anionic quencher in a cationic reversed micelle (48)

If one of the reactants completely covers the surface (e.g. the counterion of the surfactant is a quencher (49)) of the spherical micelle with radius R while the other is diffusing freely in the interior of the micelle <k> is given by equation 87

$$\langle k_q \rangle = 15D/R^2 \qquad [87]$$

If one of the reaction partners is confined to the center of a spherical micelle of radius R and the other is allowed to diffuse freely in the micelle this rate constant is given by:

$$\langle k_q \rangle = \frac{3DR_{AB}[1-(R_{AB}/R)^3]}{R^3[1-9R_{AB}/5R+(R_{AB}/R)^3-(R_{AB}/R)^6/5]} \qquad [88]$$

R_{AB}: reaction distance

D: mutual diffusion coefficient of probe and quencher

As R_{AB} is between 5 and 8 Å for most quenching processes while R approaches 20 Å for most surfactants R_{AB}/R will be intermediate between 0.25 an 0.33. Therefore in equation 80 the higher powers of R_{AB}/R can be neglected versus unity. This leads to following simplification of equation 88:

$$\langle k_q \rangle = \frac{3DR_{AB}}{R^3[1-9R_{AB}/5R]} \qquad [89]$$

For the same values of D and R and for R_{AB}/R between 0.25 and 0.33 equation 87 leads to values of $\langle k_q \rangle$ that are fifteen to twenty times larger than those obtained by equation 88 or 89. This could be the case for the quenching of dimethylindolocarbazole solubilized in $Ni(LS)_2$ micelles (49). However the small value obtained for $\langle k_q \rangle$ could suggest that the quenching process is in this case slower than diffusion controlled.

In principle also situation described by equation 88 and 89 could be applied to the quenching of pyrene tetrasulfonic acid by cations in anionic reversed micelles (44,48) if modifications to the mean first passage time due to electrical potential differences inside the reversed micelles would be taken into account. However A numerical solution of the Poisson Boltzmann equation in reversed micelles, followed by a calculation of the mean first passage time is necessary to conclude whether the situation of equation 85 or that of equation 88 prevails.

If both reaction partners are present on the surface of a spherical micelle equation 87 should be modified to:

$$\langle k \rangle = \frac{D[1-(R_{AB}/2R)^2]}{R^2\{\ln[(2R/R_{AB})^2]-[1-(R_{AB}/2R)^2]\}} \qquad [90]$$

The situation described by equation 90, which prevails (5,11,20,36b) in aqueous micelles can also be encountered in reversed micelles if the probe and the quencher have an opposite charge (50,46) of the surfactant or if both species are contain a sufficiently hydrophobic (51) moiety (pyrene sulfonic acid).

If the probe and the quencher are present on a cylinder surface with radius R and length L $<k_q>$ is given (52,54) by equation 91

$$<k_q> = \frac{3hD}{(L/2-R_{AB})[3+h(L/2-R_{AB})]}$$ [91]

where h is given by:

$$h = \frac{3D\pi RR_{AB}}{(\pi R-R_{AB})^3}$$ [92]

If the equations 87 and 90-92 are compared one can observe that $<k_q>$ is always given as a product of the mutual diffusion coefficient, a factor depending sublinear on the ratio R_{AB}/R and the inverse of the area of the surface of the micelle

Equation 87 and 90-92 are also to a first approximation useful when the probe and/or the quencher are distributed over a certain radial distance. In that case the rate constants obtained by the expressions mentioned above must be multiplied by a factor κ, giving the probability that the difference between the radial position of the probe and the quencher is less than R_{AB}. Using this approximation equation 87 can also be applied if one of the reaction partners is confined to and diffusing very rapidly over the micellar surface while the other reaction partner diffuses in the micelle interior. In that case $<k_q>$ is given (11) by equation 93

$$<k_q> = 15\sigma D/R^2$$ [93]

where σ is the probability that the difference in the azimuthal angle of the probe and the quencher is less than $\arcsin(R_{AB}/R)$.

For a reaction controlled quenching process k_q is always given by equation 94

$$k_q = \sigma\kappa k_r$$ [94]

where k_r is the rate constant of the reaction in the encounter process. Equation 94 should e.g. be applied for the quenching by iodide ions in solutions of cationic surfactants (20,21). The expressions for κ and σ were calculated for cylindrical and spherical micelles.

The validity of the first order kinetics for an intramicellar diffusion controlled quenching process was investigated by several authors(5,6). For spherical micelles with a ratio of R_{AB}/R between 0.1 and 0.4 only a small fraction of the quenching occurred in the nonstationnary regime (5,6,52,54). Furthermore for values of D/A, D/LR or D/R^2 allowing efficient quenching by one or two quenchers during the fluorescence decay time of the probe all diffusion transients finished to contribute to the fluorescence decay after a time short compared to the fluorescence decay time of the probe. This indicates that in those systems diffusion transients only become important when the number of quenchers per micelle becomes very large. Only for very large values of R_{AB}/R the amplitude of the diffusion transient becomes important. This problem could e.g. arise in case of quenching by Förster energy transfer (53).

Distortions due to diffusion transients (54) become however increasingly important for when D/A or D/R^3 becomes too small to obtain important quenching during the fluorescence decay of the probe in an aggregate containing a small number of quenchers. This situation is equivalent to a situation where the diffusion length of the excited probe becomes smaller than the largest dimension of the micelle. This situation can e.g. occur in large cylindrical micelles whose aggregation number can in principle increase without limit. In this case a large number of quenchers per micelle is necessary to obtain a significant quenching of the fluorescence. This situation has been analyzed for a probe and a quencher present on the surface of a spherical micelle. g(t) is now given by equation 95 .

$$gt) = (2h^2L^2) \ \Sigma \ \frac{\exp(-4\Gamma_n^2 Dt/L^2)}{\Gamma_n^2(L^2h^2+2Lh+4\Gamma_n^2)} \qquad [95]$$

where Γ_n is the solution of equation 96

$$\tan(\Gamma_n) = hL/2\Gamma_n \qquad [96]$$

where h is a parameter determined by R_{AB} and R and equals $3\pi R R_{AB}/(\pi R - R_{AB})^3$

8.   Conclusion

In this contribution an attempt was made to review the different kinetic expressions to which fluorescence quenching in reversed micellar systems can give rise. If one should try to use those expressions to extract information from reversed micellar systems one should always bear in mind that in a real experimental situation several of the complications (8,42,43,44) that were treated separately in this contribution can arise in the same system. Therefore a very careful selection of the probe and the quencher will always be necessary.

Furthermore the use of fluorescence quenching to determine the physicochemical properties of reversed micelles would become a more powerful technique if the global simultaneous analysis (31) of fluorescence decays obtained at different quencher concentrations following problems would be used. The use of this technique would allow to shine light on the structure of the reversed micellar systems at high R values where the intramicellar quenching (k_q) becomes slow and exchange processes become fast (8,44). Also the drastic decrease of k_q at large values of R is still a not completely resolved problem. To solve this problem it would be necessary to solve the Poisson Boltzman equation (18, 19) for different values of R and to calculate the mean first passage time (55) for an excited probe to reach a quencher or for a quencher to reach an excited probe.

Acknowledgments

M. Van der Auweraer is a "Research Associate" of the Belgian National Science Foundation. The authors are indebted to the FKFO and to the Belgian Ministery of Scientific Programmation for financial support.

References

1. a) S.E. Webber, J. Phys. Chem., 87. (1983) 347-351
   b) M. Tachiya, J. Chem. Phys., 76 (1982) 340-348
   c) M.D. Hatlee and J.J. Kozak, J.Chem. Phys., 72 (1980) 4358-4367
2. a) M. Eigen, Z. Phys. Chem. (Frankfurt am Main) 1 (1954) 176-200
   b) D. Rehm and A. Weller, Ber Bunsenges. Phys. Chem.,73 (1969) 834-
   839
3. a) M. Almgren and J.-E. Löfroth, J. Chem. Phys 76 (1982) 2734-2743
   b) G.G. Warr and F.Grieser, J. Chem. Soc. Far. Trans.I, 82 (1986)
   1813-1828 and 1829-1838
4. a) P.P. Infelta, M. Grätzel and J.K. Thomas J. Phys. Chem. 78 (1974)
   190-195
   b) M. Tachiya, Chem. Phys. Lett. 33 (1975) 289-292
   c) P.P Infelta Chem. Phys. Lett. 61 (1979) 88-91
5. a) M.Almgren and Grieser, J. Am. Chem. Soc., 101 (1979) 279-291
   b) H. Sano and M. Tachiya, J. Chem. Phys. 75 (1981) 2870-2878
   c) M. Van der Auweraer, J.C. Dederen, E. Geladè and F.C. De
   Schryver, J. Phys., Chem. 74 (1981) 1140-1147
   d) M.D. Hatlee, J.J. Kozak, G. Rothenberger, P.P. Infelta and M.
   Grätzel, J. Phys. Chem. 84 (1980) 1508-1519
6. a)Tachiya,Chem. Phys. Lett. 69 (1980) 605-607
   b) U. Gösele U.K.A. Klein and M. Hauser and , Chem. Phys. Lett.68
   (1979) 291-295
   c) Sano and M.Tachiya,J. Chem. Phys 71 (1979) 1276-1282
7. a) A. Yekta, M. Aikawa and N. Turro, Chem. Phys. Lett., 63 (1979)
   543-548
   b) R.C. Dorrance and T.F. Hunter J.Chem. Soc. Far. Trans. I
   68,(1972) 1312-1321
8. a) S.S. Atik and J.K. Thomas, J. Am. Chem. Soc. 103 (1981) 4367-4371
   and 7403-7406
   b) E. Geladè and F.C. De Schryver, J. Photochem., 18 (1982) 223-230
   c) P. Brochette and M.P. Pileni, Nouv. J. Chim., 9 (1985) 551-555
9. a) A. Henglein and Th.Proske, Ber Bunsenges Phys. Chem., 82 (1978)
   471-476
   b) A. Henglein Th. Proske and W. Scheerer ., Ber Bunsenges Phys.
   Chem., 82 (1978) 956-962
   c) A. Henglein and Th. Proske, Ber Bunsenges Phys. Chem., 82 (1978)
   1107-1112
10.a) F.M. Menger, J.A. Donohue and R.F. Williams, J. Am. Chem. Soc., 95
   (1973) 286-288
   b) H.F. Eicke, J.C.W. Shepherd and A. Steinmann, J. Coll. Interfac.
   Sci., 56 (1976) 168-176
11.M. Tachiya, in: G.R. Freeman (ed.), Kinetics of Inhomogeneous
   Processes, J. Wiley & Sons Inc., 1987 pp. 575-650
12.a) M. Maestri, P.P Infelta M. Grätzel, J. Chem. Phys. 69 (1978)
   1522-1526
   b) J.K. Thomas, F. Grieser and M. Wong, Ber. Bunsenges. Phys. Chem.
   82 (1978) 937-949
   c) B. Katusim Razem, M. Wong and J.K. Thomas, J.Am.Chem.Soc., 100
   (1978) 1679-1686
   d) M. Rodgers, M.E. Da Silva e Wheeler Chem. Phys. Lett. 43 (1976)
   587; ibidem 53 (1978) 165-169
   e) A.J. Frank, M. Grätzel, A. Henglein and E. Janata Ber. Bunsenges.

Phys. Chem. 80 (1976) 294-300

f) H. Masuhara, K. Kaji and N. Mataga, Bull. Chem. Soc. Japan, 50 (1977) 2084-2087

13.a) D. J. Miller, U.K.A. Klein and M. Hauser, Ber Bunsenges. Phys. Chem. 84 (1980) 1135-1140

b) K.Kalyanasundaram Chem. Soc. Rev. 7 (1978) 453-472

c) A.R. Watkins and B.K. Selinger Chem. Phys. Lett. 64 (1979) 250-254

d) P.P Infelta and M.Grätzel J. Chem. Phys. 70 (1979) 179-186

e) N. Turro, M. Aikawa and A. Yekta, J. Am. Chem. Soc. 101 (1979) 772-774

f) S.S. Atik, M. Nam and L.A.Singer, Chem. Phys. Lett, 67 (1979) 75-80

14.a) R.C. Dorrance and T.F. Hunter, J. Chem. Soc. Faraday Trans. I,70 (1974) 1572-1580

b) R.C. Dorrance and T.F. Hunter Chem. Phys., Lett., 75 (1980) 152-155

15.P. Lianos R. Zana, Lang J., Cazabat A.M. in: L. Mittal (Ed.), Proceedings of the 5th Symposium of Surfactants in Solution, Bordeaux France 1984, Plenum Press, 1987.

16.G. Rothenberger, P.P. Infelta, M. Grätzel, J. Phys. Chem., 83 (1979) 1871-1876

17.a) C. Tanford, The Hydrophobic Effect, Wiley, New York, 1973

b) G.S. Hartley J. Chem. Soc., (1938) 1968-1975

c) P. Mukerjee and A. Mysels ACS Symp. Ser., 9 (1975) 239

d) G.S. Hartley, Kolloid Z., 88 (1938) 22-40

18.M. Almgren, P.Linse., M. Van der Auweraer F.C. De Schryver, E. Geladè and Y. Croonen, J. Phys. Chem., 88 (1984) 289-295

19.a) R.A. Goldstein nd J.J. Kozak, J. Chem. Phys. 62 (1975) 276-291

b) S. Dieckmann and J.Frahm, Ber Bunsenges Phys. Chem., 82 (1978) 1013-1015

c) Y. Gur, I. Ravina A.J. Babchin J. Coll. Interfac. Sci. 64 (1978) 326-333 and 333-341

20.M. Van der Auweraer, E Roelants, A. Verbeeck and F.C., De Schryver, in: L. Mittal (Ed.), Proceedings 6th Symposium on Surfactants in Solution, New Dehli India 1986, in press

21.E. Roelants, E. Geladé, J. Smid and F.C. De Schryver, J. Coll Interfac. Sci., 107 (1985) 337-341

22.P.D.I. Fletcher, A.M. Howe and B.H. Robinson, J.C.S. Faraday Trans.I, 83 (1987) 985-1006

23.a) J. Lang, N. Zana,and A. Malliaris, J. Phys. Chem. in press

b) M.-P. Pileni, T. Zemḅ and C.Petit, J. Photochemistry, 28 (1985) 273-283

c) M.-P. Pileni, B. Lerebours, B . Brochette and Y. Chevalier, Chem. Phys. Lett., 118 (1985) 414-420

24.E. Geladé, N. Boens and F.C. De Schryver, J.Am.Chem.Soc., 104 (1982) 6288-6292

25.a) C. Mathew, P.K. Patanjali, A. Nabi and A.N. Maitra, Colloids Surfaces, 30 (1988) 253

b) S.J. Chen, D.F. Evans, B.W. Ninham, D.J. Mitchell, F.D. Blum and S. Pickup, J.Phys. Chem., 90 (1986) 842-847

c) A.M. Ganz and B.E. Boeger, J. Coll. Interfac. Sci., 109 (1986) 504-507

26.a) B.K. Robinson D.C. Steytler and R.D. Tack, J. Chem. Soc. Faraday
     Transactions I, 75 (1979) 481-496
    b) M. Almgren J. Van Stam, J. Swarup and J.-E. Löfroth, Langmuir, 2
     (1986) 432-438
27.N.J. Bridge and P.D.I. Fletcher, J. Chem. Soc. 79 (1983) 2161-2167
28.P.D.I Fletcher, A.M. Howe and B.R. Robinson, J. Chem. Soc. Faraday
     Transactions I, 83 (1987) 1007-1027
29.A.S. Kertes and H. Gutmann in: E. Matijevic (Ed.), Surface and
     Colloid Science. Surfactants in Organic Solvents, John Wiley & Sons
     Inc., New.York., 1976, Vol. 8 pp. 193-295
30.A. Malliaris, J. Lang, J. Sturm and N. Zana, J. Phys. Chem., 91
     (1987) 1475-1481
31.a) N. Boens, A. Malliaris, M. Van der Auweraer, H. Luo and F.C. De
     Schryver, Chem. Phys., 121 (1988) 199-209
    b) N. Boens, H. Luo, M. Van der Auweraer, S. Reekmans,F.C. De
     Schryver, A. Malliaris, Chem. Phys. Lett., 146 (1988) 337-342
    c) H. Luo, N. Boens, M. Van der Auweraer, F.C. De Schryver and A.
     Malliaris, submitted
32.a)J.B. Birks, Photophysics of Aromatic Molecules, Wiley-
     Interscience), London, 1970, ch. 7
    b) M.H. Hui and W.R Ware, J.Am.Chem.Soc., 98 (1976) 4718-4727
    c) H. Knibbe, K. Röllig, P. Schäfer, and A. Weller, J.Chem. Phys.,
     47 (1976) 1184-1185
33.a) Y.Waka, K. Hamamoto and N. Mataga, Chem. Phys. Lett. 62 (1979)
     364-367
    b) Y.Waka, N. Hamamoto and N. Mataga, Photochemistry and
     Photobiology, 32 (1980) 27-35
    c) S.S. Atik and J.K. Thomas, J. Am. Chem. Soc. 103 (1981) 3543-3550
34.a) T.F. Hunter and A.I. Younis, J.C.S. Faraday Trans. I, 75 (1979)
     550-560
    b) D.J. Miller, U.W.A. Klein and M. Hauser, Z. Naturf., 32a (1977)
     1030-1035
35 a) M. Kotlarchyk, R.B. Stephen and J.S. Huang, J. Phys. Chem.,92
     (1988) 1533-1538
    b) R.A. Day, B.H. Robinson,an d J.H. Clarke, J.C.S. Faraday Trans.
     I, 75 (1979) 132-139
    c) J.S. Huang and M.W. Kim, Phys. Rev. Lett., 47 (1981) 1462
36.M. Wong and J.K. Thomas, in: L. Mittal (Ed.), Micellization,
     Solubilization and Microemulsions, Plenum Press, New.York., 1977,
     Vol.2. pp.647
37.M. Tachiya and M. Almgren, J. Phys. Chem., 75 (1981) 865-870
38.J.-C. Dederen, M. Van der Auweraer and F.C. De Schryver, Chem. Phys.
     Lett., 68 (1979) 451-454
39.J.-C. Dederen, M. Van der Auweraer and F.C. De Schryver, J. Chem.
     Phys. , 85 (1981) 1198-1202
40.a) E. Alruin , F.Lissi, N. Blanchin, L. Miola and F.H. Quina, J.
     Phys. Chem., 87 (1983) 5166-5172
    b) F. H. Quina and V.G. Toscan, J. Phys. Chem.,81 (1977) 1750-1754
41.a) T. Nakamura and, A. Kira and M. Iwamura, J. Phys. Chem., 86 (1982)
     3559-3363
    b) T. Nakamura and, A. Kira and M. Iwamura, J.Phys. Chem., 87 (1983)
     3122-3125
42.a) E. Geladé, Ph. FD. Thesis, Katholieke Universiteit Leuven, Leuven
     (1983)

b) E. Geladé, A. Verbeeck, F.C. De Schryver, in: L. Mittal (Ed.), Proceedings 5th Symposium on Surfactants in Solution Bordeaux France 1984, Plenum Press, 1987, Vol. 5, pp. 565-579

43.M. Almgren, J.-E. Löfroth and J. Van Stam, J. Phys. Chem., 90 (1986) 4431-4437

44.A. Verbeeck and F.C. De Schryver, Langmuir, 3 (1987) 494-500

45.G.D. Correll, R.N. III Cheser, F. Nome and J.H. Fendler, J. Am. Chem. Soc., 100 (1978) 1254-1262

46.J.H. Fendler, Acc. Chem. Res., 9 (1976) 153-161

47. a) P. Brochette, T. Zemb, P. Mathis and M.-P. Pileni, J. Phys. Chem., 61, 1987, 1444-1450
b) C. Petit, P. Brochette and M.-P. Pileni, J. Phys. Chem. , 90 (1986) 6517-6521
c) J. Politi and H. Chaimovich, J. Phys. Chem., 90 (1986) 282-287
d) E. Bardez, E. Monnier and B. Valeur, J. Phys. Chem., 89 (1985) 5031-5036
e) M. Wong, J.K. Thomas and J.K. Nowak, J. Am. Chem. Soc.,99 (1976) 4730-4736

48.a) G.Voortmans, A. Verbeeck and F.C. De Schryver, submitted
b) P. Lianos and N. Zana, in: L. Mittal (Ed.), Proceedings 5th Symposium on Surfactants in Solution Bordeaux France 1984, Plenum Press, 1987

49.M.H. Abdel Kader and A. Braun, J.C.S. Faraday Trans.I, 81 (1985) 245-253

50.M.A.J. Rodgers and J.C. Becker, J. Phys. Chem., 84 (1980) 2762-2768

51.M.-P. Pileni, J.M. Furois and B.Hickel, in: L. Mittal (Ed.), Proceedings 4th Symposium on Surfactants in Solution, Lund Sweden 1982, Plenum Press, 1985, vol. 3, pp. 1471-1481

52.M. Van der Auweraer and F.C. De Schryver, Chem. Phys., 111 (1987) 105-112

53.a) P.K.F. Kogler, D.J. Miller, J. Steinwandel and J.M. Hauser, J. Phys. Chem., 85 (1981) 2363-2366
b) T. Matsuo, Y. Aso, K. Kano, Ber Bunsenges Phys. Chem., 84 (1980) 146-152
c) M.A.J. Rodgers, Chem. Phys. Lett., 78 (1981) 509-514
d) M.D. Eddiger, R.M. Domingue and M.D. Fayer, J. Chem. Phys., 80 (1984) 1246
e) H. Sato, M. Kawasaki and K. Kasatani, J. Chem. Phys., 87 (1983) 3759-3769
f) N. Tamai, T. Yamazaki, I. Yamazaki, R. Mizuma , N. Mataga, J. Phys. Chem.,, 91 (1987) 3503-3508

54.M. Van der Auweraer, S.Reekmans, N. Boens and F.C. De Schryver, submitted

55 A. Szabo, K.Schulten and Z. Schulten, J. Chem. Phys., 72 (1980) 4350-4357

PROTON TRANSFER IN REVERSE MICELLES AND CHARACTERIZATION OF THE ACIDITY IN THE WATER POOLS

B. VALEUR and E. BARDEZ

1. INTRODUCTION

Water molecules included in the core of reversed micelles form water pools which are extensively used as "tailored-to-size microreactors" for chemical (whether organic or inorganic), photochemical, biological, electron-transfer reactions, etc...(1-6). The aqueous core as a microsolvent exhibits peculiar properties due to structural changes with respect to bulk water. A thorough knowledge of these properties is of major importance for the understanding of studies on reactivity, because reactivity in any reaction medium depends not only on its chemical nature but also on its structure. The structural features of the encased water molecules, and consequently their solvation ability and their reactivity, depend on the extent of hydration of the micelle (i.e. water content) and on their localization in the aqueous core.

The particular question of "acidity" of the water pool is of fundamental interest for the study of enzymatic activity and some chemical reactions (e.g. hydrolysis). The most interesting range of water contents corresponds to rather small water pools (water-to-surfactant ratio of 3 to 10) in which the peculiar properties of water cause the largest changes in the behavior of substrates as compared to their behavior in bulk water. For instance, it is in this range that the activity and conformational properties of solubilized enzymes are most different from those observed in bulk water (3). Since the water of such aqueous cores has different properties from those of bulk water, the classical concept of pH is not readily transposable. The problem of acidity is indeed relevant to the acid-base reactivity of water molecules which depends on their structural characteristics, and therefore **acidity should be considered in the most general terms of proton transfer.** With this in mind,this chapter will be devoted first to the study of proton transfer involving water molecules in the aqueous core by using various acidic and basic probes of different localization. Then the difficulties arising from the use of the concept of pH will be discussed in light of the different attempts of determination.

2. ACIDIC PROBES IN REVERSE MICELLES: DYNAMICS OF PROTON TRANSFER

Fluorescent probes which are much more acidic in the excited state than in the ground state, tend to undergo deprotonation as soon as they are excited and are thus well suited to explore the ability of an unknown medium to accept a proton. The use of time-dependent techniques allows one to determine the excited-state rate constants for deprotonation and back recombination.

2.1 Methodology

Excited-state proton transfer can be described according to the following kinetic scheme:

$$
\begin{array}{ccc}
AH^* + H_2O & \underset{k_{-1}^*}{\overset{k_1^*}{\rightleftharpoons}} & A^{-*} + H_3O^+ \\[2mm]
{\Big\updownarrow}\, k_r + k_{nr} = 1/\tau_0 & & {\Big\downarrow}\, k_r' + k_{nr}' = 1/\tau_0' \\[2mm]
AH + H_2O & \underset{k_{-1}}{\overset{k_1}{\rightleftharpoons}} & A^- + H_3O^+
\end{array}
$$

k_1 and k_{-1} are the ground-state rate constants for deprotonation and reprotonation, respectively; k_1^* and k_{-1}^* are the corresponding rate constants in the excited state; k_f ($= k_r + k_{nr}$) and k_f' ($= k_r' + k_{nr}'$) are the emission rate constants of AH^* and A^{-*}; they are equal to the reciprocal of the lifetimes τ_0 and τ_0' in the absence of an excited-state reaction. The rate constants can be determined either by pulse fluorometry (excitation by a pulse of light) or by phase fluorometry (excitation by sinusoidally modulated light). We have chosen the latter technique which permits straightforward determination of the rate constants thanks to the differential measurements (7). As a matter of fact, if the acidic form is the only form that exists in the ground state, the phase shift $\Delta\Phi$ between the emission of the basic form and that of the acid form is given by a simple expression only involving the rate constants for the disappearance of A^{-*}:

$$
\tan \Delta\Phi = \frac{\omega}{k_{-1}^*[H_3O^+] + 1/\tau_0'}
$$

$\Delta\Phi$ can be measured directly by moving back and forth the appropriate optical filters which select the emission of AH^* and A^{-*}. The second measurement concerns the phase shift Φ_{A^-} of the basic form A^{-*} with respect to the exciting light. Then, τ_0 and τ_0' being measured separately, the values of k_1^* and $k_{-1}^*[H_3O^+]$ can be easily calculated from $\Delta\Phi$ and Φ_{A^-} (7). The problem is even simpler in the case of irreversible proton transfer ($k_{-1}* \tau_0'[H_3O^+] \ll 1$) (8).

2.2 Choice of probes

In an investigation aiming at examining the properties of water molecules according to their localization, the probes should be chosen so that they can explore different regions of the water pool: interfacial area, vicinity of the interface, and center of the pool. With this in mind, the probes should bear either no charge or only charges of the same sign as that of the polar head groups; the larger the number of charges, the higher the efficiency of electrostatic repulsion, and the deeper the localization in the water pool. If a probe bears no charge, care should be taken as regards the possible partition between the organic phase and the aqueous microphase.

TABLE 1
Acid fluorescent probes and their acidity constants in water .

name	formula	pK$_\alpha$	pK*
2-naphthol (NOH)		9.30	2.8
2-naphthol- 6-sulfonate (NSOH)		9.12	1.66
2-naphthol-6,8-disulfonate (NDSOH)		9.33	< 1
pyranine (PyOH)		7.2	0.5

In our investigations, an anionic surfactant (Aerosol-OT or AOT: sodium bis(2-ethylhexyl) sulfosuccinate) was used (solutions 0.1 M in heptane); therefore, neutral and negatively charged probes have been chosen: pyranine (PyOH), 2-naphthol (NOH), sodium 2-naphthol-6-sulfonate (NSOH), potassium 2-naphthol-

6,8-disulfonate (NDSOH). The formula of these probes are given in Table 1 together with their pK and pK* in water indicating that they are much more acidic in the excited state and thus,starting from the acid form in the ground state, they tend to undergo deprotonation as soon as they are excited. However, in AOT reverse micelles, the micro-environment of the probes is different from bulk water ; the pK* values of Table 1 are thus no longer valid and they depend on the site of solubilization. Therefore, the behaviors of the probes cannot be compared in terms of pK*.

2.3 Excited-state deprotonation of the probes according to their sites of solubilization

We now turn our attention to the solubilization sites of the probes. These sites can be assessed with the help of spectroscopic data in conjunction with the kinetics of proton-transfer in the excited state.

NOH requires particular attention because of its solubility in heptane and water; several experimental evidences (absorption and emission spectra, lifetime, and emission anisotropy) have shown that this probe is localized at the interface (where it is hydrogen-bonded to AOT polar headgroups) (9). According to the binding constant (335±10), NOH is fully bound at AOT concentrations greater than 8×10^{-2} M $^{(*)}$. Whatever the water content, no deprotonation occurs, as revealed by the emission spectrum which never exhibits the characteristic band of the naphtholate ion, whereas this band appears in bulk water. This means that hydration does not cause the proton to be displaced from the hydrogen bond established with the AOT polar head groups, and that the proton is not transferred to a water molecule because the resulting ion pair (NO^{-*}, H_3O^+) cannot be stabilized by a medium which is not polarizable enough (8,9).

In contrast to NOH, the three other probes are not soluble in heptane and they undergo deprotonation in AOT reverse micelles. The results of the dynamic experiments are shown in Figure 1. In the expression of tan $\Delta\Phi$, $k_{-1}*[H_3O^+]$ has been replaced by the apparent first-order rate constant $k'_{-1}*$ for reprotonation because writing $k_{-1}*[H_3O^+]$ implies that the recombination process is diffusion-controlled, which is not true in the present case (vide infra). As regards NSOH, an amount of water of $w = [H_2O]/[AOT] \cong 40$ is required for observing the same deprotonation rate as in bulk water, but even at such high water content, this molecule does not behave as in bulk water regarding absorption and emission

---

$^{(*)}$ It is interesting to note that, as the water content increases, the experimental data are consistent with an interface reorganization (in the domain of water-to-surfactant molar ratio <10) with a concomitant water penetration along the surfactant head groups (9).

spectra; the changes in absorption spectra as a function of water content are similar to those of NOH anchored at the interface and no isosbestic point appears. Moreover, the rate constant for proton recombination does not recover at high water content the value observed in bulk water. For these reasons, NSOH is likely to be solubilized in the vicinity of the interface rather than in equilibrium between two sites of solubilization, i.e. at the interface and in free water(8).

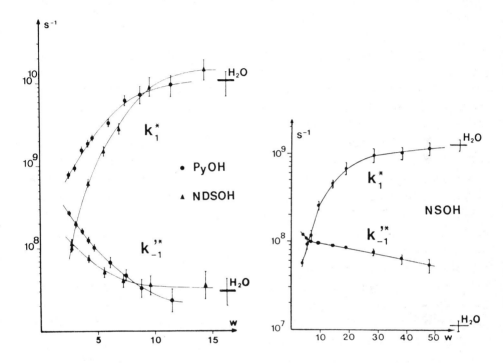

Fig. 1. Variations in rate constants of deprotonation (k_1^*) and back recombination (k'_{-1}^*) vs. water content for NSOH, NDSOH and pyranine. For comparison, the values obtained in bulk water are indicated.

Figure 1 shows a remarkable similarity of the behaviors of NDSOH (8) and pyranine (7) (the results obtained for pyranine have been confirmed by Politi et al.(31)): they both recover at w ≅ 10-12 the same spectral properties and the same kinetic behavior as in bulk water. Beyond w ≅ 10-12, as hydration of surfactant and Na^+ becomes complete, "apparently free water" can exist in the midst of the aqueous core (10,11); NDSOH and pyranine are pushed away towards this region owing to the electrostatic repulsion between the anionic heads of the surfactant and the negative charges of the probes. A localization around the center of the water pool was also observed for sodium tetracarboxylate which bears four negative charges (12). Therefore, it appears that a localization around the center can be assessed when the substrate bears more than one negative charge. Figure 2 illustrates the average residence sites of the four probes.

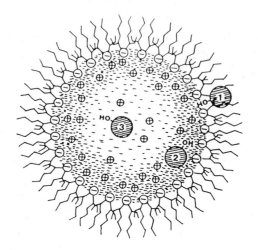

Fig. 2. Schematic illustration of the average residence sites of the probes in AOT reverse micelles: 1, NOH; 2, NSOH; 3, NDSOH and pyranine. Length of the surfactant, 11 Å. Diameter of the water pool, 18 Å at w = 3; 36 Å at w = 9. Largest dimension of the naphthol derivatives, ≈ 9 Å; hydrodynamic radius of pyranine, 4.7Å.

Before examining the information contained in the kinetic data of Figure 1 in terms of the ability of water molecules to accept a proton, let us consider first the protolysis of acids in aqueous solutions. The stepwise mechanisms I and II have been proposed. The existence of a hydrogen bridge between AH and a water molecule is a basic prerequisite. The nature of the intermediate is to be discussed as if proton transfer takes place directly (Scheme I) or via water molecules (Scheme II) (13-15).

$$A-H + \underset{\underset{H}{|}}{O}-H \rightleftharpoons A^{\ominus}\cdots H-\underset{\underset{H}{|}}{\overset{\oplus}{O}}-H \rightleftharpoons A^{\ominus} + H-\underset{\underset{H}{|}}{\overset{\oplus}{O}}-H \quad (I)$$

$$\quad\quad (a) \quad\quad\quad\quad\quad\quad (b) \quad\quad\quad\quad\quad\quad (c)$$

$$A-H + \underset{\underset{H}{|}}{O}-H + \underset{\underset{H}{|}}{O}-H \rightleftharpoons A^{\ominus}\cdots H-\underset{\underset{H}{|}}{O}\cdots H-\underset{\underset{H}{|}}{\overset{\oplus}{O}}-H \rightleftharpoons A^{\ominus} + H-\underset{\underset{H}{|}}{O} + H-\underset{\underset{H}{|}}{\overset{\oplus}{O}}-H \quad (II)$$

$$\quad\quad (a) \quad\quad\quad\quad\quad\quad\quad\quad (b') \quad\quad\quad\quad\quad\quad\quad\quad (c)$$

Indeed, oxygen acids, such as phenols, form strong hydrogen bonds with water which, because of its amphoteric nature, can act as a bifunctional link, and calculations have shown that simultaneous proton transfer with one or two intervening water molecules will occur without supplementary energetic barrier (16). Nevertheless, a water-separated ion-pair structure for the intermediate (b') implies the three dimensional hydrogen-bond network usually present in bulk water.

The backward diffusion-controlled rate constant for steps (c) —->(b) or (c) —->(b') is usually of the order of 10^{10}-10^{11} M^{-1} s^{-1} (15b); consequently, an acidic medium ([H_3O^+] $\cong$ 10^{-2}-10^{-3} M) is required for the (diffusion-controlled) pseudo-first-order rate constant (k_D[H_3O^+]) to be competitive with the rate constant for emission of species whose lifetimes are, in the present case, about 10 ns for the naphthol derivatives ($k_{em} = 1/\tau = 10^8$ s^{-1}), and 5 ns for pyranine ($k_{em} = 2 \times 10^8$ s^{-1}). In a neutral or weakly acidic medium, the diffusion-controlled rate for recombination is slow with respect to the rate of emission and therefore it cannot be measured by fluorescence experiments. Consequently, the measured rate constant k'^*_{-1} is the rate for geminate recombination, (b) or (b')——> (a).

In AOT reverse micelles at water contents larger than w $\approx$ 10-12, the kinetic behavior of NDSOH and pyranine, residing around the the center, indicates that the surrounding water is bulklike water regarding protolysis. A type-II mechanism with water-separated ion pairs may thus be proposed. On the other hand, even with large amounts of water, the recombination rate of NSOH, i.e. its geminate recombination, is still much faster than in bulk water which means that the microenvironment of the probe is not able to solvate the ion-pair structure (b) or (b'). In the aqueous core, a region which is expected to behave in this way is the vicinity of the surfactant monolayer where the water molecules are involved in the hydration of polar heads and sodium ions (10).

As the amounts of water are decreased (w < 10), a decrease in the deprotonation rate is observed with a concomitant increase in the reprotonation rate with respect to bulk water, for all the probes (NOH not being considered). A decrease in rate constant for proton ejection reflects a decrease in the probability of proton transfer from the donor (the probe) to the acceptor, i. e. a water cluster ; recent studies have shown that these clusters are likely to involve four water molecules, so that the specific entity representing the hydrated proton is $H_9O_4^+$ (17-19). Under conditions where the number of available water molecules is reduced, and/or where the tridimensional H-bonded struture of water is partially broken, the probability of transfer is reduced (as observed upon addition of an alcohol (20-21) or in biological environments (22-23)). As a matter of fact, in small micelles, the hydration of the polar heads and counterions is incomplete so that the whole water present is highly bound and oriented in solvation shells (10) and NMR experiments (24-26) and fluorescence lifetime studies on xanthene dyes (27) have shown that the water contains less hydrogen bonding than in normal bulk water. The water molecules being mainly oriented in the solvation shells of sulfonate and sodium ions, their mobility is reduced; such kind of water is thus unable to hydrate a proton, because it is now clearly established that the rotations of water molecules play a predominant role in the formation of the acceptor cluster for either proton transfer or electron transfer (*). Hindered rotations reduce the acceptor character of water and affect its thermodynamic and kinetic properties (19).

Consequently, it is expected that the reduction of protolysis results from structural features: (i) Available water molecules for proton transfer must be removed from the inner hydration shells, which requires subtantial amounts of work and accounts for the enhancement of the activation energy of the reaction; (ii) Because of the lack of complete hydration, the initial formation of the hydrogen-bonded complex between AH and H2O may be considered as an additional step in the overall mechanism and may significantly affect the kinetics of deprotonation; (iii) In a reduced water pool, proton transfer leads to the formation of contact ion pairs (b) which cannot be efficiently stabilized in a medium of low polarizability, as observed in the vicinity of the interface of larger micelles. Enhancement of the rate for back recombination is thus to be expected.

The above results clearly show that **the ability of water molecules in the water pool to accept a proton depends on their localization because this ability is related to the H-bonded water structure which gradually changes as a function of the distance with respect to the interface from hydration water to a normal three-dimensional water network.**

Politi and Chaimovich (35) have used the same probes and have interpreted their results in terms of distribution of the probes between the aqueous interfacial region (outer water pool) and the inner water pool according to Zinsli's biphasic model (29); this model has also been used by Kondo and co-workers (30) to explain their results on pyranine. In contrast, we believe that there is a gradual change in the structural properties of water molecules with the distance from the interface and that a time-averaged preferential location of the probe is a more realistic explanation. Such a gradual change of properties is consistent with a continuous variation of the electric field and of the distribution of the counterions within the water pool (36).

2.4 Activity of water

Gutman et al. (23) have established that the probability of proton transfer in aqueous solutions is a direct function of the chemical activity of water, the rate constant k_1* is given by the empirical relation

---

(*) The analogy between proton hydration and electron hydration should be mentioned: as a matter of fact, studies of hydrated electrons in reverse micelles (28) have shown that the absorption spectrum and the lifetime of the hydrated electron are dependent on the pool size and are almost identical with those of bulk water only for w ≥ 15.

$$k_1^* = k^*{}_1{}^0 (a_{H_2O})^n$$

where $k^*{}_1{}^\circ$ is the rate constant in pure water (reference state); for pyranine in concentrated electrolyte solutions, n is independent of the electrolyte (LiBr, LiCl, KCl, MgCl$_2$) and was empirically determined as being 6.9. Considering the analogy between the addition of water in a reverse micelle and the dilution of a concentrated electrolyte solution, the above equation can be used for the calculation of water activity around the center of the aqueous core from the kinetic measurements obtained with pyranine. The water activity as a function of the solubilized amount of water is shown in Figure 3. One should notice the good agreement of this activity

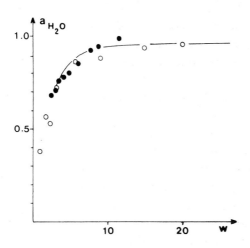

Fig. 3. Activity of water as a function of water content: theoretical function (solid line) (33), vapor pressure determination (o) (32) and values of the present work (●) (7).

plot with both the activities yielded by measurements of partial pressure of water vapor (32,37) and those calculated from the electrostatic model of Jönsson and Wennerström (33). The decrease in water activity as amounts of solubilized water are decreased reflects the progressive concentration of the aqueous core with sodium ions, constituting the environment of pyranine.

The same method can be used to evaluate the water activity of the microenvironments probed by the naphthol derivatives NSOH and NDSOH (34,35). From the rate constants of proton transfer and calibration curves obtained in homogeneous electrolytic solutions, it is possible to get numerical values of water activity. In contrast to the results of Politi and Chaimovich (35), we have observed that the water activity derived from the data obtained with NDSOH (34) recovers the value in bulk water at w $\approx$ 10-12, like pyranine; this is consistent with the preferential location of both these probes around the center of the water pool. However, the values obtained at lower water contents are significantly different for pyranine and NDSOH. Moreover, the behavior of NSOH is different. The difficulty of interpretation of these results and others (38) is to be discussed.

Then the question arises as to the validity of the method because thermodynamic equilibrium implies that the chemical potential of water is the same throughout the micellar solution, and therefore, for a given state of reference, the activity of water molecules is the same whatever their localization. Let us consider a reverse micelle at high water content: the water molecules around the center of the water pool have properties very close to those of bulk water and their activity is thus close to unity; consequently, the water molecules in the vicinity of the interface should have the same activity. At first sight, the fact that, at the same water content, probes of different localizations do not lead to the same value of activity is not in accordance with thermodynamics. However, one should keep in mind that, when using acid probes, the values of activity are derived from the determination of deprotonation rates; a proton is thus involved, the activity of which is not constant throughout the water pool (for a charged species, this is the electrochemical potential which is constant). Therefore, probing methods aiming at the determination of water activity must be used with caution.

3. BASIC PROBES IN REVERSE MICELLES

Acidic probes have allowed the study of the ability of water molecules to accept a proton. In the same way, basic probes behavior is expected to reflect the ability of the same water molecules to donate a proton, i.e. their "acidic" character.

In fact, in the case of AOT reverse micelles, the reactivity of basic compounds is a quite crucial question, and several difficulties crop up as soon as one attempts to introduce and study bases in the aqueous core.

The first obstacle stems from the fact that AOT samples may contain acidic impurities resulting from both the manufacturing processes (synthesis and purification) and the self-hydrolysis of the ester function. Increased attention has been paid in the last few years to this problem (3,39,40). The importance of even traces of acidic water-soluble impurities must be borne in mind : for instance, an impurity present at the 0.05 % level (mole per cent) leads, in a solution containing an [H_2O]/[AOT] ratio of 20, to an impurity concentration in the water microphase of 1.4×10^{-3} M. When the water/detergent molar ratio is reduced to, say 5, the concentration is multiplied by four. In homogeneous aqueous solutions, such acid concentrations should correspond to pH values lying between 2 and 3 for a strong acid, or between 3.0-3.5 for a weak acid of pK around 4.

Thus the purity of the surfactant appears to be essential when dealing with basic compounds,which is demonstrated by the following experiment : two hydrophilic very weak bases, benzimidazole (pK = 5.30) and 6-methoxy-5-sulfoquinoline (pK = 4.16), were incorporated in AOT reverse micelles, using successively two different commercial AOT samples, one purchased from Fluka and the other from Sigma. The U.V. absorption spectra allowed us to determine the degree of protonation of the bases and to observe that both bases displayed completely opposite behavior according to the AOT sample: whatever the water content, both were under protonated form in the AOT from Fluka, but under neutral form in the AOT from Sigma. As a matter of fact, further potentiometric titrations indicated 0.5 % acidic impurities in the former AOT, whereas no acidic impurity was detectable in the latter one.

This simple experiment provides fairly compelling evidence of the risk of wrong interpretations, whenever the quality of the material has not been carefully checked.

The second difficulty is relevant to the chemical nature of the surfactant itself. As it has been already emphasized (39), the ester function may undergo an hydroxide-promoted hydrolysis, even in the aqueous core of the micelles.

When the medium is free of acidic impurities the dissolution and protonation of a basic compound B generate hydroxide ions, according to the equilibrium :

$$B + H_2O \rightleftharpoons BH^+ + OH^-$$

and the more so, the stronger the basic function.

However, every time hydroxide ions appear, hydrolysis may occur at a noticeable rate, because of the great excess of AOT over hydroxide : for an $[H_2O]/[AOT]$ ratio of 5, the "effective" ester concentration in the aqueous microphase is 11 M, and it decreases to 2.8 M when $[H_2O]/[AOT] = 20$. As OH^- ions are consumed, the above equilibrium is displaced, leading to more and more protonation. Consequently, a time-dependent increase of BH^+ is likely to be observed, and no conclusions can be drawn on the ability of water to protonate B.

The following experiment substantiates this contention. When adenine (pK = 9.80) or sodium naphtholate (pK = 9.55) are added in their solid form to a 0.1 M AOT/water/heptane solution (free of acidic impurities), the water-soluble protonated form BH^+ appears according to the sequence :

$$
\begin{array}{ccc}
& H_2O & \\
B\ \text{solid} \longrightarrow & B_{aq} \longrightarrow & BH^+{}_{aq} + OH^-{}_{aq} \\
\text{in excess} & \underbrace{\hspace{5cm}} & \\
\text{(insoluble in heptane)} & \text{aqueous core} &
\end{array}
$$

As expected if OH- ions further react, spectrophotometrical measurements show that the amount of BH+ increase with time, owing to the progressive solubilization of the solid. For instance, when [H2O]/[AOT] = 9, we observed that it took about one hour for each of the compounds to double its concentration under the BH+aq form.

As a result, in such conditions, protonation of the base is screened by hydrolysis of AOT, which is unavoidable in basic medium. Thus, the outstanding affinity of "proton sponge" for AOT w/o microemulsions previously reported (41) may be partly due to this phenomenon.

At that point of the discussion, it appears evident that the only way to overcome the difficulty is to use dynamic methods, so that the hydrolysis velocity becomes negligible during the time scale of the measurement.

Transposition of the aforementioned dynamic studies to the protonation of fluorescent basic probes in the excited state implies to find a compound able to exist under the neutral form in the ground state and to undergo protonation in the excited state. Hence, besides pK constraints of pK $\leqslant$ 5 and pK* $\geqslant$ 11, the probe must have a lifetime long enough to be protonated in its excited state.

Unluckily, very few basic compounds are suitable for such an investigation. Amongst quinoline derivatives, 6-methoxy-quinoline can be mentioned (pK = 5.2 ; pK* = 11.8) (42) but it turned out that this compound was only soluble in the heptane phase, showing no affinity either for the water core or for the interfacial region. Another endeavor was to synthesize 5-sulfo-6-methoxy-quinoline, which, on the contrary, proved to be soluble only in water. However, despite favorable pK values (pK = 4.16 ; pK* $\approx$ 11.5) very little excited-state protonation was observed in neutral medium because the lifetime of the basic form (0.38 ns) revealed to be too short to allow significant protonation. Consequently, no accurate kinetic measurements were possible with that compound. In fact, discovering a suitable fluorescence basic probe to make use of dynamic techniques in AOT reverse micelles remains quite a challenging problem.

Lastly, even if the lack of dynamic results does not provide parallel results to those obtained with acidic probes, it must be pointed out that **water included in AOT reverse micelles does not reveal a peculiar acidity towards basic compounds**, as long as acidic impurities are removed. As a matter of fact, a compound like benzimidazole (pK = 5.30) is not protonated in acid-free AOT, even at low water contents (see above). Furthermore, every time that an increased solubilization under the protonated form is observed, the eventuality of AOT hydrolysis must be kept in mind, because it occurs at the expense of the basic form of the compound, via hydroxide ions.

4. CHARACTERIZATION OF THE ACIDITY IN THE AQUEOUS CORE

Such a characterization is important as soon as ionizable compounds are to be solubilized in the water pool. The case of enzymes is of particular interest. It is tempting to use the classical concepts of pH and pK, but several difficulties arise when transposing these concepts to confined water in reverse micelles, vesicles or biological systems.

4.1 Critical review of various attempts

Three types of methods have been proposed to evaluate the acidity in the aqueous core: direct determination using a glass electrode, ^{31}P-NMR spectrometry, use of pH indicators.

Menger and Yamada (43) have reported pH measurements in solutions of AOT in heptane by means of a glass electrode in the presence of solubilized phosphate and borate buffers. At large water contents (w>25), the measurement yields a value of pH similar to that of the buffer before injection, which means that the amount of water is sufficient to cover the electrodes. At lower water contents, the interpretation of the data in terms of micellar effect is difficult owing to problems of calibration and junction potential.

Another method of evaluation of pH in water pools is provided by ^{31}P-NMR. The chemical shift of the phosphorus atom of some compounds depends on pH, a property which has been used for the measurement of local pH inside and outside cellular membranes. This method can be applied to anionic reverse micelles (sodium octanoate, AOT) by comparing the chemical shifts of phosphate buffers in the aqueous core and in homogeneous solutions, in order to define an empirical acidity scale (44,45). However, it is necessary to assume that the pK values of the buffers remain unchanged in the water pools, the anions $H_2PO_4^-$, HPO_4^{2-} and PO_4^{3-} undergoing no interaction with the polar headgroups from which they are repelled. It should be noted that the values of pH that are determined by this method, are found to be generally within 0.4 pH unit of the value for the buffer solution before injection, which means that the values measured in this way reflect the pH around the center of the water pool where the buffer is efficient. This objection (3) reduces the interest of the method.

The third possibility is the use of acid-base indicators whose absorption spectra are pH-dependent. An extension of this method has been proposed with the use of fluorescent compounds (especially for biological applications): the proportions of the acid and basic forms in the ground state are determined from the analysis of the emission as a function of the excitation wavelength.

In many investigations, the determination of the micellar pH, or the apparent pK (pK_{app}) of the indicator in the aqueous core, is based on the Henderson-Hasselbach equation written in the following form:

$$pH = pK_{app} + \log \frac{[Ind]}{[HInd]} \tag{1}$$

where [Ind]/[HInd] represents the ratio of the concentrations of the basic and acid forms of the indicator as obtained from the absorption spectrum. This method allows one to calculate the pH assuming that the value of pK_{app} is equal to the value in bulk solutions (46). Alternatively, the value of pK_{app} can be calculated if it is assumed that the value of pH is identical either with the pH of a buffer solution injected in the micellar core (47-50), or with the value derived from ^{31}P-NMR data (44,45) (vide supra). The results show that hydrophilic indicators, with negatively charged acid and basic forms , undergo only slight variations in pK_{app} (<1). On the other hand, if one or both forms of the indicator is hydrophobic, a large increase of pK_{app} with respect to bulk solutions is observed ; for instance, the value found for paranitrophenol anchored at the interface is 11.5 instead of 7.14 in water (47). In several cases, the values of pK_{app} of the same hydrophobic indicator depend on the nature and concentration on the buffer, and on the water content of the micelle (44,50a,50d). These observations reveal the existence of interactions between the dye, the micellar components and those of the buffer solution.

In the case of indicators located at the interface or in its vicinity, the assumption that the pH is equal to that of the bulk buffer solution is not valid. El Seoud (50) has calculated the interfacial pH on the basis of an ion-exchange equilibrium occuring between sodium and hydrogen ions :

$$Na^+_b + H^+_f \rightleftharpoons Na^+_f + H^+_b$$

The subscripts f and b refer to the free and interfacially bound species, respectively. In such a model, hydrogen ions are accumulated in the water involved in the hydration of the polar heads. From the concentration of H^+_b calculated for the total volume of the solubilized water, a value of pH_b can be obtained, which is 0.5 to 1.5 units lower than pH_f (in the center of the waterpool). Such an estimation of interfacial pH is interesting, but the model suffers from stringent assumptions:
- the aqueous core is assumed to be biphasic with two types of hydrogen ions: (i) bound hydrogen ions, corresponding to pH_b (but does a proton involved in an ion pair have a character of free acidity?), (ii) free hydrogen ions, corresponding to pH_f. However, such an image of the aqueous core is not in agreement with the gradual change of the properties of the water pool from the interface to the center; in particular, the probability of presence of an hydrogen ion should gradually decrease with increasing distance from the interface, as for sodium ions (36).
- several aspects of the calculation are questionable: the equilibrium constant for exchange is taken equal to that of sodium dodecyl sulfate, concentrations in volume

are used for interfacially bound species, the activity of water is not taken into account, etc...

Moreover, this calculation does not account for variations in pK_{app} of 4 to 5 units, as observed in some cases.

4.2 Thermodynamic aspects

In an aqueous environment, the definition of pH is related to the thermodynamic activity of free hydrogen ions (51,52):

$$pH = - \log a_{H+} \qquad [2]$$

The relation between between activity and concentration (expressed in molarity) of hydrogen ions is

$$a_{H+} = f_{H+} [H^+] \qquad [3]$$

where f_{H+} is the activity coefficient relative to molarity.

The equilibrium constant for autoprotolysis of water, which is the same throughout the water pool, should be written with activities:

$$K_w = \frac{a_{H+} . a_{OH-}}{a_{H2O}} = 10^{-14} \qquad \text{(at 25°C)} \qquad [4]$$

It must be emphasized that, in a reverse micellar solution at equilibrium, the electrochemical potential of H^+ is constant throughout the water pool. Let us recall that the electrochemical potential of a charged particle is the sum of its chemical potential and its potential energy at a given electrical potential. This electrical potential is obviously not constant in the aqueous core of reverse micelles (36), and therefore neither is the chemical potential of H^+ (and OH^-). Thus it follows that **the activity of H^+ is not constant throughout the water pool, and only a local pH can be defined**.

When a pH indicator is used, the acidity constant should be written with activities:

$$K = \frac{a_{H+} . a_{Ind}}{a_{HInd}} \qquad [5]$$

hence,

$$pH = pK + \log \frac{[Ind]}{[HInd]} + \log \frac{f_{Ind}}{f_{HInd}} \qquad [6]$$

By comparison with Eq. 1, we obtain:

$$pK_{app} = pK + \log \frac{f_{Ind}}{f_{HInd}} \qquad [7]$$

Therefore, pK_{app} depends on the factors able to modify the activity coefficients:
- ionic strength (if Ind and/or HInd are ionic);
- specific interactions depending on the chemical nature of the indicator and the surrounding species (e.g. buffer components);
- structural changes of the medium.

In reverse micelles, these effects may all be involved but to a different extent according to the localization of the indicator, the size of the micelle and the nature of the buffer. In the aqueous core of AOT reverse micelles, the concentrations of sodium ions can be very high, and thus the effect of ionic strength requires further attention.

4.3 Ionic strength effects

It is worth examining the influence of ionic strength on the pK_{app} of a dye in the absence of surfactant. We have chosen pyranine (pK= 7.2) because it should be particularly sensitive to the ionic strength owing to the number of negative charges it bears, i.e. three in the acidic form and four in the basic form. The absorption spectrum was recorded as a function of ionic strength (addition of KCl) in a buffer at pH=7.2. The evolution of the absorption spectrum indicates that the ratio [PyO⁻]/[PyOH] is changed. Since the pK is a constant and the pH is kept constant by the buffer, Eq.6 shows that the variations in [PyO⁻]/[PyOH] compensate for the variations in the ratio of activity coefficients f_{Ind}/f_{HInd} as a function of ionic strength. Once the ratio [PyO⁻]/[PyOH] is determined, pK_{app} can be calculated by means of the following equation:

$$pK_{app} = pH - \log \frac{[PyO^-]}{[PyOH]} \qquad [8]$$

- at ionic strengths I < 1 M: the ratio [PyO⁻]/[PyOH] increases with increasing I; the resulting decrease in f_{Ind}/f_{HInd} is in agreement with the Debye-Hückel theory: the activity coefficient of PyO⁻ (valence 4) is indeed expected to decrease more than that of PyOH (valence 3). At a ionic strength of 1 M, pK_{app} is equal to 6.7, i.e. 0.5 unit lower than the value in pure water.
- at ionic strengths I > 1 M: the spectra do not exhibit isosbestic points any more and an increase, instead of a decrease, in pK_{app} with increasing I is now observed. In this range, the specific interactions and the structural changes are predominant.

This experiment shows that the ionic strength induces a change in pK_{app} of no more than 0.5 units. For less charged indicators, the variation is even smaller; e.g. for 2,3-dinitrophenol, pK_{app} = 3.80 at I = 0.5 M (KCl) instead of 4.1 in water (53).

It is worth noting that the variations in pK_{app} due to an ionic strength effect in homogeneous solutions, are of the same order of magnitude as those observed for an indicator located around the center of the micellar aqueous core: for instance, pK_{app} = 1.65 for maleic acid in AOT/heptane solutions, instead of 1.95 in aqueous solutions (50). This coincidence leads us to suggest that the effect of ionic strength is mainly responsible for the variations in pK_{app} observed for hydrophilic indicators residing around the center, whereas the effects of specific interactions and structural changes might be negligible.

This statement is confirmed by our dynamic studies on fluorescent probes located around the center (see Section 2.3) which show that beyond $w \approx 10\text{-}12$, there is no further micellar effect on deprotonation because water molecules surrounding the probe behave as normal water with respect to protolysis. Therefore, at lower water contents, the reduction of pK_{app} is likely to result from an effect of "ionic strength": an accumulation of sodium ions in the vicinity of the negatively-charged basic group slows down the recombination rate of the basic form with a proton.

In contrast, these considerations cannot explain the large increase in pK_{app} observed for indicators localized at the interface or in its vicinity: 4.4 units for paranitrophenol (47), 1.7 unit for 2,4-dinitrophenol (48) (this value is to be compared with the opposite and smaller variation in pK reported above for 2,3-dinitrophenol as a function of ionic strength). It should be noted that 2,4-dinitrophenol is less hydrophobic than paranitrophenol, and the difference in ΔpK can be explained in terms of different localizations of these probes.

5. CONCLUDING REMARKS

All the experiments aiming at characterizing the acidity in the aqueous core of reverse micelles, show that in the center of water pools large enough to contain a sufficient amount of free water $(w > 10)$, the classical concepts of pH and pK can be used; the variations in apparent pK of acid-base indicator can be interpreted in terms of ionic strength effect, and the deprotonation and reprotonation rates are close to those observed in bulk water.

On the other hand, in small water pools $(w < 10)$, or at the interface (even for larger micelles), the problem becomes extremely difficult, and in our opinion, attempts to get direct or indirect information on acidity in terms of hydrogen ion concentration should be abandoned not only because activity of hydrogen ion is to be considered instead of concentration, but also because this activity is not constant throughout the aqueous core; the large difference between the vicinity of the interface and the center precludes the estimation of an average value. In addition, such an average value would have no meaning for solubilized enzymes that bears various acido-basic groups experiencing different environments that are able or unable to protonate or deprotonate them.

In small water pools or at the interface, acidity must be rather viewed in terms of the tendency of water molecules to donate or accept protons. This tendency can be directly estimated by dynamic measurements of deprotonation and/or protonation rates of suitable probes with special attention to their site of solubilization. Examples of such an approach are described in Section 2 of this chapter.

Acknowledgments. The authors are indebted to Prof. M. Chemla for helpful discussions on thermodynamic aspects.

REFERENCES

1   J.H. Fendler and E.J. Fendler, "Catalysis in Micellar and Macromolecular Systems", Academic Press: New York, 1975.
2   Reverse Micelles, P.L. Luisi and B.E. Straub, Eds., Plenum Press, New York,1984.
3   P.L. Luisi and L.J. Magid, CRC Crit. Rev. Biochem. **20**, 409 (1986).
4   K. Martinek, A.V. Levashov, N. Klyachko, Y.L. Khmelnitski and I. V. Berezin, Eur.J.Biochem. **155**, 453 (1986).
5   M. Wacks, Proteins: Structure, Function, and Genetics, **1**, 4 (1986).
6   P.L. Luisi, M. Giomini, M.P. Pileni, and B.H. Robinson, Biochim. Biophys. Acta,**947**,209 (1988).
7   E. Bardez, B.T. Goguillon, E. Keh, and B. Valeur, J. Phys. Chem.,**88**, 1909(1984).
8   E. Bardez, E. Monnier, and B. Valeur, J. Phys. Chem.,**89**, 5031 (1985).
9   E. Bardez, E. Monnier, and B. Valeur, J. Colloid Interface Sci., **112**,200(1986).
10   M. Zulauf and H.F. Eicke, J. Phys. Chem., **83**, 480(1979).
11   H.F. Eicke, Chimia, **36**, 241 (1982).
12   a. B. Valeur and E. Keh, J. Phys. Chem., **83**, 3305(1979).
     b.E. Keh and B. Valeur, J. Colloid Interface Sci., **79**, 465 (1981).
13   A. Weller, Prog. React. Kinet., **1**,187 (1961).
14   E. Grunwald and D. Eustache in "Proton Transfer Reactions", E. Caldin and V. Gold , Eds., Chapman and Hall, London, 1975, Chapter 4.
15   a. M. Eigen and L. de Maeyer, Proc. R. Soc. London, Ser. A, **247**,505 (1958).
     b. M. Eigen, Angew. Chem.,Int. Ed. Engl., **3**,1 (1964).
16   J. Busch and J.R. de la Vega, Chem. Phys. Lett., **26**, 61 (1974).
17   H. Shizuka, T. Ogiwara, A. Narita, M. Sumitani and K. Yoshihara, J. Phys.Chem., **90**, 6708 (1986).
18   J. Lee, R.D. Griffin and G.W. Robinson, J. Chem. Phys., **82**, 4920 (1985).
19   G.W. Robinson, P.J. Thistlethwaite and J. Lee, J. Phys. Chem.,**90**, 4224 (1986).
20   D. Huppert and Kolodney, Chem. Phys.**63**, 401 (1981).
21   M. Kaschke, A. Granesz, and J. Kleinschmidt, Laser Chem., **7**, 41 (1987).
22   M. Gutman, E. Nachliel and D. Huppert, Eur. J. Biochem., **125**, 175 (1982).
23   M. Gutman, D. Huppert and E. Nachliel, Eur. J. Biochem., **121**, 637 (1982).
24   M. Wong, J.K. Thomas and T. Nowak, J. Am. Chem. Soc., **99**, 4730 (1977).
25   A. Llor and Rigny, J. Am. Chem. Soc., **108**, 7533 (1986)

26   J. Rouvière, J. M. Couret, M. Lindheimer, J.L. Dejardin and R. Marrony, J.Chim. Phys., **76**, 289 (1979).
27   M.A.J. Rodgers, J. Phys. Chem., **85**, 3372 (1981).
28   M.P. Pileni, B. Hickel, C. Ferradini and J. Pucheault, Chem. Phys. Lett., **92**,308, (1982).
29   P.E. Zinsli, J. Phys. Chem., **83**, 32223 (1979).
30   H.Kondo, I. Miwa, and J. Sunamoto, J. Phys. Chem., **86**, 4826 (1982).
31   M.J. Politi, O. Brandt, and J.H. Fendler, J. Phys. Chem., **89**,2345 (1985).
32   R. Kubik, H.F. Heicke and B. Jönsson, Helv. Chim. Acta, **65**,170 (1985).
33   B. Jönsson and H. Wennerström, J. Colloid Interface Sci., **80**,428 (1982).
34   E. Bardez and B. Valeur, unpublished results.
35   M. J. Politi and H. Chaimovich, J. Phys. Chem., **90**, 282 (1986).
36   F. Akoum and O. Parodi, J. Physique, **46**, 1675 (1985).
37   W.I. Higuchi and J. Misra, J. Pharmaceut. Sci., **51**, 455 (1962).
38   K. Galvin J.A. Mc Donald, B.H. Robinson, and W. Knoche, Colloid Surf., **25**, 195 (1987).
39   P.D.I. Fletcher, N.M.Perrins, B.H. Robinson and C. Toprakcioglu, in "Reverse Micelles", P.L. Luisi and B.E.Straub, Eds,Plenum Press, New York, 1984, p.69.
40   P.L. Luisi, P. Meier, V.E. Imre and A. Pande, in "Reverse Micelles", P.L. Luisi and B.E.Straub, Eds,Plenum Press, New York, 1984, p.323.
41   E. Bardez and B. Valeur, Chem. Phys. Letters,**141**, 261 (1987).
42   E. Pines, D. Huppert, M. Gutmann, N. Nachiel and M.Fishman, J. Phys. Chem.,90, 6366 (1986).
43   F.M. Menger and K. Yamada, J. Am. Chem. Soc., **101**, 6731 (1979).
44   R. E. Smith and P. L. Luisi, Helv. Chim. Acta, **63**, 2302 (1980).
45   a. H. Fujii, T. Kawai and H. Nishikawa, Bull. Chem. Soc. Jpn., **52**, 2051 (1979).
      b.H. Fujii, T. Kawai, H. Nishikawa and G. Ebert, Colloid Polym. Sci.,**260**, 697 (1982).
46   F. Nome, S.A.Chang and J.H. Fendler, J. Chem. Soc., Far. Trans. 1, **72**, 296 (1976).
47   F.M. Menger and G. Saïto, J. Am. Chem. Soc., **100**, 4376 (1978).
48   A. Levashov, V.I. Pantin and K. Martinek, Kolloid Zh.,**41**, 453 (1979).
49   A.T. Terpko, R.J. Serafin and M.L. Bucholtz, J. Colloid Interface Sci.,**84**, 202 (1981).
50   a. O.A. El Seoud, A.M. Chinelatto and M.R. Shimizu, J. Colloid Interface Sci.,**88**, 420 (1982).
      b. O.A. El Seoud and M.R. Shimizu, Colloid Polymer Sci., **260**, 794 (1982).
      c. O.A. El Seoud and R.C. Vieira, J. Colloid Interface Sci., **93**, 289 (1983).
      d. O.A. El Seoud and A.M. Chinelatto, J. Colloid Interface Sci., **95**,163 (1983).
      e. O.A. El Seoud, in "Reverse Micelles", P.L. Luisi and B.E.Straub, Eds,Plenum Press, New York, 1984, p.81.
51   R. G. Bates, "Determination of pH. Theory and Practice", Wiley, New York,(1973).
52   "Definition of pH scales, standard reference values, measurement of pH and related terminology (Recommendations 1984)", Pure Appl. Chem., **57**, 533 (1985).

53   "Treatise on Analytical Chemistry", Kolthoff and Elving, Eds, Wiley, New York, 1975, Part I, Vol.11, p. 6996.

FEMTOSECOND REACTIVITY OF EXCESS ELECTRON IN REVERSED MICELLES WATER POOL

Y. GAUDUEL, S. POMMERET, A. ANTONETTI

1.   INTRODUCTION

The use of organized assemblies incorporating atomic or molecular cluster is extensively encouraged to develop novel area of chemistry and biochemistry (1-5). In this way the mixture of alkyl chain surfactant with water in oil permit to obtain microemulsions (W/O) with specific properties which cannot be obtained in homogeneous polar or non polar media : compartimentalization of molecules, reduction of reaction dimensionallity, membrane mimetic effects, polydispersion of aqueous spheres at the nanometer scale.

In a ternary phase diagram for surfactant (Aerosol OT) hydrocarbon-water systems, the L_2 phase corresponds to reversed micelles, i.e. stable isotropic solutions (oil/surfactant/water system). The third component (water molecules) are solubilized as water droplets (nanometer scale) by a monolayer of surfactant in a bulk organic solvent (hydrocarbon). The figure 1 shows that in the L_2 phase diagram, the partition of sodium bis (2-ethylhexyl) sulfosuccinate is exclusively defined at the interface in which the polar lead groups are directed towards the continuous hydrocarbon phase (n-heptane). In this configuration, AOT reversed micelles can dissolve large amount of water ($H_2O < 10\%$ v/v). One of the most fondamental parameter used to define the reserved micelles size is the water to surfactant molar ratio ($W = [H_2O]/[AOT]$).

Numerous theoretical works and experimental investigations involving small-angle neutron-scattering (6-9), ultracentrifugation (10,11), photon correlation spectroscopy (12), time-resolved fluorescence probing (13-17), small angle X ray scattering, NMR spectroscopy (18-20), fluorescence polarization technique (21,22) and Raman scattering (23) have substantially improved the understanding of the structure of reversed micellized systems, the state and properties of sequestered water bubbles.

The best available experiments in term of the microorganization of the water molecules in AOT reversed micelles are those of neutron and light scattering or NMR spectroscopy (19,22,24-26). In particular, the ^{1}H and ^{23}Na NMR spectroscopy studies of the state of sequestered water clusters in AOT reversed micelles have shown that at low [H2O] / [AOT] ratio (W 5) all the

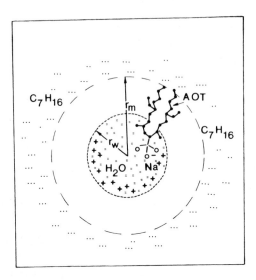

Fig. 1.A. Schematic representation of an AOT reversed micelle showing the non polar phase (n-heptane), the interfacial region with surfactant tails and the sequestred water pool. r_m and r_w represent the radius of reversed micelle and water droplet respectively.

solubilized water molecules ($n \cong 300$/micelle) are expected to interact with the polar head groups of surfactant. These molecules participate in the solvation of counterion (Na^+) through strong ion-dipole interactions between the counterion and the water. This specific sequestered interfacial water is characterized by a high activation energy for the ^{23}Na relaxation process (6.85 Kcal/mol. with W 5). These ion-dipole interactions lead to a specific organization of water molecules around the sodium ion and provide favourable distribution of ionic clusters [$Na(H_2O)^+_n$] within the inner aqueous core. Raman scattering studies provide a detailed description of the structure of water in reversed micelles. The bonded water near the polar surface suffer a loss of structure linked to the interaction with counterions (23). In this more disordered phase, the regular tetrahedral structure of water is not observed. The NMR investigations on the rotationnal correlation time T_c obtained from T_1 clearly demonstrate that the amount of free water tends to zero when the water molecules are immobilized by strong binding of the sodium ions (W 5). Similar interations are confirmed by the effect of the water pool size (W) on the chemical shift of H_2O within reversed micelles or the IR spectral data of water in AOT reversed micelles. NMR and spectroscopic data on the structure and state of water become similar to those of water bulk when size of water pool is increasing (W > 20). However even at high [H$_2$O]/[AOT] ratio (W 50) theoretical and experimental investigation of constrained water in reversed micelles underline a two-state model for the description of the water bubble : an aqueous larger corresponding to very viscous water close to the interface and an inner pool with properties similar to bulk water (19,28-30).

The microstructure and states of sequestered water bubbles in reversed micellar systems offer the opportunity to investigate specific reactions (charge separation, monoelectronic transfer) in unique aqueous media stabilized by interfacial layer. Photochemical and radiochemical techniques have allowed the investigation of chemical reactivities occurring in monodisperse micropocket which mimic biological interfacial water or membrane functions. Indeed, the flash photolysis of solubilized probes (phenothiazine, methylphenothiazine) or pulse radiolysis of nonpolar fluid (iso-octane, cyclohexane, n-heptane) have permitted to obtain information on the behaviour of primary species (hydrated electron) following energy deposition in reversed micelles (31-35) or on photoelectron transfer from compartimentalized donor (m-methylphenothiazine, phenothiazine, chlorophyll) to acceptor (methylviologen, cytochrome C or propylviologensulfonate) (36-38). In these experiments, the electron is used as microprobe of local dynamics of molecular organization induced by the presence of an excess electron. The dynamics of formation of the solvated electron is of considerable interest for the understanding of the early events involved in a monoelectronic transfer in the condensed matter. Moreover, the energetics and the time dependence of the electron-medium interaction play an important role in the formation of this exceptional radical. Recent investigations on the capture of excess electron by water pool of reversed micelles have demonstrated that electron attachement [e$^-$ + n(H$_2$O)] at high W occurs with a rate that exceeds a diffusion controlled process (33). However, the elucidation of the primary steps of charge separation and electron capture by water pool (electron thermalization, localization and solvation) have always been limited by the instrumental resolution.

The improvements in ultrashort pulse generation made possible new photochemical advances in the elucidation of detailed mechanisms of the primary events of the electron transfer in micellized water clusters. In the section 2 we focus on the specific technical points involved in the femtosecond spectroscopy of reversed micelles. The section 3 will deal with recent data on the dynamics of the primary events following the interaction of ionizing radiation with a chromophore in AOT reversed micelles. Picosecond free radical reactions occurring in water clusters between sequestered oxidized coenzyme and solvated electron within water clusters will be discussed in the section 4.

2. FEMTOSECOND SPECTROSCOPY OF AOT REVERSED MICELLES

The topic of generating subpicosecond pulses has been discussed at length in previous paper (39,40). However, in this section we describe the specific points used for femtosecond investigation of electron reactivity in reversed micelles. Spectral and kinetic studies have been carried out using femtosecond pump-probe method.

The figure 2 represents the general scheme for generation of femtosecond pulses, amplification and time-resolved spectroscopy. The generation of high power femtosecond laser is based on colliding pulse mode-locked dye laser (CPM) containing two following jets : one jet of rhodamine 6G and one very thin jet of saturable absorber. The pulses generated in this ring cavity are as shorter $80 \cdot 10^{-15}$s and the wavelength is centered around 620 nm. The presence of four prisms in the cavity permit to compensate the group velocity dispersion. The amplification of the femtosecond pulses (60 angströms wide spectrum) due to the group velocity dispersion and compensated by means of an ajustable delay line composed of a grating pair (GP) or a four prisms arrangment. The amplification chain allows output pulses of energy above 1 mJ and typically 100 fs duration (10^{-13}s) when using a pumping energy of 300 mJ at 520 nm with a pump duration of 6 ns at a 10 Hz repetition rate.

The obtaining of high intensity femtosecond pulses are fundamental for spectroscopic purpose in that they can generate white light pulses of comparable duration (so called "continuum"). This continuum light is used as a weak delayed pulse beam while the intense pulse, required for perturbing the analyzed micellar system remains in the initial fundamental pulse or with less energy in its second harmonic using a thick KDP crystal (20 μJ at 310 nm). In our experimental conditions, this ultraviolet light constitutes the pump pulse used to initiate the primary photochemical reaction i.e. photoionization of micellized phenothiazine. Previously, papers have described the analysis of induced absorption kinetics measured over the whole visible spectrum and in the near infrared up to 1.25 μm (41).

For the essential, the absorption change induced by a pulse intensity $I_p(t)$ on a test pulse of intensity $I_T(t)$ with a temporal delay between the two pulses (as determined by the variable delay line) is given in our small signal conditions by :

$$\Delta\alpha(\tau) = \int_{-\infty}^{+\infty} A(\tau - \tau') \int_{-\infty}^{+\infty} I_p(t)\, I_T(t + \tau')\, dt\, d\tau' \qquad [1]$$

In case of instantaneous molecular response A(t) is a constant while for a non null molecular response time T it becomes $A(t) = 1 - \exp(-t/T)$ for $t > 0$.

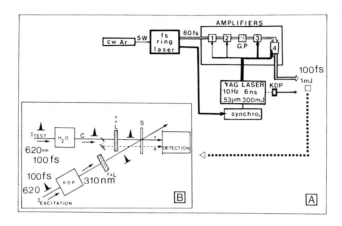

Fig. 2-A. General scheme for generation and amplification of femtosecond pulses including the CPM laser, the four amplifier stages and the gratting pair (GP) used for pulses compression. The insert B shows the principle of femtosecond spectroscopy of reversed micelles (F : filter, L : lens, T : test pulse, R : reference pulse, I : pump pulse, S : sample, C : continuum).

3.   ELECTRON LOCALIZATION AND SOLVATION IN AOT REVERSED MICELLES

The preparation of AOT inverted micelles and the investigation of the primary events of excess electron capture by water clusters have been performed at ambient temperature (21 +/- 1°C). The method of preparation of these microemulsions have been described in previous paper (42). The photochemical generation of excess electrons in AOT reversed micelles has been performed by resonant two photon ionization of a polycyclic chromophore (phenothiazine) having a low ionization potential (I_g 6.5 eV).

The dynamics of electron capture by water pool of the PTH/n-heptane AOT/H_2O system is obtained through infrared and visible femtosecond spectroscopic investigations. The interaction of intense UV laser pulses with PTH leads to an ultrafast charge separation in the hydrophobic region of reversed micelles. To account for this ionization process, a likely photochemical reaction may be considered. Ultrashort powerful laser UV radiation ($I > 10^8$ w/cm^2) can be absorbed through a non linear process, namely two photon absorption. In the specific case of PTH, it is suggested that the absorption of a 4 eV (310 nm) would lead to a vibrationally excited level in the first excited singlet state (PTH S_1) i.e. 2.9 eV above the ground state. In a two step process, a second photon will be absorbed by the singlet state S_1 of PTH giving an highly excited state which may dissociate in a thermally actived process to subsequently yield a charge separation. In this hypothesis the excess electron (e^-_{qf}) is characterized by an average kinetics energy of about 1 eV.

$$\text{PTH} \xrightarrow{h\nu} [\text{PTH}^*] \rightarrow [\text{PTH}^+...e^-] \rightarrow \text{PTH}^+ + e^-_{qf} \qquad [2]$$

The picosecond absorption changes obtained following the femtosecond photoionization of micellized PTH in AOT reversed micelles (W 50) exhibit a structureless and unsymmetric band extending a high energy tail above 2.5 eV and peaking around 1.7 eV (43). From the available data literature on the radiation chemistry of aqueous solution (44) this visible band is assigned to an hydrated state of the electron. If we assume an absorption coefficient of 1.85 10^4 M^{-1} cm^{-1} for the hydrated electron at 720 nm (45,46) the estimate of the concentration of this radical after each femtosecond ultraviolet pulse is about 4.2 +/- 0.2 μM. This concentration decreases below 2 μM in reversed micelles with very low H$_2$O/AOT ratio (W 5). This result indicates that the efficiency of the femtosecond capture and trapping by water bubbles decreases with decreasing the water pool size. In agreement with previous pulse radiolysis or photolysis experiments (32,34,45) the spectroscopy properties of the hydrated electron generated by photochemistry depends on the weigh-in amount of sequestered water (W). The existence of a significant blue shift ($\Delta E \cong 0.21$ eV) at low W (W 5) is assigned to a high concentration of counterion (Na$^+$) within the inner aqueous cluster. Similar effects of high ionic strength on spectroscopic properties of hydrated electron have been described through nanosecond radiolysis of concentrated ionic solution and have been assigned, in this range of concentrations, to a change of the electron hydration energy (47). The time-resolved investigations of electron capture by water pool of reversed micelles have been performed according to configuration of the figure 2. The main results are summarized in figure 3 and table 1. From these results, it can be observed that the apparent risetime of the induced absorption (T_{obs}) in the red spectral region ($\lambda_t = 720$ nm) is dependent on the weighed-in amount of water i.e. the size of the water microdroplet. More precisely, there is a decrease of the $K_{obs} = 1/T_{obs}$ when the number of sequestered water molecules is increased in AOT reversed micellar system. The changes of the [H$_2$O] / [AOT)] ratio from 50 to 5 modify significantly the efficiency of an ultrafast excess electron capture by AOT reversed micelles water pool. A specific analysis of time resolved data from different size of water clusters will be discussed at length in a next paper (43).

TABLE 1

Effect of weighed-in amount of water (W = [H2O] / [AOT] on the risetime of 720 nm induced absorption following femtosecond photoionization of PTH in AOT reversed micellar (AOT 0.1M/ n-heptane/water).

$[H_2O] / [AOT]$	$n\ H_2O^*$	$T_{obs.}$ (10^{-15} s)
W = 5	$0.2\ 10^3$	90 +/- 10
W = 30	$18\ 10^3$	130 +/- 10
W = 40	$40\ 10^3$	150 +/- 10
W = 50	$70\ 10^3$	180 +/- 10

* n : average aggregation number of water molecules per micelle. These values are estimated from data of Eicke and Rehak (10).

In this paper we will discuss only some of the recent data obtained with large aqueous clusters (W 50). The concerned results are summarized in the figure 4. In AOT reversed micelles without water droplet (W 0), the risetime of the induced absorption at 720 nm can be well fitted to a one build up time of 80+/-10 fs. This time constant taking into account the position of the zero time delay as previously shown (31) is assigned to the formation of a trapped state of electron in the hydrocarbon phase of reversed micelles ($e^-_{n-heptane}$). The time-revolved spectroscopy data obtained following laser photolysis of phenothiazine in reversed micelles (W 50) are shown in the part A of figure 4. The 720 nm risetime showns a real time delay in comparison with control curve (W 0). Indeed, in reversed micelles containing large amount of water molecules, the electron transfer from an excited state of PTH leads to a subsequent reactivity with the entrapped water cluster.

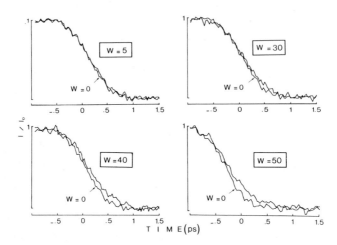

Fig. 3 Femtosecond date on the dynamics of photoionization of phenothiazine
($4\,10^{-4}$ M) in AOT reversed micellar systems (AOT : 0.1 M in n-heptane) with
various molar ratio of water and AOT (W = [H_2O] / [AOT]).

The kinetic data obtained in the infrared and visible spectral region (figure 4)
have been analyzed with a dynamical model wich takes into account the
contribution of two populations of electrons : one population is assigned to a
trapped state in the non polar phase (n-heptane) and the other to a hydrated state
within the inner core of reversed micelles (43). The risetime of the induced
absorption at 720 nm (W 50) can be fitted to the convolution of the pulse profile
and the extended signal rise by use of combination of two monoexponential laws
($1 - \exp(-t/T_X)$). The first time constant $T_{X1} = 80+/-10$ fs corresponds to the
appeerence of localized electron in the hydrocarbon phase (figure 5). The second
time constant (T_{X2}) has been determined from experiments performed in the
infrared (figure 4B). This absorption spectroscopy kinetics has been analyzed by
a kinetical model which takes into account the existence of a transient state of the
electron before solvation in water microdroplet. The infrared kinetics is well
fitted to the equation 3 giving an appearance time T_1 of 140 +/- 10 fs and a
lifetime T_2 of 270 +/- 10 fs.

$$A_{(t)} = A_o[1/(T_2 - T_1)] \, [\exp(-t/T_2) - \exp(-t/T_1)] \tag{3}$$

The incomplete recovery of the signal at 1000 nm is due to the contribution of the e^-_{trap} in n-heptane and e^-_{sol} in an aqueous microdroplet as shown at 720 nm. Indeed at 720 nm, the induced absorption T_{x1} and T_{x2} is perfectly fitted to the convolution of the pulse profile and the expected signal rise (equation 4) by use of the same values of T_1 and T_2 as determined at 1000 nm.

$$A_{(t)} = A_o \, [1 - 1/(T_2 - T_1)] \, [T_2 \exp(-t/T_2) - T_1 \exp(-t/T_1)] \tag{4}$$

These femtosecond spectroscopic investigations on AOT reversed micellar systems clearly demonstrate that the primary event occurring after the photoionization of PTH correspond to an ultrafast excess electron capture by the aqueous microdroplets with a rate that exceeds the estimates of a simple diffusion-controlled reaction. Previous experiments have established that the rate of attachment of exces electron to a micelle water pool k ($e^- + [H_2O]_m$) was faster than a diffusion controlled rate (10^{15} $M^{-1}s^{-1}$) for large water concentration (W 30) (33). In large water droplets (W 50, H_2O 6%) the excess electron capture in the water bubble occurs with a pseudo first order rate of 7.1 10^{12} s^{-1}. In these conditions, we can estimate that the rate constant $k[e^- +(H_2O)_n]$ is 7 10^{17} $M^{-1}s^{-1}$. Consequently, the femtosecond electron attachment ($e^- + nH_2O$) with water cluster cannot be interpreted in term of simple diffusion controlled reaction involving spherical species (electron and water clusters). From the static absorption spectrum of PTH in AOT reversed micelles and in agreement with previous experimental works showing that the solubility of PTH in n-heptane is significantly increased by alkyl chain AOT, we concluded that this chromophore is preferentially localized near the interfacial water. This localization of PTH would be favourable to an ultrafast electron capture by surfactant solubilized water pool following femtosecond photoionization of PTH (figure 5).

The present experiments have permit to apprehend the dynamics of the early events which initiate the hydrated electron formation in water clusters of reversed micelles. The mechanisms of femtosecond charge separation leading to an electron capture can be understood in term of a tunneling process between an excited state of phenothiazine and unoccupied levels of the system aq/e^-_{aq} inside the aqueous inner core. One of the main result obtained in the present study is the direct observation of a transient infrared species following the electron transfer from phenothiazine to large water cluster. From the femtosecond data available in pure liquid water (50), this transient infrared species is assigned to a precursor of hydrated electron in the polar inner core of reversed micelles. Consequently, it is

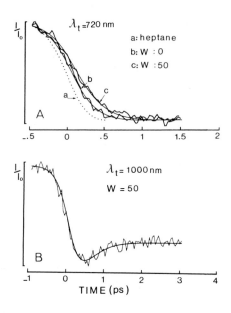

Fig. 4. A. Time resolved induced absorption following femtosecond UV excitation of phenothiazine (4 10^{-4} M) in AOT reversed micelles (AOT 0.15 M-n-heptane). The lines b, represent computed best fits giving appearance time constant of 90 fs (W 0) and 180 fs (W 50). The insert B shows the time resolved spectroscopy data at 1000 nm following femtosecond photoionization of PTH (4 10^{-4}M) in AOT reversed micelles (W 50). The lifetime of the transient species (e^-_{presol}) is assumed to be 270 + 20 fs.

clear that the photoelectron solvation in the water bubble does not proceed through a single step. The risetime T_1 of this infrared tail (140 +/- 10 fs) includes both the electron ejection from the chromophore, the electron thermalization and localization inside the aqueous micellized cluster. Previously, it has been estimated that the thermalization distance of an epithermal electron with an excess kinetic energy of 1 eV is about 40 angströms (49). Such a distance is smaller than the mean radius (r_W) of the aqueous microphase measured from small angle aqueous neutron scattering of X ray scattering : r_W = 60 A for W 50 (10). This would permit to suggest that when a photoelectron is preferentially ejected through the charged interface and captured by bulk like water cluster, its solvation likely occurs in the inner polar phase of water droplet (figure 5).

n‒HEPTANE $\Big/$ AOT $\Big/$ H$_2$O $\Big/$ AOT PTH n‒HEPTANE

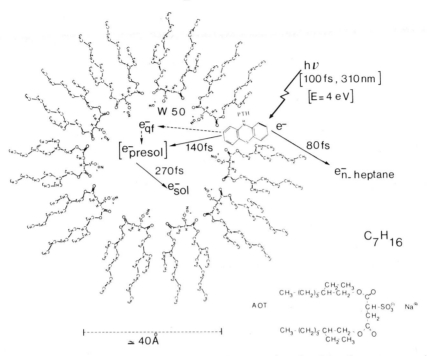

Fig. 5. Sequence of primary photochemical events involved in charge separation and electron capture by reversed micelle water pool. e^-_{qf} : quasi free electron, $e^-_{n\text{-heptane}}$: localized electron in nonpolar phase, e^-_{presol} : infrared prehydrated electron, e^-_{sol} : hydrated electron.

Once the electron gets trapped (e^-_{presol}), the final stabililization of the localized electron towards a fully solvated state occurs in 270 fs. The kinetics reported here demonstrate that the initial negatively charged cluster $[(H_2O)_n^-]$ i.e. (e^-_{presol}) is a well chemical entity spectroscopically characterized on a femtosecond time scale. These studies provide unique experimental basis for testing recent theoretical approaches on excess electron coupling in liquid water. The very fast appearance of the presolvated electron is at a time short compared to any nuclear motion, solvent dipole orientation or thermal motion of water molecules (Frank Condon principle) and its implies that efficient mecanism involved in the localization process of electron do not require large molecular and dynamical reorganization. These femtosecond investigations do not permit to distinguish precisely if the electron creates its own trapping site (self trapping mechanism) or searchs pre-existing shallow traps. In agreement with plausible picture of electron trapping in pure water (53,54), our IR data are compatible with a distribution of pre-existing sites which behaves monotonically in energy down to -1.4 eV. In the appearance of a presolvated state (e^-_{presol}), the efficient down to -1.4 eV. In the appearance of a presolvated state (e⁻$_{presol}$), the efficient role of shallow traps i.e. (V$_0$ 0.58) for a cross section around 20 A^2, would correspond to a spatially extended electron-trap of about 4. This size is greater than the radius of the fully solvated electron (r = 2.3-2.8 angströms). The time resolved appearance of the initially non fully hydrated electron (140 fs in reversed micelle with large W) rules out the assumption of a predominent role of high density of pre-existing deep traps into which the electron falls directly after photoejection. Our femtosecond data showing the existence of a precursor of solvated electron (e⁻$_{presol}$) demonstrate that the presence of counterions Na$^+$ in water bubble does not provide more favourable spatial distribution of pre-existing deep traps that in pure liquid water. It is interesting to notice that in large water droplet, the appearance time and the lifetime of this primary species (prehydrated electron) are similar to values obtained in pure liquid water (50). This would permit to suggest that when a photoelectron is ejected through the charged interface of reversed micelles and captured by the bulk like water cluster, its early localization likely occurs in the inner polar phase containing free water molecules. NMR spectroscopy of AOT micellized water cluster have demonstrated that i) the counterion sodium is preferentially localized near the micellar solubilized water interface (70%), ii) the unbounded water molecules remain localized in the inner core of aqueous clusters (19). Additionnal investigations by time-resolved fluorescence (22) agree for an inhomogeneous polar interior of AOT reversed micelles including a viscous polar boundary layer of about 6 A and an unbounded water droplet (free water molecules). In agreement with these NMR studies on the organization of water clusters in AOT/n-heptane reversed micelles, the femtosecond spectroscopic and dynamic data on electron capture by large water clusters permit to establish that the early events involved in the localization and solvation are not influenced by the properties of the sequestered water bubble and remained similar to those in large water bulk (50). Owing to the fact that theoretical models which takes into account the role of solvent fluctuations in determining the rate of adiabatic electrontransfer in liquids tend to demonstrate that molecular rearrangement near a point charged may induce equilibrium of the binding energy at the subpicosecond time scale (51,52), the energy distribution of pre-existing traps in water bubbles and their specific role in the excess electron capture and solvation are still debatable questions.

4.   ELECTRON TRANSFER WITH OXIDIZED PYRIDINE NUCLEOTIDE

Pulse radiolysis investigations of quenching reactions with dilute solutions of scavenger (NO$_3^-$, cytochrome C, chymotrypsine, ribonuclease) have permit to obtain information on the distribution of entrapped molecules and kinetical data on a monoelectronic transfer in reversed micellar systems (37,38). In this section IV we discuss about photochemical investigations of reversed micelles containing concentrated sequestered solution of oxidized coenzyme (NAD$^+$, nicotinamide adenine dinucleotide). The aim of these studies is to investigate the one electron transfer process with NAD$^+$ in water clusters following micellar capture of excess electron. The stabilization of oxidized coenzyme in sequestered aqueous microdroplet will provide useful redox system to investigate free radical reaction (univalent reduction) in interfacial water.

$$NAD^+ + e^- \xrightarrow{K} NAD^\circ \text{ (Pyridinyl radical)} \tag{5}$$

The possibility that a primary reaction involves a one-electron transfer between phenothiazine and an oxidized coenzyme entrapped in the aqueous phase of AOT reversed micelle has been investigated by femtosecond absorption spectroscopy. Nicotinamide adenine dinucleotide, a component of the mithochondrial electron chain transport is implicated in important biochemical oxidation-reduction reactions. Much of the interest in oxidized NAD derives from its function in the electron transfer reaction mechanisms in redox systems (55,56).

Using the micellar system phenothiazine/AOT interface/water-oxidized coenzyme as a simple model for redox reaction we have investigated the kinetics of one-electron transfer between the donor located in the hydrophobic core of the micelle (PTH) and the acceptor sequestered in the water bubble. The solubilization of an oxidized coenzyme within the aqueous core of a reversed micelle is of interest for the development of biomimetic ractions. NAD$^+$ was purchased from Sigma Chemical Corporation and was dissolved at different concentration in the water pool of reversed micelles.

To assess the structural integrity of the oxidized coenzyme, absorption spectra were taken for different [NAD$^+$] concentration and W. Static spectral characterization of micellar samples was run out on a dual-beam Varian 2300 spectrophotometer in the UV and visible spectral region. A spectrum of a normal AOT reversed micelles systems is used as reference. The figure 6 summarizes the results obtained for [NAD$^+$ = 10.5 mM] in the water pool.

By comparison with spectrum of NAD$^+$ in aqueous solution, the spectrum of sequestered NAD$^+$ in reversed micelles shows a characteristic band centered around 320 nm. When W is increased from 20 to 60 the insert of figure 6A demonstrates that this additionnal band becomes undistinguishable. For low W (5<W<20), the band centered at 320 nm increases significantly. These spectral observations suggest that a fraction of the oxidized coenzyme (NAD$^+$) interact with the interfacial layer giving a specific complex characterized by this

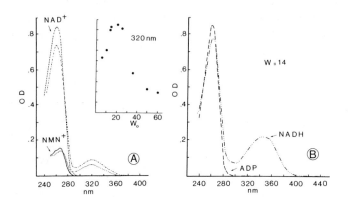

Fig. 6 Absorption spectra of nucleotide (10.5 mM in water pool) sequestered in AOT reversed micelle systems (AOT 0.10 M, n-heptane-H$_2$O). A : Spectral data for NAD$^+$ and NMN$^+$ in bulk water (-----, —), in reversed micelles W 14 (---,---, ...). The insert represents the influence of the molar ratio [H$_2$O]/[AOT] on the intensity of the visible band centered atound 320. B : Absorption spectra of ADP (10.5 mM) and NADH (10.5 nM) in AOT reversed micelles (AOT 0.1 M, n-heptane, W 14).

absorption band around 320 nm. It is interesting to notice that a similar band is observed with the oxidized mononucleotide (NMN) but neither with adenine mononucleotide (ADP) nor reduced form of dinucleotide (NADH). This would suggest that i) this band originates from an oxidized pyridinyl cycle, ii) the interaction depends on the properties of the water microdroplet and the hydration degree of the micelle. It can be also suggested that the negatively charged sulfogroup of AOT molecule can form complex with positively charged pyridinil cycle but not with neutral pyridium cycle of NADH. These conclusions are in agreement with previous micellar enzymology investigations which demonstrate that the catalytic activity of solubilized enzymes is dependent on the degree of hydration of reversed micelles i.e. [H$_2$O]/[AOT] ratio (4,5).

Picosecond kinetics of solvated electron with oxidized pyridine nucleotide was studied using femtosecond ultraviolet pulse to initiate charge transfer. The time resolved data performed in the red spectral region (λ_t=720 nm) are summarized in the figure 7. The upper part of this figure shows that in reversed micelle without electron acceptor there is no significant relaxation of the induced absorption in the first 200 ps following laser photoionization. These results are in agreement with the estimate of the subnanosecond lifetime of solvated electron in AOT reversed micellar systems (32,38). The rate constant for the electron transfer with pyridine dinucleotide was determined by following the rate of disappearance of the hydrated electron absorption at 720 nm. The curves of the figure 7 represent a typical time-resolved transfer at pH 2.7. NAD^+ reduces the lifetime of this radical. Notice also a first early relaxation in the first picosecond following the excitation. This relaxation is attibuted to an excited state of NAD^+ (57). The rate of the univalent reaction between hydrated electron and oxidized coenzyme in the smallest surfactant solubilized water pool (W<20) is significantly depressed in comparison with high water pool size. This result is in agreement with radiochemical reaction occurring between solvated electron and diluted concentration of acceptor (31). In the conditions where $[e^-_{sol}] \ll [NAD^+]$, the pseudo-first order condition prevails and the slower decay signal can be fitted to a monoexponential (exp -t/T) giving a bimolecular rate constant of 2.8 $M^{-1}s^{-1}$ and 1.5 10 M^{-1} s^{-1} for W 50 and 20 respectively.

Whatever is the water pool size, the rate constant for the reaction of hydrated of hydrated electron withpyridine nucleotide (k 10 $M^{-1}s^{-1}$) is indicative of a diffusion-controlled process where practically every encounter yields a product (pyridinyl radical $NAD^°$). However, the bimolecular rate constant at the picosecond time scale is lower than values obtained in aqueous biomicelles (57). This would indicate that in AOT reversed micelles, specific factors may influence the dynamics of the univalent reduction occurring in large water bulk of water-restricted environment (effect of hydrogen bound, dielectric properties, ionic strength).

The variation of the bimolecular rate constant with W can be interpreted, in our biomicellar system by a significant effect of the polar interface on the localization of the oxidized coenzyme (NAD^+) inside the water bubble.

These femtosecond spectroscopic investigations demonstrate that in reversed micellar aggregates the oxidized coenzyme NAD^+ maintains its biological capacity to mediate redox reactions. More precisely, the one-electron reduction of NAD^+ triggered by a photochemical electron transfer leads to the formation of the pyridinyl radical ($NAD^°$). The reactivity of the hydrated electron with the oxidized coenzyme is dependent on the properties and state of water bubbles in reversed micelles. Numerous theoretical and experimental studies performed in micellar systems have shown that the reactivity of reactant with solvated electron are dependent on i) the location of the acceptor inside the aqueous phase of reversed micelles ii) the weighed-in amount of water (58-60). In the conditions where the electron acceptor concentration is low (micromolar scale)

138

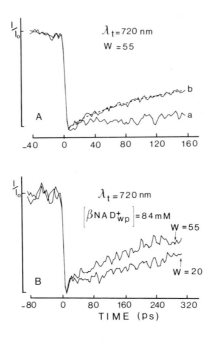

Fig. 7. A : Time-resolved absorption following the femtosecond photolysis of phenothiazine (4 10^{-4} M) in AOT reversed micelles (AOT 0.1M, n-heptane) a : reversed micelles without electron acceptor, b : reversed micelle containing NAD$^+$ (68 10^{-3} M). B : Effect of the water pool size on the dynamics of absorption decay.

the bimolecular rate constant is expected to vary as $1/w^3$ or $1/w^2$ depending on whether acceptor is located at the interface or in the water pool. In our experimental conditions (high concentration of electron acceptor) the absence of the classical dependence of the quenching rate constant of hydrated electron with the molar ratio W would suggest that the Poisson distribution of the probe in the reversed micelles is not verified. Moreover, at the picosecond time scale, the influence of a water pool collision or an exchange process in the bimolecular reaction is unconsiderable. In return, from picosecond data obtained in aqueous micelles (57) the contribution of several factors may regulate the time-dependence of the bimolecular rate constant k[e$^-_{sol}$ + NAD$^+$] : electrostatic factors, transient gradient, ionic strength. Investigations in a scheme of a diffusion-controlled reaction including these factors are in progress. The picosecond studies of the initial one-electron transfer reaction with oxidized coenzyme sequestered in water clusters is useful to elucidate an ultrafast free-radical reaction close to the one occurring in biological membranes active micropocket enzyme.

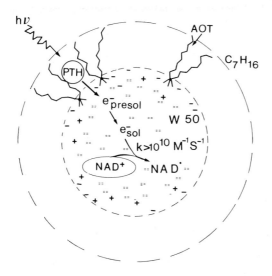

Fig. 8. Representation of the primary photochemical events occurring in a reconstitued redox system containing an electron donor (PTH 4 10^{-4} M), an electron acceptor in the sequestered aqueous micropocket. The reversed biomicellar system contains AOT 0.1 M, n-heptane, PTH 4 10^{-4} M, sequestered water (W 50) and NAD.

Conclusion

In AOT reversed micelles containing large amount of sequestered water molecules (W 50, 5 10^4 water molecules per micelle) the charge separation following femtosecond photolysis of phenothiazine (PTH$^+$ ...e$^-$) leads to a fast electron capture by aqueous clusters and a subsequent trapping for which the surrounding molecules are not in equilibrium configuration. The capture efficiency is dependent on the water pool size. The infrared absorption spectroscopy data demonstrate the existence of an early transient i.e. a precursor of the fully equilibrated state of solvated electron with a lifetime of 270 fs. In AOT reversed micelles containing concentrated solution of oxidized NAD$^+$, the femtosecond initiation of a monoelectronic transfer leads to a subsequent free-radical reaction involting e$^-_{sol}$. The attachment of a solvated electron to a micellized oxidized coenzyme corresponds to a bimolecular diffusion-limited process. Further investigations in the infrared are in progress to study the kinetics of free-radical reaction which compete with electron capture and solvation in water clusters.

Acknowledgments
The authors are thankful to N. Yamada for technical assistance.

REFERENCES

1   J.H. Fendler, J. Phys. Chem. 84 (1980) 1485-1491.
2   N.J. Turro, M. Gratzël, A.M. Brown, Angew. Chem. Int. Ed. Engl. 19
     (1980) 679-696.
3   C.J. O'Connor, T.D. Lomax, R.E. Ramage, Ad. Colloïd. Interf. Sci. 20
     (1984) and references therein.
4   K. Martinek, A.V. Levashov, N.L. Klyachko, V.I. Pantin, L.V. Berezin,
     Bioch. Biophys. Acta 657 (1981) 277-294.
5   S. Barbaric, P. Luisi, J. Am. Chem. Soc. 103 (1981) 4239-4244.
6   C. Cabos, P. delord, J. Appl. Cryst. 12 (1979) 502-510.
7   M. Kotlarchyk, J.S. Huang, S.H.J. Chen, J. Phys. Chem. 89 (1985) 4382.
8   C. Cabos, J. Marignan, J. Phys. Lett. 46 (1985) 1267.
9   B.H. Robinson, C. Toprakcioglu, J.C. Dore, J. Chem. Soc. Farad. Trans. 1
80   (1984) 13-27.
10   F.F. Eicke, J. Rehak, Helv. Chim. Acta 59 (1976) 2883.
11   B.H. Robinson, D.C. Stegler, R.D.J. Tack, J. Chem. Soc. Faraday Trans. 1
     75 (1979) 481.
12   M. Zulauf, H.F. Eicke, J. Phys. Chem. 83 (1979) 480-486.
13   N.J. Bridge, P.D.I. Fletcher, J. chem. Soc. Faraday Trans. 1 79 (1983)
     2161.
14   H.F. Eicke, P.E. Zinsli, J. Coll. Int. Sci. 65 (1978) 131-140.
15   J. Lang, A. Jada, A. Malliaris, J. Phys. Chem. 92 (1988) 1946-1953.
16   E. Keh, B. Valeur, J. Coll. Int. Sci. 79 (1981) 465-478.
17   E. Bardez, E. Mounier, B. Valen., J. Phys. Chem., 89 (1985) 5031-5036.
18   S.G. Frank, G. Zografi, J. Colloïd. Interface Sci. 28 (1968) 66.
19   M. Wong, J.K. Thomas,T. Nowak, J. Am. Chem. Soc. 99 (1977)
     4730-4736.
20   A. Maitra, J. Phys. Chem. 88 (1984) 5122.
21   M. Wong, J.K. Thomas, M. Gratzël, J. Am. Chem. Soc. 98 (1976) 2391.
22   P. E. Zinsli, J. Phys. Chem., 83 (1979) 3223-3232.
23   F. Mallamace, P. Migliardo, C. Vasi, F. Wanderlingh, Phys. Chem. Liq. 11
     (1981) 47-58
24   D. Balasubramanian, J. Indian Chem. Soc. 58 (1981) 633.
25   K.F Thomson, L.M. Gierasch, J. Am. Chem. Soc. 106 (1984) 3648-3652.
26   C.A. Martin, L.J. Magid, J. Phys. Chem., 85 (1981) 3938-3944.

27   P. Stilbs, B. Lindman, J. Colloïd. Interface Sci. 99 (1984) 290-...

28   M.J. Politi, H. Chaimouich, J. Phys. Chem., 90 (1986) 282-287.

29   P. Grigoloni and M. Maestro, Chem. Phys. Lett., 127 (1986) 248-252.

30   C.A. Boicelli, M. Giomini, A.M. Giuliani, Appl. Spectro. 38 (1984) 537.

31   A.J.W.G. Visser, J.H. Fendler, J. Phys. chem. 86 (1982) 945-950.

32   V. Calvo Pevez, G.S. Beddard, J.H. Fendler, J. Phys. Chem. 85 (1981) 2316-2319.

33   G. Bakale, G. Beck, J.K. Thomazs, J. Phys. Chem. 85 (1981) 1062.

34   M.P. Pileni, B. Hickel, C. Ferradini, J. Pucheault, Chem. Phys. Lett. 92 (1982) 308.

35   M.P. Pileni, Chem. Phys. Lett. 81 (1981) 603-605.

36   M.P. Pileni, B. Lerebours, P. Brochette, Y. Chevalier, J. Photochem. 28 (1985) 273-283.

37   M.P. Pileni, T. Zemb, C. Petit, Chem. Phys. Lett. 118 (1985) 414.

38   M.P. Pileni, P. Brochette, B. Hickel, B. Lerebours, J. Colloïd. Interface Sci. 98. (1984) 549-554.

39   R.L. Fork, C.V. Shank, R.T. Yen, Appl. Phys. Lett. 41, 223.

40   A. Migus, A. Antonetti, J. Etchepare, D. Hulin, A. Orszag, J. Opt. Soc. Am. B2 (1985) 584.

41   Y. Gauduel, A. Migus, J.L. Martin, Y. Lecarpentier, A. Antonetti, Ber. Bunsenges. Phys. Chem. 89 (1985) 218-222.

42   Y. Gauduel, A. Migus, J.L. Martin, A. Antonetti, Chem. Phys. Lett. 108 (1984) 319-322.

43   Y. Gauduel, S. Pommeret, N. Yamada, A. Migus, A. Antonetti, submitted to publication.

44   M. Anbar, E.J. Hart, in "The hydrated electron", New-York (1970).

45   G.E. Hall, G.A. Kenney-Wallace, Chem. Phys. 32 (1978) 313.

46   D.N. Nikogosyan, A.A. Oraevsky, V.I. Rupasov, Chem. Phys. 77 (1983) 131-143.

47   I.V. Kreitus, J. Phys. Chem. 89 (1985) 1987-1990.

48   M. Wong, M. Gratzël, J.K. Thomas, Chem. Phys. Lett. 30 (1975) 329.

49   H. Neff, J.K. Sass, H.J. Lewerenz, H. Ibach, J. Phys. Chem. 84 (1980) 1350.

50   Y. Gauduel, J.L. Martin, A. Migus, N. Yamada, A. Antonetti "In Ultrafast Phenomena V" Chemical Physics, G.R. Fleming, A.E. Siegman, Ed., Springer Verlag. 308-311. A. Migus, Y. Gauduel, J.L. Martin, A. Antonetti, Phys. Rev. Lett. 58 (1987) 1559-1562.

51   D.F. Ca   lef, P.G. Wolynes, J. Physd. Chem. 87 (1983) 3387.

52   J. Gaathon, J. Jortner, Can. J. Chem. 55 (1977) 1801.

53   A. Mozumder, Radiat. Phys. Chem. (1988) (forthcoming).

54   J. Schnitker, P.J. Rossky, G.A. Kenney Wallace, J. Chem. Phys. 85 (1986) 2986.

55   E.J. Land, A. Swallow, Biochem. Biophys. Acta 162 (1968) 327-337.

56   B.H.J. Bielski, P.C. Chan, J. Am. Chem. Soc. 102 (1980) 1713-1717.

57   Y. Gauduel, S. Berrod, A. Migus, N. Yamada, A. Antonetti, Biochemistry 27 (1988) 2509-2518.

58   C. Petit, P. Brochette, M.P. Pileni, J. Phys. Chem. 90 (1986)) 6517-6521.

59   M. van der Auwersen, J.C. Dederen, E. Gelade, J. de Schryver, Chem. Phys. 74 (1981) 1140.

60   U. Gösele, U.K. Aklein, M. Hauser, Chem. Phys. Lett. 68 (1979) 291.

PHOTOCHEMISTRY IN REVERSE MICELLES

M.P.PILENI

1.INTRODUCTION

Few photochemical reactions have been investigated in reverse micelles in comparison to those published for micellar solutions or for vesicular systems.In this paper, the data obtained photochemically using semiconductors or small metallic particles as a sensitizer,are reviewed.The photochemical reaction occuring in a magnetic field,and the photoinduced reactions are reviewed in other chapters of this book. Several groups have studied the quenching of an excited state of a sensitizer by a quencher or an electron acceptor. From their work, the distribution law of quenchers in reverse micelles and some structural data for the reverse micelles have been obtained. Photoelectron transfers employing dyes as a sensitizer and electron acceptors such as "small molecules" or proteins have been studied. In several cases, the forward electron transfer reaction from the singlet or triplet state is followed by a very fast back reaction.When sensitizer and electron acceptor are on opposite sides of the interface of the reverse micelles, the back electron transfer process is greatly delayed . Some studies showing electron transport from the water layer to the organic layer are presented. The formation of singlet oxygen in reverse micelles is followed directly or by using a chemical reagent. Reverse micelles are able to control the reaction role by restricting the mobility of the substrate in a specific reaction field. These phenomena are shown in studies using spiropyran or photodimerisation processes .

2. PHOTOELECTRON TRANSFER

Light is used as an electron pump,promoting charge transfer from donor to acceptor.The reaction studied is the following:

$$D^* + A \longrightarrow D^+ + A^-$$

where D is the photosensitizer and A the electron acceptor. The photosensitizer used are ruthenium salts such as $Ru(bpy)_3^{2+}$ (where (bpy) is 2,2' bipyridyl) , metal porphyrin derivatives and phenothiazine derivatives. The electron acceptors are ferricynanide, copper salts, quinones and viologens or protein (1-13).The microheterogeneity of the system can retard the back reaction and allows the utilization of the reduced acceptor in chemical processes.

The study of the photoelectron transfer reaction is employed to learn what processes may occur in such systems. It is reasonable to assume that the photoelectron transfer reaction occurs when the donor and the acceptor are located in the same droplet.

However, it has been observed by several authors(14-16) that exchange of material between reverse micelles is takes place easily. This induces a delay in the forward electron transfer process.

2.1. Forward photoelectron transfer:

The forward reaction of the photoelectron transfer is studied by following the decay of the sensitizer excited state at various electron acceptor concentrations. The kinetics observed are strongly dependent on the location of the electron acceptor and the sensitizer. If the probe is associated with the micelles two-component decays are observed. The fast decay is due to intramicellar quenching and is characterized by a quenching rate constant, k_q. When the lifetime of the sensitizer in its excited state is shorter than the time for intermicellar exchange, the slow transient component is due to the decay of the sensitizer in its excited state located in micelles containing no quenchers. If the lifetime of the sensitizer in its excited state is longer than the intermicellar exchange process, the slope of the slow-component decay increases with quencher concentration, These decays are determined by a Poisson distribution of the donor and acceptor molecules among the water pools,. With this assumption the following equation, describes the complex decay observed(17-29):

$$\{S\}(t) = \{S\}(0) \ exp \ \{- C_1 \ t - C_2.t\{1-exp(- C_3.t)\}\} \qquad -1-$$

where : $\{S\}(t)$ and $\{S\}(0)$ are respectively the concentration of the sensitizer in its excited state at time t and at zero time

$$C_1 = k_o + k_q \cdot k_e\{Q\} / \{ k_q + k_e \{WP\} \qquad -2-$$
$$C_2 = k_q \cdot n_q / \{ k_q + k_e\{WP\} \qquad -3-$$
$$C_3 = \{ k_q + k_e\{WP\} \qquad -4-$$

$\{WP\}$ is the concentration of the water pool, $\{Q\}$ is the concentration of the quencher, n_q is the average number of quenchers per micelle and is equal to $\{Q\}/\{WP\}$, k_o the first order rate constant governing the decay of the excited state in the absence of quencher, k_q is the intramicellar quenching rate constant, k_e is the quencher exchange rate constant involving collisions between water pools.

A non-linear least squares computation gives the estimations of C_1, C_2 and C_3 for the experimental curves.

From such kinetic treatments, three quantities k_q, k_e and n are determined.

The kinetic laws are, as is expected (equation 1), biexponential. However in some cases, the kinetic quenching appears to be governed by homogeneous kinetics. The steady state Stern-Volmer plot exhibited a linear dependence on the quencher concentration, the fact that the kinetics are homogeneous rather than heterogeneous is explained (25, 30) in terms of the existence of water pools with fast micellar processes with transfer of the electron acceptor from one water droplet to another within the excited state decay time of the sensitizer. It has been shown that the kinetic equation is:

$$\{S\}(t)/\{S\}(0) = 1/\{1 \ / \ k_e/ \ k_0\{Q\}\} \qquad\qquad -5-$$

Several photosensitizers and electron acceptors have been used to determine these quenching and micellar exchange rate constants. The most popular systems were $RuII\text{-}Fe(CN)_6^{3-}$ or porphyrin-viologen. With RuII the fluorescence techniques are utilized whereas with porphyrin laser photolysis techniques are used, to investigate the nature of the microemulsions or reverse micelles.

For all the systems used, the quenching decay rate constant decreases with increasing the water content (5,6,9,13).It has been shown that it is associated with diffusion phenomena occurring in the water pool because of the increase of the droplet radius on increasing the water content.From the change of the quenching rate constant with the water content, the location of the reactants can be deduced(28). The quenching rate constants determined in various microemulsions or reverse micelles systems are strongly dependent on the nature of the microemulsion (5,6,9,13) and on that of the reactants ((5,6,9,13),30,31). This indicates that quenching rate constant is strongly dependent on the location of the sensitizer and the electron acceptor : when both are located in the water pool the forward electron transfer is very fast whereas when the sensitizer is located at the interface and the electron donor in the water pool or vice versa, the electron transfer is slower. The micellar exchange rate constant is not strongly dependent on the water content (13,15,20,30,31). However the micellar exchange process is strongly dependent on the aggregate: in oleate microemulsions by replacing hexanol by pentanol or hexadecane by dodecane, the biexponential kinetic decay is changed into a monoexponential decay which is attributed to an increase in the micellar exchange rate constant.The presence of several additives induces considerable changes in the micellar exchange rate constant (31) . From the average number of quenchers per micelle, the water pool concentration {WP} and the aggregation number are deduced.

Assuming the droplet is spherical and that the water pool volume results from the total volume of water molecules, the water pool radius can be obtained.

2.2:Back reaction:

The microheterogeneity of the system retards or speeds up the back reaction. This depends mainly on the location of the photosensitizer and on the electron acceptor. The delay in the back reaction allows the use of the reduced acceptor in the chemical process.

2.2.1 The reactants are both located in the water pool:

In several systems when both reactants are located in the water pool, no reduced forms are obtained(10). With a second electron donor added to the system, which is able to reduce the oxidized photosensitizer,the reduced electron acceptor is stabilized. This indicates that the excited state quenching observed is due to an electron transfer and the forward reaction is followed by a very fast back reaction. This is attributed to the close proximity of the reactants.

2.2.2 The reactants are separated by the interfacial wall:

This photoelectron transfer has been obtained using tetraphenylporphyrin or chlorophyll as the sensitizer and viologens as electron acceptors (13). In these cases, a large delay in the back electron transfer is observed. With chlorophyll quantitative data are obtained. The back reaction kinetic decay takes place on a millisecond time scale whereas the forward reaction takes place on the microsecond time scale.The kinetic decay depends on the water content and shows two distinct parts (figure 1).

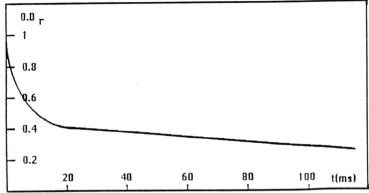

Figure 1: Kinetic decay of the chlorophyll cation formed from photoelectron transfer to chlorophyll to viologen in reverse micelles.

The fast component is attributed to the intramicellar back reaction whereas the long lived component is due to the micellar exchange processes. A schematic representation of this kinetic model is shown on figure 2. According to this model there is a competition between the back reaction (rate constant k_b) and the parent ion separation reaction (rate constant k_s). The joining up rate constant is given by k_j. The direct recombination (without separation) is believed to be responsible for the rapid part of the decay, whereas the delayed recombination (involving separation followed by joining up of the parent ions) is assumed to be responsible for the long-term decay. A mathematical treatment of the model yields a system of differential equations which cannot be solved analytically:

$$\frac{d}{dt}\frac{[\,M^+\,]_t}{[\,M^{+/-}\,]_{t=0}} = -B\left\{\frac{[\,M^+\,]_t}{[\,M^{+/-}\,]_{t=0}}\right\}^2 + A\frac{[\,M^{+/-}\,]_t}{[\,M^{+/-}\,]_{t=0}}$$

$$\frac{d}{d}\frac{[\,M^{+/-}\,]_t}{[\,M^{+/-}\,]_{t=0}} = (C+A)\frac{[\,M^{+/-}\,]_t}{[\,M^{+/-}\,]_{t=0}} + B\left\{\frac{[\,M^+\,]_t}{[\,M^{+/-}\,]_{-t=0}}\right\}^2$$

where $\{M^+\}$ and $\{M^{+/-}\}$ are the concentrations of the droplets containing respectively the oxidized photosensitizer and both the oxidized and the reduced forms of the reactants. $\{M\}_T$ is the total droplet concentration.

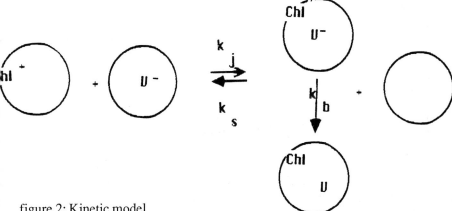

figure 2: Kinetic model

The rate constants for the separation reaction, k_s and the joining up reaction, k_j, can be estimated `since the total concentration of water pools, $\{M\}_T$, as a function of w is known and $\{M^{+/-}\}_{t=0}$ can easily be approximated from the initial optical density. The ratio k_j/k_s of the joining-up to the separation rate constants is less than or equal to 100 whereas the ratio k_q/k_b of the forward to the back electron transfer rate constants is about 1000. This indicates on one hand that the parent ions are prevented from separating and on the other hand that the back electron transfer is strongly inhibited compared with the forward reaction. It is concluded that there is a specific interaction between the chlorophyll cation and reduced viologens and in spite of the proximity of the parent ions, the back reaction is inefficient.`The interface may take a prominent part in the ion pair stabilization. The slowness of the back electron transfer is probably due to the interaction between the parent ions.

2.3 Electron transport from the water layer to the organic layer (32, 33):

The photochemical redox reactions involving the water molecules as electron donors and some quinones as electron carriers have been studied with a view to approaching artificial photosynthesis. Anthraquinone-2-sulfonate, AQ, irradiated in a reverse micellar solution composed of benzylhexadecyldimethylammonium hydroxide (CDBAH) -water-benzonitrile (as an oil) is photoreduced by an electron transfer from hydroxide ion and equilibrated with water, to produce the corresponding radical anion (AQ^-) as a stable product.

After sufficient formation of the radical anion (AQ^-) is observed, the selective irradiation (>450nm) of the radical anion by visible light causes a rapid disappearance of the AQ^- itself. The electron is transfered from the radical anion to benzonitrile step by step. The presence of trace amounts of dihydrodimer and trimer of benzonitrile is observed. This dimer and trimer may be produced from the radical anion of benzonitrile which suggests that an effective irreversible electron transport is takes place from the water layer incorporated in the reverse micelles to the organic layer benzonitrile as solvent. The mechanism is:

$$
\begin{array}{lcl}
AQ & \xrightarrow{\text{450nm}} & AQ^* \\
AQ^* + OH^- & \longrightarrow & AQ^{\cdot-} + OH^{\cdot} \\
2\ OH^{\cdot} & \longrightarrow & H_2O_2
\end{array}
$$

$$AQ^{-\cdot} \qquad \underline{\quad 365nm \quad} \rightarrow \qquad AQ^{-\cdot *}$$

$$AQ^{-\cdot *} + C_7H_5N \quad \longrightarrow \qquad AQ + C_7H_5N^{-\cdot}$$

$$x\ C_7H_5N^{-\cdot} \qquad \longrightarrow \qquad dimer,\ trimer$$

The reaction in reverse micelles is investigated in which the bulk phase solvent itself acts as an acceptor at high concentration. By 365nm irradiation, the radical $AQ^{-\cdot}$ again appears and disappeared with visible light (450nm).

The succesive visible light irradiation of the resultant radical anion $AQ^{-\cdot}$ increases the reductive power of $AQ^{-\cdot}$ itself and this photo redox system may cover a wide range of redox potential by stepwise two-photon excitation. The photoreduction by an electron transfer through the cationic reverse micelle surface takes place and an effective irreversible electron transport from the water layer to the organic layer is attained. Anthraquinone-2-sulfonate acted as an electron carrier at the micellar interface by the stepwise two-photon excitation.

2. 4. Photochemical reactions using proteins as an electron acceptor or as a catalyst:

Reverse micelles are able to solubilize enzymes or proteins and protect them from the unfavorable action of organic solvents by means of surfactants. Hence the study of catalytic properties of enzymes can be extended to organic media. Two differents experiments have been performed using photochemical processes:

2.4.1- A protein plays the role of an electron acceptor in the photoelectron transfer or a photoionisation process:

N methylphenothiazine (MPTH) is used as the photoactive compound, which reduces cytochrome c (8). With a second electron donor in the system, which is able to reduce the oxidized MPTH, irreversibly reduced cytochrome c is observed under continuous irradiation . Reduction of cytochrome c and c_3 by pyrene photoionization occurs (7). Reversible reduction of the proteins was observed which means that the proteins have maintained their original configuration within the micelles

2.4.2 An enzyme plays the role of catalyst in the reverse micelles:

Hydrogenase (34) from desulfovibrio vulgaris is encapsulated in reverse micelles with cetyltrimethylammonium bromide as surfactant and chloroform/octane mixture as solvent. Reducing equivalents for hydrogenase catalyzed hydrogen production are provided by vectorial photosensitized electron transfer from a donor (thiophenol) in the organic phase through a surfactant Ru^{2+} sensitizer located in the interface to methylviologen concentrated in the aqueous core of the reverse micelle. The result shows that reverse micelles provide a microenvironment that stabilizes hydrogenase against inactivation and and allows an efficient vectorial photosensitized electron and proton flow from the organic phase to hydrogenase in the aqueous phase

3.Singlet oxygen produced photochemically (35-38):

Reverse micelles are frequently used as models for complex biological systems. An example of this is the study of the behaviour of singlet oxygen microheterogeneous systems as a model for the activity of singlet oxygen in the photodynamic damage in biological systems.

Singlet molecular oxygen is generated by collisions with a photosensitizer in its excited state such as rose bengal or acetophenone. Singlet oxygen is observed by its own luminescence decay at 1270nm or by an indirect chemical method monitoring O_2 in fluid systems. In these methods oxidizable substrates such as rubrene, diphenylbenzofuran (DPBF) or anthracenepropionic (ADPA) are used. The sensitizer is either located in the bulk hydrocarbon phase (2 acetonaphtone) or in the aqueous interior phase (xanthene dye rose bengal). Similarly the quencher can be solubilized in the bulk phase (ß carotene) or in the water pool (Ni^{2+}, azide ions or tryptophane). From the study of the quenching of singlet oxygen by quencher located in the interior aqueous phase or in the bulk it is deduced that singlet oxygen is not restricted to one of the different phases but can diffuse freely through the whole system.

A model based on lifetime measurements of singlet oxygen, velocity and equilibrium constants for the exchange of this monitor was developed . The only components present prior to the 10ns excitation pulse are the solvent, micellar constituents, water soluble sensitizer and oxygen. In less than 1μs after the pulse the only difference is that some of the oxygen has been converted to $O_2(^1\Delta_g)$. The kinetic model used schematically represents the transfer of $O_2(^1\Delta_g)$ between an exterior and an interior phase. k^+ and k^- are the first order rate constants of the entrance and of the exit to and from the water pool. The model simplified is given figure 3 Where Δ and Σ represent excited and ground state oxygen species respectively and the subscripts "bulk" and "wp" refer to the bulk phase and to the water pool of the reverse micelles. k_b and k_{wp} are the decay rate constants of singlet oxygen in the bulk and in the water pool , k_q is the quenching rate constant of singlet oxygen in the water pool. From the kinetic analysis it is deduced that singlet molecular oxygen (Δ) can either migrate from the heptane medium into the water pools or from the water pool to the bulk organic phase. The rate constants for the entrance and exit of molecular oxygen to and from the water pool reverse micelles and the decay rate constants of singlet oxygen in the bulk phase and in the water pool for two different systems are given in references 35 and 38.

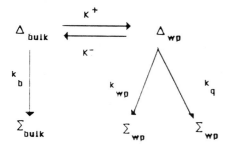

Figure 3: various processes of oxygen

The oxygen singlet ($^1\Delta_g$) is 9 times more soluble in heptane than in the interior aqueous phase. The H bonding capacity of the entrapped water can be monitored via the fluorescent lifetimes of xanthene dyes like rose bengal (RB). In protic solvents like water and alcohols these dyes have a very short lifetime, probably due to an efficient intersystem crossing from excited singlet and triplet states. The smaller the H bonding capacity of the solvent the longer are the lifetimes of these labels. This monitor will be anchored in the interface with its apolar tail. The results show no difference between RB and $C_{16}RB$ so it can be safely assumed that RB is also associated with the interface. At w=0, the H bonding capacity of the entrapped water is of the same order of magnitude as in t-butanol and increases gradually until a limiting value is reached at w>20. At w ≤20, the lifetime of RB is still much longer than in bulk water. The strong correlation between this quenching rate constant and the water content (and hence the size of the water pools) observed clearly points out the gradual change in the mobility and structuredness of the water molecules in the AOT reverse micelles as the size of the water pool varies. Quenching kinetics thus offer a feasible method for monitoring water structure in reverse micelles. The decay of singlet molecular oxygen depends on the water content of the water pool. For a given water content the decay of singlet molecular oxygen depends on the quencher concentration. The quenching rate constant varies with the water content. Quenching studies with water soluble molecules yield information about the water structure in the micelles. When the azide ion was used as quencher a maximum constant quenching rate constant is reached at w =30. This rate is about 80% of the value in bulk water, indicating that the entrapped water under these circumstances behaves almost like normal water. Tryptophan behaves differently. The tryptophan molecule is probably associated with the interface. In this case the quenching rate is also constant at w > 3 but the value is only 50% of that in the bulk water. These results clearly indicate that there is a large difference between water in the neighbourhood of the interface and in the center of the micelle.

4: Rate control by restricting mobility of the substrate in specific reaction fields:

Reverse micelles are investigated in order to evaluate the effect of the reverse micelles in controlling the reaction rates or pathways by restricting the mobility of the substrates situated in a specific reaction field.

4.1 The photochromism of a water soluble spiropyran (39)

The photochromism study involves on one hand a colorless spiropyran which is a closed spirostructure "1" and on other hand the open form "2" which has a red or orange color strongly dependent on pH and solvent polarity at a given temperature.There is an equilibrium between the colorless and the colored forms of spiropyrans. The reaction in either direction can occur both photochemically and thermally. Irradiation of compound **2** with visible light induces the formation of the compound "1". Keeping the latter colorless solution (compound "1") in the dark induces the reappearance of the color of the compound"2". The present system shows a typical thermocoloration-photobleaching cycle. The monitor shows a negative photochromism in polar solvents such as water, methanol or ethanol as well as in the AOT reverse micelles.The thermocoloration rate of the spiropyran of "1" decreases in terms of stabilization in the ground state of the neutral closed ring species "1" and /or destabilisation of the polar transition state toward the ionic product by solvation with less polar solvents. Nevertheless, the microscopic polarity in the AOT micelles is comparable to that in methanol . The conversion rate from "1" to"2" in the AOT micelles was decelerated by 10-20 times compared with that in methanol. This is explained in terms of the restriction in the internal rotation of the molecule during the thermocoloration : Molecular structure considerations suggests to us that in order for"1" to open the pyran ring and take a possibly conjugated and plane conformation like the merocyanine dye structure"2", two a bonds which originally were linked to the olefinic double bond,must rotate with the cis-trans isomerisation. Most probably, the rotative movement is controlled by the rigidity of the field. It may be reasonably concluded, thus, that the deceleration rate of the thermocoloration of"1" in the reverse micellar core must be caused by the restriction of the mobility and internal rotation effects on the thermocoloration process. It is found that the photobleached species being at a higher energy level has a longer lifetime in the restricted field provided by reverse micelles than in the homogeneous regular solutions which suggests that in the specific reaction field as provided by the reverse micelles, it may be possible to hold the labile substrate at the higher energy level by restricting the freedom of molecular motion.

4.2 photodimerisation of stibene derivative(40):

In homogeneous solution, at low concentration, trans-4- stibazolium cations show only monomer absorption and fluorescence and a trans to cis photoisomerization is obtained by direct irradiation.By increasing the concentration an excimer fluorescence is observed with a slight broadening of the absorption spectrum. The yield of cis stibazolium cation photoproduct decreases and concurrent with the appearance of excimer fluorescence the head-tail photodimer is chiefly produced

In AOT reverse micelles, at high water content, the excimer fluorescence appears, at a lower concentration than that obtained in homogeneous solution. There is a decrease in the production of cis stibazolium cations and of head-tail photodimer accompanying the onset of excimer emission. Concurrently the head-head dimer appears, which is not produced in homegeneous solution and by decreasing the water content, w, the yields of cis stibazolium cations and of head-tail photodimer decrease whereas that of head-head dimer increases.Such a remarkable change in the photodimer ratio with the water pool content indicates that the photophysics and photochemistry of cis stibazolium cations are modified by both the concentrating and orienting properties of RM. Thus at relatively high w values the micellar concentration decreases which induces an increase in the local concentration of stibene derivatives and then favours the normal excimer fluorescence and the head-tail photodimer as is observed in aqueous solutions. As w decreases so that little "free" water is present, the orientation of stibazolium cations with respect to the charged interface must become more constrained so that the predominant excimer formed fluoresces at shorter wavelengths and decays preferentially to head-head dimer . The significant result of this study is the finding that efficient and selective formation of the syn head to head photodimer is correlated with a ground state association that can be controlled by stilbazolium-surfactant and surfactant-water ratios. These results indicate that for small water pools the charged interface of the reverse micelles presents an organized surface that can lead to a topological control not readily attainable in other fluid media.

154

REFERENCES:

-1- M.Wong, J.K.Thomas and M.Gratzel J.Am.Chem.Soc., 98, 2391, (1976)

-2- M.A.J.Rodgers and Da Silva E. Wheeler, Chem.Phys.lett., 53, 165, (1978)

-3- S.S. Atik, and L.A Singer, Chem. Phys. Lett., 59, 519 (1978)

-4- M.P. Pileni, and M. Gratzel J. Phys. Chem., 84, 1822 (1980)

-5- S.S Atik and J.K Thomas J. Am. Chem. Soc., 103, 2550, (1981)

-6- S.S Atik and Thomas, J.K., J. Am. Chem. Soc., 103, 3553,(1981)

-7- V.Calvos-Perez, G.S. Beddard, and J.H.,Fendler, J. Phys.Chem., 85, 23126 (1981).

-8- M.P.Pileni Chem Phys Let. 81, 603, (1981)

-9- J.M.Furois,P.Brochette and M.P.Pileni, J. Colloid Interfac Sci., 97, 552 (1984).

-10- P.Brochette and M.P.Pileni Nouveau J.Chim., 9, 551,(1985)

-11- S.M.B.Costa and R.L.Brookfield J.Chem.Soc., Faraday Trans,2, 82, 991, 1986

-12- P. Brochette,J.Milhaud and M.P.Pileni "Surfactant in solution" (1987) ed. P.Bothorel and K. Mittal

-13- P.Brochette, T.Zemb, P.Mathis and M.P.Pileni J.Phys.Chem.,91, 1444, (1987).

-14- H.F.Eicke, J.C.W. Shepherd and A. Steinemann J.Coll.Int.Sci., 56,168, (1976)

-15- P.D.I.Fletcher and B.H.Robinson Ber.Bunsenges.Phys.Chem., 85, 863, (1981)

-17- P.P.Infelta. M. Gratzel and J.K.Thomas J. Phys. Chem., 78, 190 (1974).

-18- M.Tachiya Chem. Phys. Lett., 33, 289 (1978).

-19- A.Yekta, M.Aikawa and N.J.Turro Chem. Phys. Lett., 63, 543 (1979)

-20- J.C. Dederen, Van der Auweraer, and F.C.de Schryver, Chem. Phys. Lett., 63, 543 (1979).

-21- G.Rothenberger, P.P.Infelta and M. Gratzel J. Phys. Chem., 83, 1871 (1979).

-23- M.Tachiya Chem.Phys.Lett., 68, 291, (1979)

-24- M.Tachiya J. Chem. Phys., 76, 340 (1982)

-25 -E.Geladé and F.C.de Schryver J.photochem., 18, 223 (1982)

-26- M.P.Pileni, B. Hickel and J. Puchault, Chem. Phys Lett. 92, 308 (1982).

-27- M.P.Pileni, P.Brochette, B. Hickel, and B.Lerebours, J. Colloid Interface Sci., 98, 549 (1984).

-28- C.Petit, P.Brochette and M.P.Pileni J.Phys.Chem., 90, 6517,(1986)

-29- M.P.Pileni,J.Mihaud, T.Zemb, P.Brochette and B.Hickel, Surfactant in solution" ed.P.Bothorel and K.Mittal (1987).

-30- S.S.Atik and J.K.Thomas J.Phys.Chem.n, 85, 3921, (1981)

-31- S.S.Atik and J.K.Thomas J.Amer.Chem.Soc., 103, 7403, (1981)

-32- H.Inoue and M.HidaBull.Chem.Soc.Jpn., 55, 1880 (1982)

-33- J.Sunamoto, K.Iwamoto, M.Akutagawa, M.Nagase and H.Kon J.Am.Chem.Soc., 104, 4904 (1982)

-34- R.Hilhorst, C.Laane and C.Veerger Proc.Natl.Acad.Sci., USA,79, 3927,

(1982)

-35- M.A.J.Rodjers J.Phys. Chem., 1981, 85, 337

-36- I.B.C Matheson and M.A.J.Rodgers J.Phys.Chem.,86,884, 1982

-37- P.C.Lee and M.A.J.Rodgers J.Phys.Chem., 1983, 87, 4894

-38- M.A.J.Rodgers and P.C.Lee J.Phys.Chem., 1984, 88, 3480

-39- J. Sunamoto, K.Iwamoto, M.Akutogawa, M.Nagase and H.Kondo
J.Am.Chem.Soc., 104, 4904, (1982)

-40- K.Takagi, B.R.Suddaby, S.L.Vadas, C.A.Backer and D.G.Whitten
J.Am.Chem.Soc., 108, 7865, (1986)

MAGNETIC-FIELD DEPENDENT REACTIONS IN REVERSE MICELLAR SYSTEMS

U.E.STEINER

1.   INTRODUCTION

In reverse micelles surfactant molecules enclose a polar core, where considerable amounts of water may be solubilized. For water/surfactant molar ratios (denoted as w) exceeding a value of about 10, the intramicellar water approaches the character of a thermodynamic microphase (1) distributed in fairly monodisperse droplets of up to several tens of nanometers radius. For these the term nanodroplets has been coined by Eicke (2). Since the water nanodroplets constitute potential microreactors, the radius of which may be easily controlled by w, we found them very attractive for investigating the recombination kinetics of spin-correlated radical pairs (RPs), which in homogeneous solution would predominantly undergo non-correlated second order recombination.

Studies of the magnetic-field dependence of kinetics and product yields of reactions involving radical pair intermediates have revealed the important aspect of electron spin correlation in radical pairs (3) during the period of time where diffusional geminate re-encounters take place. Such magnetokinetic effects are of great interest as a source of detailed information on the dynamics of radical pair reaction kinetics and how it is influenced by the medium. To understand the basis of magnetokinetic effects with radical pairs it is essential to recall that these 'remember' the spin alignment of their precursor state (singlet or triplet) for some time, even when they may have become separated to a distance where the exchange interaction between the unpaired radical electrons is zero. The loss of spin correlation in the radical pair which we will denote as 'spin evolution' may be due to two types of mechanisms: a coherent one, based on the difference in frequency and orientation of the axis of electron spin precession which the unpaired electron spins undergo in the different isotropic hyperfine fields of the two radicals, or an incoherent one which is due to stochastic motion of the two electron spins caused by fluctuating anisotropic magnetic interactions, the same type of mechanisms which are also responsible for the linewidth of hyperfine lines in the esr spectra of individual radicals. The degree of spin evolution may be quantified in terms of time-dependent singlet or triplet character (p_S or p_T) of the radical pair, which will approach the values of 1/4 or 3/4, respectively, at times which depend on the strength of the interactions mentioned and on the strength of an external magnetic field, but typically will not exceed a few microseconds. At any time t after the radical pair has been created the value of p_S or p_T will be directly reflected by the recombination probability of the radical pair on a subsequent encounter at t. This probability is generally close to the value of $p_S(t)$ since radical pairs usually can only recombine to form products in their singlet ground states and do so at a rate which is usually diffusion-controlled. The yield of geminate recombination (so-called cage recombination), or conversely, the yield of free radicals undergoing so-called escape reactions, will in general depend on the initial spin multiplicity, the spin-independent probability of geminate re-encounters, the degree of spin evolution between successive re-encounters and hence on the average time elapsing between these and on the strength of an applied magnetic field. From the details of the magnetic-field dependence of the radical pair kinetics one can draw conclusions about the mechanism responsible for the spin evolution. In addition, rate parameters characteristic of the various reaction channels of the radical pair may be evaluated. The principles outlined above have been studied for many examples of radical pairs in homogeneous solutions, where the geminate re-encounter probability decays within several nanoseconds to tens of nanoseconds. In theses cases spin evolution is only probed during a correspondingly short interval of time. On the other hand, much longer periods of geminate re-encounters are found in micellar solutions where the micelles, with typical radii of about 20 Å, act as kind of supercages (4) for the radical pairs generated within them. When choosing microemulsion nanodroplets as media for studying magnetic-field dependent radical kinetics, it was our intention to take advantage of the polar interior of these aggregates for assisting electron transfer reactions and of the ease of size variation of the nanodroplets which can be extended much beyond the size of normal micelles, so that a wider range of time intervals between geminate re-encounters can be realized.

This allowed us to extend the study of radical pair spin evolution to longer times than previously accessible.

2. MAGNETIC FIELD EFFECTS WITH TRIPLET RADICAL PAIRS

In polar media triplet radical pairs may be conveniently generated by using photo electron transfer reactions with dye triplets. Various aspects of such reactions, especially the dependence of free radical yield on solvent polarity, solvent viscosity and specific properties of electron donors have been studied in our laboratory using thionine and related dyes as excited triplet electron acceptors (5-8). One of the main characteristics of our findings concerning the yield of free radicals and its magnetic-field dependence in homogeneous solutions was, that geminate recombination in radical pairs of the type $^3(A^. D^{.+})$ is only of minor importance. However, the yield of free radicals is significantly affected by heavy atom substituents, an effect which has been attributed to the enhancement of intersystem crossing (ISC) in a triplet exciplex $^3(AD)^+$ assumed to be a precursor of the solvent-shared radical pair. The lack of geminate radical pair effects in homogeneous solution must be attributed to the absence of Coulombic attraction in radical pairs of the 0/+ charge type, which renders geminate re-encounter probabilities too low, thus precluding radical pair spin evolution to become kinetically apparent.

The reverse micellar system including the w/o microemulsion region of w we utilized for the study of geminate radical pair recombination kinetics and spin effects was made up of cetyldimethylbenzylammonium chloride (CDBA) solutions (0.04 M) in benzene with various amounts w of water added, wherein the cationic dye had been dissolved. Dye concentrations were such that the average population of a nanodroplet was of the order of a few percent, so that production of more than one RP per nanodroplet was negligible. The H_2O/CDBA/benzene reverse micellar system was first introduced by Klein, Hauser and Miller (9). According to their results aggregation numbers increase approximately proportional to w^2 for w>10, a behaviour which is consistent with the nanodroplet model of microemulsions (1). In this region of w the nanodroplet radius varies approximately as 1.9 w Å.

2.1 Radical pair kinetics in nanodroplets of variable size

The influence of nanodroplet size on geminate RP recombination kinetics and its magnetic-field dependence was studied in detail for RPs produced by quenching thionine triplet with aniline as electron donor (10). Although aniline is solvated mainly in the unpolar bulk phase its access to the dye triplets in the polar interior of the surfactant/water aggregates is not inconveniently reduced as compared to the kinetics in homogeneous solution. For w=15 the bimolecular quenching rate constant was found to be

$$H_2\overset{\oplus}{N}\underset{N}{\diagdown}X\diagdown NH_2 \qquad X= S \text{ thionine } (TH^+)$$
$$X= O \text{ oxonine } (OxH^+)$$

$1.4 \times 10^9 M^{-1} s^{-1}$, dropping to about $6 \times 10^8 M^{-1} s^{-1}$ for w=32.5, which is to be compared with a value of $3.4 \times 10^9 M^{-1} s^{-1}$ for methanolic solution. Thus, with an aniline concentration of 0.05 M triplet decay is fairly complete after 30 ns, even for the highest w applied. In the quenching process an electron is transferred from aniline to the dye triplet, giving rise to the typical dye semiquinone spectrum, which in the reverse micellar system resembles very much that in methanol. The kinetics of semithionine decay was followed by nanosecond laser flash spectroscopy using an excimer-laser-pumped dye laser for excitation. Characteristic signals observed with various values of w are shown in Figure 1. Kinetically the decay proceeds in two steps. The faster one, occuring in the 200 - 1000 ns time regime, is unparalleled in homogeneous solution and must be attributed to geminate intra-nanodroplet radical pair recombination, whereas the slower kinetic component which extends to the millisecond region, belongs to the non-correlated inter-nanodroplet radical recombination, corresponding to the homogeneous second order rate process observed in homogeneous solution. The quantitative kinetic analysis of the geminate process was based on the following kinetic scheme (I), according to which the dye radical concentration $c_{TH^.}$ decays as given in eq.[1]:

$$c_{TH^.}(t) =$$

$$c_{TH^.}(0) [k_{esc}/(k_{esc}+k_{rec})]\{1+(k_{rec}/k_{esc})\exp[-(k_{esc}+k_{rec})t]\} \quad [1]$$

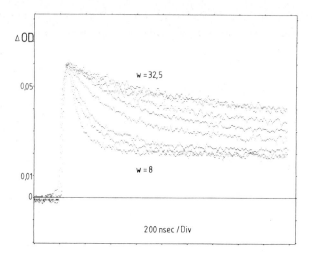

Fig.1. Radical pair recombination kinetics observed at zero field
with RPs obtained from thionine triplet quenching by aniline in
water/CDBA/benzene microemulsions of various w (8, 11.5, 15,
18.5, 22, 25, 5, 29, 32.5). The absorption of thionine semiquinone radical was monitored at 420 nm using the technique of laser
flash spectroscopy (25).

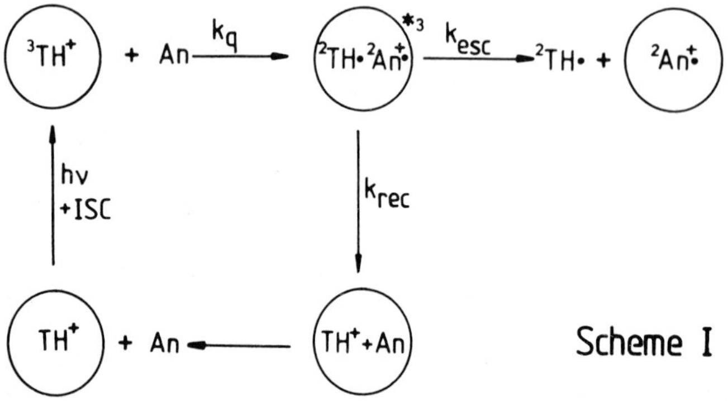

Scheme I

Using the first-order rate constants k_{rec} for geminate recombination and k_{esc} for escape of a radical from the nanodroplet of its
generation the efficiency of radical pair escape η_{esc} is given by

$$\eta_{esc} = k_{esc}/(k_{esc} + k_{rec}) \qquad [2]$$

The rate constants obtained for various w by applying kinetic
fits based on eq.[1] to the observed radical decay curves are
plotted versus w in the diagrams shown in Figure 2.

The values of k_{rec} are strongly w-dependent, varying approximately as w^{-3} for w>15, which implies an inverse proportionality to the volume of the nanodroplets, as would be expected for a diffusion-controlled process if the reactants were free to move in the volume of the intramicellar water core. Also indicated in the diagram are the lines to be expected theoretically for diffusion controlled pair reactions in homogeneous spheres with radii corresponding to those of the nanodroplets investigated and assuming a viscosity of 1 cP (like that in bulk water) or 10 cP, respectively. Since the observed values of k_{rec} are clearly lower, it is not very likely that the radical pairs undergo unrestricted motion in the water pool, unless the water there is assumed to have a much higher viscosity than free bulk water.

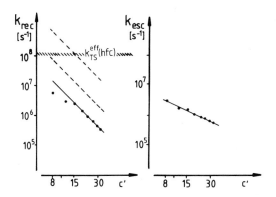

Fig.2. Rate constants k_{rec} and k_{esc} of Scheme I evaluated as a function of w from transient signals as shown in Figure 1.

Most probably the radicals will be associated with the Stern layer at the water/surfactant interface for most of the time, being only occasionally released for free diffusional trips into the interior of the water pool. Possibly these trips are the main transport mechanism by which geminate radical pair re-encounters are brought about.

Since the radical pairs are produced with triplet spin alignment they cannot recombine before having undergone a triplet-to-singlet spin change. In Figure 2a is also indicated the effective rate constant at which such processes should occur at zero magnetic field. This value of about $10^8 s^{-1}$ was derived from

the theory of spin motion as induced by isotropic hyperfine
coupling (11,12) in the radical pairs considered here. Comparing
this rate constant with the observed rate constant of radical
pair recombination it becomes clear that in zero field the spin
process cannot be the rate determining step of the recombination
process. Under the prevailing conditions the radical pair should
recombine as if in spin equilibrium, i.e. show no spin memory for
the precursor multiplicity.

Compared to k_{rec} the w dependence of k_{esc} is much weaker and
is approximately proportional to w^{-1}, i.e. to r_{nd}^{-1}. The r_{nd}
dependence of k_{esc} is the same as that of the surface-to-volume
ratio in a sphere and could be interpreted as indicating a
uniform distribution of the escaping radical species over the
nanodroplet volume. However, as pointed out above, this interpretation would be at variance with the low absolute value of k_{rec}.

Whereas for the reaction system considered in this section
there is no indication of spin memory effects on the radical
recombination kinetics at zero field, such effects become clearly
apparent if magnetic fields are applied (cf. Figure 3). As the
field becomes higher the geminate process is slowed down and the
yield of escaping radicals is concomitantly increased. A kinetic
analysis of the decay curves according to eq [1] showed that only
the rate constant k_{rec} is significantly affected by a magnetic

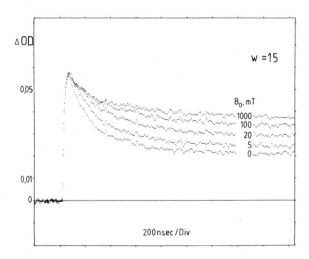

Fig.3. Magnetic-field dependence of transient kinetics with
[3](TH· An·+) RPs in CDBA-enclosed water nanodroplets in benzene
(adapted from ref 10). Conditions of observation as in Figure 1.

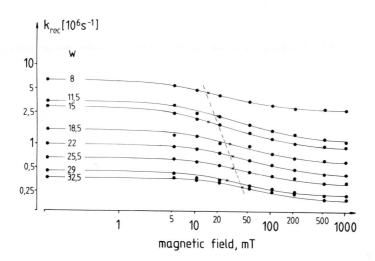

Fig.4 Magnetic-field dependence of rate constant k_{rec} of gemi-nate recombination of $^3(TH\cdot\ An\cdot^+)$ RPs in water/CDBA/benzene microemulsions for various w values corresponding to nanodroplet radii between 15 Å and 60 Å (10).

field (10). Its field dependence in nanodroplets of various size is depicted in Figure 4.

Although the absolute values of k_{rec} vary by a factor of about 20 over the range of w-variation, the shapes of the curves representing the magnetic-field dependence are essentially unchanged. The field dependence seems to approach saturation at fields of about 1 Tesla, where k_{rec} is reduced by a factor of two to three with respect to its corresponding zero-field values. On each curve is marked the interpolated position where

$$k_{rec}(B_{1/2}) = k_{rec}(B=0\ T) + 0.5\{k_{rec}(B=1\ T) - k_{rec}(B=0\ T)\} \qquad [3]$$

which defines a corresponding $B_{1/2}$ value for each w. Despite the scatter displayed by these $B_{1/2}$ values, there is a clear trend for $B_{1/2}$ to increase with the nanodroplet radius ($B_{1/2} \approx 14$ mT for w = 8 to $B_{1/2} \approx 40$ mT for w = 32.5). The slope indicated corresponds to a value of -2, implying a relation:

$$k_{rec}(B_{1/2}) \sim B_{1/2}^{-2} \qquad [4]$$

A qualitative explanation of the magnetic field dependence found

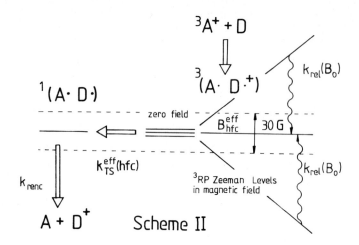

Scheme II

for k_{rec} can be given in terms of Scheme II. In order that a RP be able to recombine, its total electron spin has to acquire some singlet character. As explained above, this cannot be a rate determining process in zero field since $k_{TS}^{eff}(hfc)$ » k_{renc}. However, as the field increases, the increase of the Zeeman splitting will slow down spin equilibration between the outer and central Zeeman levels and eventually render it sufficiently slow so that it may become apparent in the recombination rate constant. The facts that $B_{1/2}$ is several times larger than the effective isotropic hyperfine field of about 30 Gauss (3 mT) calculated by a relation given by Weller et al. (13), and that a definite saturation limit of the magnetokinetic effect is not attained below 1 T, clearly indicate that T-S coupling mechanisms other than isotropic hyperfine coupling (i.e. the one which is responsible for CIDNP and CIDEP effects) are still operating on the time scales of our experiments even at fields where this isotropic hfc mechanism of $T_{\pm}/T_o$,S mixing has been suppressed. These mechanisms must imply exchange of the $T_{\pm}/T_o$,S energy difference between the electron spin system and the bath, i.e. they must correspond to longitudinal magnetic (T_1-) relaxation processes. Only if these processes become slow enough (compared with the re-encounter rate characterized by a rate constant k_{renc}) a magnetic field effect will appear in the recombination kinetics. Thus we may assume that the magnetic-field dependence observed for k_{rec} reflects the magnetic-field dependence of the relaxation rate constant k_{rel}. As a consequence of this, however, we have to infer from Figure 4 that k_{rel} must strongly depend on w, namely in the same way as $k_{rec}(B=0) \approx 1/4 k_{renc}$. This would imply that electron spin relaxation in the radical pair occurs mainly during re-encounters. This is a strong indication that the dominant spin relaxation mechanism in question might be provided by the stochastic modulation of magnetic dipolar interaction between the two radical spins during a re-encounter. Estimates of the contribution of such a mechanism have been reported by Hayashi and Nagakura (14). Their results show a strong field-dependent decrease of the corresponding relaxation rate constant which is also of the appropriate order of magnitude. To see in which way the w-dependence of $B_{1/2}$ observed in our investigation can be related to this mechanism will, however, require a more detailed quantitative treatment.

2.2 Spin-orbit coupling effects on magnetokinetic behaviour of triplet RPs.

In low viscosity homogeneous solvents spin motion has no appreciable effect on the efficiency of geminate recombination for RPs of the type $^3(A^{\cdot}\ D^{\cdot +})$, because the probability of geminate re-encounters is too small. Nevertheless a very marked decrease of the free radical yield has been observed even in homogeneous solutions of low viscosity, on introducing heavy atom substituents into the electron donor (15). These effects and their magnetic-field dependence have been quantitatively accounted for by considering spin-orbit coupling in the triplet exciplex preceding the formation of a solvent-shared radical pair (16). Using microemulsion nanodroplets as a reaction medium the long periods during which geminate re-encounters take place in such aggregates made it possible to study the role of spin-orbit coupling on the spin evolution in radical pairs,too.

The data presented in Table 1 characterize radical formation and geminate recombination kinetics in several systems involving thionine triplet quenching by halogen anilines. Two main effects are borne out by the free radical yields. As in methanol, in the reverse micellar system too, the free radical yield decreases monotonically with increasing spin-orbit coupling (note the position-dependent effect of the bromine substituent, which may be also related to the effect of spin-orbit coupling, since the latter depends on the density of the unpaired electron experienced by the heavy atom nucleus in the oxidized donor radical (6)). Comparing the same donor in both solvents, the free radical yield

TABLE 1

Radical production and geminate RP decay with halogen anilines as quenchers of thionine triplet in MeOH and reverse micelles

electron donor	MeOH[a] Φ_{fr}[c]	CDBA reverse micellar system (w = 15)[b]				
		Φ_{fr}[c]	η_{esc}[d]	k_{esc}[e]	k_{rec}[e]	f_{rec}[f]
aniline	1.00	0.29	0.36	1.75	3.0	3.2
4-Cl	0.97	0.27	0.40	1.75	3.0	1.9
3-Br	0.90	0.21	0.48	2.9	3.1	1.5
2-Br	0.70	0.13	-	-	-	-
4-Br	0.51	0.10	0.45	2.9	3.4	1.0
4-I	0.13	0.06	0.38	3.1	5.1	1.0

[a] from ref 17 [b] from ref 18 [c] yield of free radicals for complete triplet quenching [d] cf. eq. [2] [e] in units of $10^6 s^{-1}$ [f] defined as $k_{rec}(B=1T)/k_{rec}(B=0T)$

is reduced in the micellar system by a factor of 2 - 5. Nanosecond time-resolved experiments have shown that in all of these systems a considerable fraction of radicals , present immediately after the triplet quenching, decays in a geminate process on the nanosecond time scale (18,25). The rate constants of geminate radical pair decay are not very sensitive to heavy atom substitution. With the help of the time-resolved measurements the yield of free radicals Φ_{fr} may be decomposed into a product of the initial yield of RPs Φ_{rp} which depends on SOC according to the exciplex mechanism and a factor η_{esc} measuring the efficiency of radical escape, which is practically 1 in homogeneous methanolic solution and between 0.36 and 0.48 in the reverse micellar system (eq [5]). In zero field the factor η_{esc} is not significantly modified by heavy atom substitution. This again emphasizes that in zero field spin evolution does not seem to be the rate determining step in geminate recombination of triplet RPs produced in such reverse micellar aggregates.

$$\Phi_{fr} = \Phi_{rp} \cdot \eta_{esc} \qquad [5]$$

The relative magnetic change of the free radical yield in the CDBA reverse micellar system was recorded as a continuous function of a magnetic field (18) with the donors listed in TABLE 1. These so-called MARY (19) spectra are displayed in Figure 5.

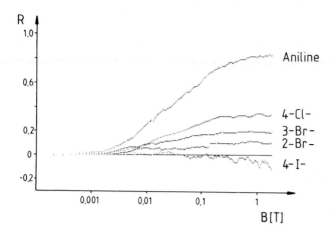

Fig.5 MARY spectra of relative magnetic field effect R on the free radical yield resulting from thionine triplet quenching by various halogen anilines in water/CDBA/benzene microemulsions. The yield of free radicals was monitored by thionine ground state depletion at 610 nm 20 μs after the laser pulse (18).

For the systems with 4-Cl-, 3-Br- and 2-Br-aniline they have practically the same shape and the same $B_{1/2}$ as observed with aniline. However, the MARY intensity is reduced as the spin-orbit coupling increases. Again, there is the 3- > 2- > 4-position dependence of the substituent SOC effect. For the 4-Br-aniline the MARY intensity is zero and not depicted in Figure 5. For the 4-I-derivative the MARY spectrum shows a negative intensity, however with a significantly higher $B_{1/2}$ than in the other cases.

Nanosecond time-resolved magnetokinetic experiments have revealed that the positive MARY intensity in Figure 5 is due to a magnetic-field dependence of k_{rec} which renders η_{esc} magnetic field dependent (18,26). However, due to the heavy atom substituents, the geminate radical pair recombination process becomes less susceptible to a magnetic field. This means that the rate of triplet-singlet transitions which is slowed down in a magnetic field, is less strongly affected if heavy atoms are present. We suggest that this may be due to an increasing contribution of the spin-rotational relaxation mechanism, which is independent of the

$$k_{rel}(\text{s-rot}) = 1/T_1 = (12\pi r^3)(\Delta g_\parallel^2 + 2\Delta g_\perp^2)kT/\eta \qquad [6]$$

magnetic field strength and increases with the square of the g-tensor anisotropy, which relates it directly to the strength of spin-orbit coupling (20). Making use of the relation in eq [6], where r is the hydrodynamic radius of the radical (here we assume a value of 3 Å) and η is the viscosity of the solvent, we expect the following values (18):

4-Cl: $2 \times 10^5 \eta^{-1} s^{-1} cP$ 3-Br: $7 \times 10^5 \eta^{-1} s^{-1} cP$

4-Br: $3 \times 10^6 \eta^{-1} s^{-1} cP$ 4-I : $1 \times 10^7 \eta^{-1} s^{-1} cP$

These rate constants will mark the lower limits to which the $T_{\pm} \to S, T_0$ relaxation rate constant may be decreased by a magnetic field. If $k_{rel}(B \to \infty) > k_{rec}(B=0)$ no magnetic field effect will be found. If $k_{rel}(B \to \infty) < k_{rec}(B=0)$ and approaches $k_{rec}(B=0)$ from below, as in the donor series investigated, the saturation value of the magnetic-field effect will go down. Under the conditions at which the MARY spectra in Figure 5 were obtained $k_{rec}(B=0) \approx 3 \times 10^6 s^{-1}$. Setting $\eta = 1$ cP (the viscosity of bulk water) the $k_{rel}(s\text{-rot})$ values in the substituent series listed above would in fact approach and exceed $k_{rec}(B=0)$ which can explain the observed reduction and finally complete disappearance of MARY intensity with these substituents. It must be pointed out, however, that the results on the w-dependence of $k_{rec}(B=0)$, as described in the last section, suggest a significantly higher effective viscosity for the <u>translational</u> motion of the radicals.

As for the negative sign of the MARY effect with 4-I-aniline, we suppose that this is due to a magnetic-field effect on Φ_{rp} and is caused by the triplet mechanism, which is also found in homogeneous solution. The corresponding effects to be expected for the other halogen derivatives investigated here are much smaller, so that they are not detectable at the level of experimental accuracy available with the low dye concentrations in the micellar systems.

3.   MAGNETIC FIELD EFFECTS WITH SINGLET RADICAL PAIRS

Since radical pairs produced with singlet spin alignment may undergo spin-allowed fast recombination to form singlet ground state products the yield of detectable, freely diffusing radicals is usally very low when singlet precursors are involved. Exceptions may be found where the singlet recombination becomes slow due to unfavourable Franck-Condon factors, which requires that the energy of the radical pair is fairly high (typically > 20000

cm^{-1}) In such cases recombination to form a locally excited triplet state may be faster if the radical pair has achieved the necessary triplet spin alignment (21,22). Another type of cases where high radical yields ensue form electron transfer reactions with an excited singlet state has been encountered by Iwa et al.(23) with singlet excited oxonine/electron-rich aromatic amine donor systems. Here the radical pair energy is far below the excited singlet state and, in fact, is already closely above the singlet ground state of the pair of reactants. Due to the chemical similarity of such a singlet reaction system with our systems for generating the triplet radical pairs, we chose the reaction between oxonine singlet and N.N-tetramethylparaphenylene diamine (TMPDA) showing a Φ_{fr} or 0.48 in methanolic solution, to investigate the magnetic-field dependent reaction kinetics of radical pairs generated with <u>antiparallel</u> spins in microemulsion nanodroplets.

Oxonine (OxH^+) and its semireduced radical ($OxH^{\cdot}$) have absorption spectra which closely resemble the corresponding ones of thionine and its semiquinone radical. Due to the replacement of the S-atom in thionine by the lighter O-atom in oxonine the spin-orbit coupling is reduced to an extent that practically no spontaneous triplet formation occurs, so that no triplet reaction will interfere and the spin alignment of the generated radicals corresponds to a well defined singlet. In the CDBA reverse micellar system photoexcited oxonine produced an appreciable background of long-lived transient absorption. Therefore the investigations were carried out in microemulsions made up of H_2O/AOT nanodroplets in isooctane. This system is very well known in the literature. The relation between the radius of the surfactant-coated nanodroplets and the water/surfactant ratio w has been experimentally determined by the method of dynamic light scattering (1). Using this method it was also confirmed by Eicke and coworkers (24) that TMPDA concentrations up to 0.1 M in the oil phase do not cause significant changes of the radii of the nanodroplets.

As in homogeneous methanolic solutions, in the AOT reverse micellar system, too, a considerable amount of $OxH^{\cdot}$ radicals are detectable after a laser flash, when $^1OxH^{+*}$ is quenched by TMPDA (26,27). The kinetic behaviour of the radicals as observed on the nanosecond time scale is shown in Figure 6 for various w and for a homogeneous solution in methanol. At the particular wavelength

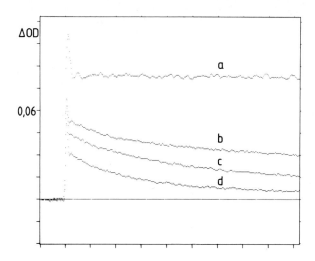

Fig.6 RP recombination kinetics observed at zero field in water/AOT/isooctane microemulsion (w= 32 (b), 16 (c), 8 (d)) and in methanol (a) with RPs resulting from 70% quenching of oxonine fluorescence by TMPDA (27). The oxonine semiquinone radical was monitored at 413 nm. The time scale is 100 ns/div for (a) and 200 ns/div for (b-d).

where the radical kinetics is recorded there is also an appreciable $S_1 \rightarrow S_n$ absorption of the dye, causing a pronounced absorption peak during the laser flash. Two main features are documented by Figure 6. Firstly, the amount of radicals detectable directly after the laser increases with w and is largest in pure methanol. Since care was taken that the same amount of laser quanta were absorbed in each experiment and the degree of fluorescence quenching, as controlled by the TMPDA concentration was the same, this result implies that there is a certain amount of very rapid geminate RP recombination taking place already during the laser excitation, the contribution of which decreases with w and has the least effect in the homogeneous solvent. Secondly, the radical decay in the microemulsion system is almost entirely geminate, whereas the radical decay in methanol extends to the millisecond region and is of second order with a rate constant of $2.8 \times 10^9 s^{-1}$. As was also found in the case of thionine/aniline RPs in CDBA based microemulsions, the rate constant of recombination to be followed on the time scale of several hundred nanoseconds decreases with w. However, the w-dependence of k_{rec}, which in the present system may be approxima-

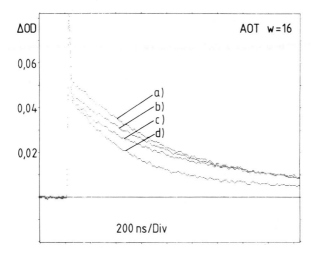

Fig.7 Magnetic-field dependence of singlet RP kinetics (26). Conditions correspond to those described in Figure 6, curve c.

ted by a proportionality to w^{-1}, is much less pronounced than in the former case. The magnetic field effect on the radical pair recombination kinetics is shown in Figure 7. Whereas there is only a very small decrease ($\approx 2\%$) of the free radical yield in methanol, in the microemulsion system the corresponding change in the amount of radicals present directly after the laser flash is about -20%. The magnetic-field effect in the microemulsion system is a twofold one. Besides the decrease of the amount of radicals detectable at the earliest times after the laser pulse, the magnetic field effect on which shows a $B_{1/2}$ of about 8 mT and saturates below 50 mT, there is also a magnetic-field effect on the geminate recombination kinetics. The effective first order rate constant k_{rec} of geminate recombination remains fairly constant up to fields of the order of 50 mT but then increases by a factor of 1.8 when the field is raised to 1 T.

An interpretation of the observed kinetic behaviour and its magnetic field dependence can be given in terms of Scheme III, according to which two different types of RPs have to be considered. One of them recombines on a time scale shorter than 10 ns, the other one on the time scale of several hundrerd nanoseconds. The species related to the latter process will be denoted as 'trapped' radical pair, whereby it is implied that both radicals are immobilized in the waterpool/surfactant inter-

172

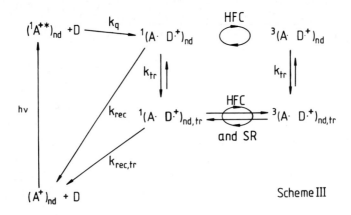

Scheme III

face layer, whereas in the other type of RP at least one of the radicals may be able to diffuse freely through the water pool until it recombines with its geminate counterpart or it becomes trapped, too. That there must be two types of RPs with largely different mobilities has already been inferred from the w-dependence of the radical concentration after the laser flash. The fact that the initial yield of this more slowly recombining type of RP may be influenced by a magnetic field leads us to assume the possibility of conversion of the fast-diffusing type of RP into the trapped one. Although it cannot be excluded that part of the trapped RPs originate directly in this situation through the photoelectron transfer process, a considerable part of them must originate as the freely diffusing RP type. With this assumption the magnetic field effect on the yield of trapped RPs can be explained as a consequence of spin evolution in the fast-diffusing RP. This spin evolution, which during the few nanoseconds of the lifetime of the latter type of RP can be brought about only by the <u>coherent</u> hfc mechanism, decreases the initial singlet character of the RP, thus decreasing also its recombination probability. However, the coherent $S \to T_\pm$ mixing process may be suppressed by quite small magnetic fields, so that the singlet character tends to be preserved for a longer time and the recombination will occur more efficiently in a magnetic field, with a concomitant decrease of trapped radicals. In fact a model calculation based on this mechanism allowed us to estimate the values of the rate constants $k_{rec} (\approx 10^9 s^{-1})$ and k_{tr} $(\approx 10^8 s^{-1})$ from the magnetic field dependence of the yield of trapped radicals (26).

When the RPs are in the trapped situation re-encounters occur much less frequently and there is enough time for spin equilibration, even in fields where the coherent $S \to T_+$ process is already suppressed. In order that the corresponding incoherent process be also slowed down below the re-encounter frequency, higher fields are necessary. If this requirement is met the recombination rate constant will increase. Assuming that on each re-encounter in zero field the trapped RPs recombine with a probability of 25%, corresponding to a spin-statistical singlet character, and in high field, where only $S \to T_o$ equilibrium can be established between two re-encounters, with a probability corresponding to a 50% singlet character, a maximum magnetokinetic effect of +100% on $k_{rec,tr}$ may be anticipated. In fact, the observed increase in $k_{rec,tr}$ of +80% at 1 T is quite close to this theoretical limit.

4.   CONCLUSION

In this contribution we have summarized and discussed the results of our investigations on several basic aspects concerning the behaviour of spin-correlated radical pairs enclosed in the nanodroplets of w/o microemulsions.

The recombination kinetics of radical pairs created with triplet spin correlation was systematically studied as a function of nanodroplet size and magnetic field. The time constants of recombination varied from 0.15 μs to 2.6 μs. They were proportional to the volume of the water nanodroplets, however the recombination was clearly slower than to be expected for a diffusion-controlled process in the interior of the nanodroplets. In zero magnetic field the decay of spin correlation due to isotropic hyperfine coupling is faster than the re-encounter process and not rate determining for recombination. In a magnetic field of 1 T the effective recombination rate constant is slowed down by a factor of 2 - 3. The field $B_{1/2}$ necessary to obtain half of the 1 Tesla effect increases from 14 mT to 40 mT as a monotonic function of the nanodroplet radius. The magnetic-field dependent recombination rate reflects the magnetic-field dependence of a spin relaxation process which depends on radical pair re-encounters and may be attributed to electron spin dipolar interaction. Application of halogen substituted radicals revealed contributions of a magnetic-field independent spin relaxation mechanism, the importance of which gradually increases with the spin-orbit coupling in the corresponding radical. This effect is most probably due to the spin-rotational relaxation mechanism.

When creating radical pairs with <u>singlet</u> <u>spin</u> <u>correlation</u> radical recombination processes and magnetic-field effects could be separately observed in two different time and field domains ($\leq$10 ns, $\leq$ 50 mT; $\geq$ 100 ns, $\geq$ 50 mT). They may be attributed to radical pairs diffusing freely through the interior of the water nanodroplet and others which are immobilized in the water/surfactant interface. A magnetic field accelerates the recombination due to interference with the (coherent) isotropic hyperfine coupling mechanism in the first case and with (incoherent) relaxation mechanisms in the second case.

In concluding we note that future developments in the application of magnetokinetic effects to micellar and microemulsion systems will need to develop quantitative models of spin relaxation. These will then provide access to specific details of radical pair motion in such media and thus will add to the basic knowledge which is required for a systematic progress in the methods of reaction control.

5.   REFERENCES

1   M.Zulauf and H.-F. Eicke, J.Phys.Chem. 83 (1979) 480-486
2   H.-F.Eicke, S.Geiger, F.A. Sauer and H.Thomas, Ber.Bunsenges.
     Phys.Chem. 90 (1986) 872-876
3   For reviews cf. K.M. Salikhov, Yu.N. Molin, R.Z. Sagdeev,
     A.L. Buchachenko (Eds): Spin Polarization and Magnetic
     Effects in Radical Reactions, Amsterdam: Elsevier 1984; U.E.
     Steiner and T. Ulrich , Chemical Reviews in press
4   N.J. Turro and G.C. Weed, J.Am.Chem.Soc. 105 (1983) 1861-1868
5   U.E. Steiner, G.Winter and H.E.A. Kramer, J.Phys.Chem. 81
     (1977) 1104-1110
6   U.E. Steiner and G. Winter, Chem.Phys.Lett. 55 (1978) 364-368
7   G.Winter and U.E. Steiner Ber.Bunsenges.Phys.Chem. 84 (1980)
     1203-1214
8   G.Winter, H.Shioyama and U.E.Steiner, Chem.Phys.Lett. 81
     (1981) 547-552
9   U.K.A. Klein and M.Hauser, Z.Physik.Chem. Frankfurt am Main,
     90 (1974) 215-220; D.Miller, Doctoral Thesis, Stuttgart
     1977.
10   T. Ulrich and U.E. Steiner, Chem.Phys.Lett. 112 (1984) 365-
     370
11   K.Schulten and P.G. Wolynes, J.Chem.Phys. 68 (1978) 3292-
     3297; E.W.Knapp and K.Schulten J.Chem.Phys. 71 (1979) 1878-
     1883
12   W.Schlenker, T.Ulrich and U.E.Steiner Chem.Phys.Lett. 103
     (1983) 118-123
13   A.Weller, F.Nolting and H.Staerk, Chem.Phys.Lett. 96 (1983)
     24-27
14   H.Hayashi and S.Nagakura, Bull.Chem.Soc.Jap. 57 (1984) 322-
     328

15   U.E.Steiner Z.Naturforsch. 34a (1979) 1093-1098
16   U.E.Steiner, Ber.Bunsenges.Physik.Chem. 85 (1981) 228-233; T.Ulrich, U.E.Steiner and R.E.Föll, J.Phys.Chem. 87 (1983) 1873-1882
17   H.-P. Waschi and U.E.Steiner, unpublished ; H.-P.Waschi, Doctoral Thesis, Stuttgart 1983.
18   T.Ulrich, U.E.Steiner and W.Schlenker, Tetrahedron 42 (1986) 6131-6142
19   W.Lersch and M.E. Michel-Beyerle, Chem.Phys. 78 (1983) 115-126
20   P.W.Atkins and D.Kivelson, J.Chem.Phys. 14 (1966) 169-174
21   K.Schulten, H.Staerk, A.Weller, H.-J.Werner and B.Nickel, Z.Physik.Chem. Frankfurt am Main 101 (1976) 371-390
22   M.E. Michel-Beyerle, R.Haberkorn, W.Bube, E.Steffens, H.Schröder, H.J. Neusser and E.W.Schlag, Chem.Phys. 17 (1976) 139-145
23   P.Iwa, U.E.Steiner, E.Vogelmann and H.E.A.Kramer, J.Phys.Chem. 86 (1982) 1277-1285
24   U.Hofmeier, M.Borkovec and H.-F. Eicke, private communication
25   T.Ulrich and U.E.Steiner, unpublished; T.Ulrich Doctoral Thesis , Konstanz 1986
26   D.Baumann, T.Ulrich and U.E.Steiner, Chem.Phys.Lett. 137 (1987) 113-120
27   D.Baumann and U.E.Steiner, unpublished; D.Baumann, Diplom Thesis , Universität Konstanz 1986

HYDRATED ELECTRONS IN REVERSE MICELLES

M.P. PILENI

1. INTRODUCTION

The present paper gives an overview of the physical properties of hydrated electrons obtained by pulse radiolysis in AOT reverse micelles(1-5). From the quenching of hydrated electrons by various probes such as "small" molecules or proteins, the distribution law of these quenchers and their locations in reverse micellar solutions are deduced. The hydrated electron is expected to be a very good probe to test the water pool of reverse micelles. Because of its small size, it is assumed that it does not perturb the micellar structure of reverse micelles. It is expected, in negatively charged reverse micelles, to be expelled by electrostatic interactions to the center of the water pool. The hydrated electron reacts with most the organic and inorganic molecules except saturated hydrocarbons and alkali metals, sulfate, persulfate and halogen ions. For all these reasons, the hydrated electron may be considered as one of the best probes of reverse micelles.

2. RESULTS AND DISCUSSIONS

2.1. Spectroscopic data for hydrated electrons in reverse micelles

Above $w = 20$, the transient optical spectrum obtained by pulse radiolysis in AOT reverse micelles is characterized by an absorption band centered at 710nm (1, 3). The absorption spectrum is close to that obtained in a homogeneous aqueous solution. The hydrated electron , e^-_{aq} , concentrations are linearly dependent on the absorbed doses. Evidence that the spectrum is due to e^-_{aq} formation is obtained by scavenging experiments. The addition of electron scavengers such as N_2O or O_2 to the reverse micellar systems decreases the lifetimes of the 710 nm absorption. The optical densities are linearly dependent on the absorbed doses. Therefore, the absorption spectrum is attributed to electrons hydrated in the micellar water pool.

By decreasing the water content, a decrease in the hydrated electron concentration is obtained. At w values above 15, the hydrated electron concentration is independent of the water content in the water pool whereas at lower values it falls with decreasing w. To explain these results we assume that excess electrons formed in the hydrocarbon are captured by the water pool. In fact by pulse radiolysis of AOT micelles in isooctane, C_8H_{18} , containing various amounts of water, electrons are produced in the bulk (isooctane) and in the water pool by the following processes:

$$C_8H_{18} \longrightarrow C_8H_{18}^+ + e^-$$

$$H_2O \longrightarrow H_2O^+ + e^-$$

$$H_2O^+ + H_2O \longrightarrow H_3O^+ + OH^\cdot$$

$$e^- + H_2O \longrightarrow e^-_{aq}$$

$$OH^\cdot + AOT \longrightarrow R$$

The hydrated electron absorption spectrum changes when the water content in the water pool decreases.

Below w = 20, the hydrated electron absorption spectra are blue shifted and less intense with decreasing w (3).The width of the absorption spectra of hydrated electrons in reverse micelles depends on the water content. At low water content the absorption spectrum is broader than that observed in aqueous solution or at w values above 15. This broadening may be attributed to changes in the properties of water in reverse micelles at low water content.

The extinction coefficient of the hydrated electron is determined by comparing the hydrated electron spectrum and that of reduced viologen. In aqueous solution these spectra are similar and the extinction coefficient of reduced propylviologen sulfonate at 610nm is close to the extinction coefficient of the hydrated electron at 650 nm(6).For all of the w values, the variation of the optical densities of reduced propylviologen sulfonate at 610 nm is similar to that of the hydrated electron at 650 nm. Because of the similarity of the absorption spectra of these two species in aqueous solution and in reverse micelles and because of the close values of their extinction coefficients at 650 nm and 610 nm respectively in aqueous solution it is concluded that the extinction coefficient of the hydrated electron in reverse micelles is close to that obtained in aqueous solution (equal to $18500 \, cm^{-1} M^{-1}$ at 710 nm) and does not differ by more than 10%.

In aqueous solution, the hydrated electron reacts with AOT with a rate constant, k_1, equal to $4.7 \times 10^7 \, M^{-1} s^{-1}$. Hence the half-life of the hydrated electron in the reverse micelle is shorter by one order of magnitude than in aqueous solution under similar conditions but it reacts with the surfactant molecules.

The decay of hydrated electrons follows first order kinetics and their lifetimes depends on the water content in the water pool. On diminishing w, the hydrated electron lifetimes decrease. However they depend on the surfactant purity and on the origin of the surfactant(6). Similar changes in the variation of the physical properties with the water content have been observed by NMR relaxation (7). However these changes occurred at w = 6 corresponding to the solvation shell of the counter ion and to a rigid micellar core. At w values lower than 5 no solvation of the electron has been obtained by different authors (7,8). We propose the following model:

- At low w values the water in the pool is highly immobilized and is not able to solvate electrons.

- When w increases, the water at the center of the pool is little affected by the interface and the probability of electron solvation also increases. The fact that the absorption spectra of hydrated electrons inside the micelle are shifted toward short wavelentghs compared to bulk water, shows that the water in the pool is different from bulk water. This could be due to the fact that the local concentration of Na^+ is high.

- When w increases further, the probability of electron capture and solvation by the water pool increases and the spectra of the hydrated electron become closer and closer to that observed in bulk water.

2.2. Determination of the distribution law of probes in RM from quenching studies of hydrated electrons by nitrate ions

In aqueous solution, hydrated electrons are strongly quenched by sodium nitrate with a rate constant equal to 3.10^9 $M^{-1}s^{-1}$. This quenching has been studied in AOT in reverse micelles having water contents between 10 and 60.

Below w = 25, for a given value of w, the first order rate constant of hydrated electrons is independent of the presence of nitrate ions in the water pool. However, the hydrated electrons yield decreases with increasing nitrate ion concentration. This can be explained by the fact that some of the hydrated electrons react rapidly with the nitrate, and are not observed while the remainder are unquenched, and are thus insensitive to the nitrate ions. The observed decay is attributed to the fraction of hydrated electrons formed in the water pool containing no quencher. The ratio of the optical density of hydrated electrons obtained in the presence and in the absence of quencher is equal to the probability of finding no quencher in a water pool, P_0.

$$P_0 = \frac{OD}{OD_0}$$

where OD and OD_0 are respectively the optical densities of hydrated electrons in the presence and in the absence of nitrate ions.

Two statistical models (9-22) describing the distribution of solubilized molecules in micellar systems were developed on the basis of their interaction with the host aggregates:

1. The distribution function corresponding to the Poisson law is believed to hold for direct and reverse micelles:

$$P_j = \frac{n^{-j} . e^{-n}}{j!}$$

where P_j is the fraction of micelles containing j molecules of quencher and n is the average number of quencher molecules per micelle($n = [NO_3^-]/[WP]$). The quencher and the water pool concentrations are $[NO_3^-]$ and [WP] respectively. The Poisson law is the mathematical decription of quencher solubilization without any cooperative effect : the adsorption of a quencher in a micelle has no effect on the probability of the adsorption of another molecule in the same micelle.

2. The second distribution proposed is geometrical which is not based on a definite model of direct micelles and is still open to discussion :

$$P_j = \frac{n^{-j}}{(1 + n)^{j+1}}$$

Using these two statistical models, the calculations of the fraction of micelles containing zero quencher are respectively :

with a Poisson distribution law : $Ln\, P_0 = n$

with a geometrical distribution : $1/P_0 = 1 + n$

Below w=40, figures 1A and B show the linearity of LnP_0 with nitrate concentration. This strongly supports a Poisson distribution law i.e. a non cooperative solubilization of nitrate ions and hydrated electrons. Figure 1B shows a severe

180

curvature of the plot of 1/Po against [NO₃⁻], indicating that the experimental precision is largely sufficient to eliminate the geometrical and related distributions.
These results are in agreement with those given in the literature (9-22). The slope of
a linear plot of LnP_0 versus[NO₃⁻] gives 1/[WP] for different w values. Thus, it
gives the water pool concentration, [WP] for different w values.

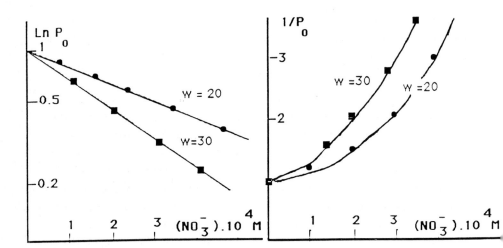

Figure 1: Variation of P_0 with [NO₃⁻] at w=20 and w=30

For w equal or up to 40 the hydrated electron decay follows a complex rate law in
the presence of NO₃⁻. To determine the hydrated electron quenching rate constant,
the kinetic treatment is deduced from a Poisson distribution (9, 19). Assuming that
the intramicellar quenching rate constant, k_q, is greater that the product of the intermicellar exchange rate constant, k_e, and the micellar concentration, the kinetic
treatment produces the following time dependence for the hydrated electron concentration :

$$Ln\, [\,e^-_{aq}\,]/[\,e^-_{aq}\,]_0 = (\,k_0 + k_e \cdot [NO_3^-]\,]\, t - n(1 - exp\,(-k_q)t) \qquad -1-$$

where $[\,e^-_{aq}\,]$ and $[\,e^-_{aq}\,]_0$ are respectively the hydrated electron concentrations at
time t and time zero, k_0 is the first order rate constant governing the hydrated elec-

tron decay in the absence of nitrate, k_e is the bimolecular exchange rate constant for water pool collisions, n is the average number of nitrate ions per water pool (n =$[NO_3^-]$/[WP]), where$[NO_3^-]$ and [WP] are respectively the nitrate ion and the water pool concentrations). Equation 1 is identical to:

$$Ln [e^-_{aq}]:[e^-_{aq}]_0 = Ln(OD/OD_0) = C_1 t - C_2(1 - exp(-C_3)t) \qquad -2-$$

where OD and OD_0 are respectively the optical densities of the hydrated electron at time t and time zero. The coefficients C_1, C_2 and C_3 were adjusted to the experimental data using equation 2 in a nonlinear least squares fitting procedure.

Figure 2 shows the experimental and simulated curves for different nitrate concentrations. The long time decay curves presented in figure 2 for various nitrate concentrations do not depend on the nitrate concentration indicating that the hydrated electron disappears before nitrate exchanges could occur. This is consistent with the exchange rate constant ($k_e = 3.10^7 M.s^{-1}$) determined by similar kinetic studies using porphyrins or chlorophyll as a probe and viologens as quenchers (21, 23, 24) and to those determined by stopped flow (25). This is also confirmed by the independence of the C_1 values determined from our calculations with nitrate concentrations.The difference between the results obtained with nitrate ions at low and high water content is due only to the fact that at low water content, the quenching rate constant is very large and quenching is not observable in the nanosecond time scale.

The quenching rate constant, k_q, decreases with increasing the water content. This is in agreement with all the quenching rate constants determined in reverse micelles (4, 17,18, 20-24).

The water pool concentration is deduced from the linear plot of $C_2 = n$ against the nitrate concentration,

Assuming that the micellar solution is monodispersed and that the water molecules are all located in the micellar water pool, the total volume occupied by N water molecules per micelle is :

$$w. (AOT). 10^{-3}/55. (WP). N$$

where N is the Avogadro number. Assuming that micelles are spherical, the radius of the water pool is derived from

$$r_w^3 = 3.w. [AOT] . 10^{-3}/55. (WP) . 4. \pi$$

Figure 3 shows the water pool radii obtained at various water contents from this calculation. These values are compared to those obtained by small-angle X-rays or neutron scattering experiments (26-30). The water pool radii determined at various water contents by small angle neutrons are in good agreement with those obtained by pulse radiolysis. Figure 3 shows the differences between the values obtained with these two techniques and by small angle X-ray scattering. This is due to the diffusion of the head polar group in the latter case. The slope of the line obtained from SAXS or from SANS or pulse radiolysis is equal to 1.5. The same value can be obtained from a geometrical calculation assuming the volume of the sphere is governed by the volume of the water molecules, $V_{aq} = 30 \, A^3$, and its surface by the surfactant of the head polar group of AOT molecules, $S_{AOT} = 60 \, A^2$, (details are given in chapter 3). It can then deduced that at $w > 15$, the water content, w, is directly related to the water pool radius.

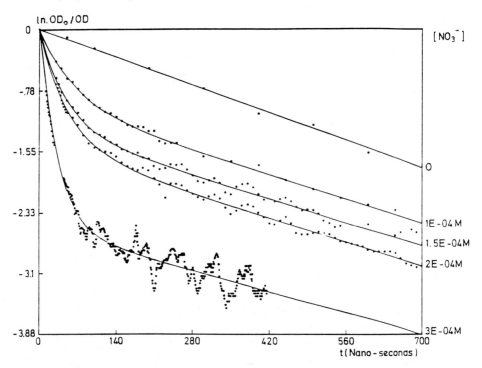

Figure 2: Kinetic decay of the hydrated electron at various ion concentrations.

2.3: <u>Determination of the location of the quenchers from the kinetic treatment</u>:

The quenching rate constants of hydrated electrons for various quenchers have been determined kinetically. The choice of the quenchers used is directly related to their locations.

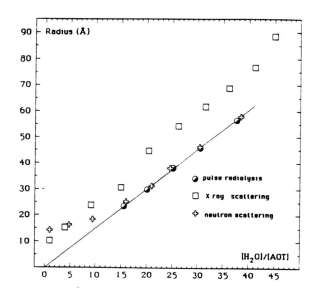

Figure 3: Variation of the water pool radius, r_w, with the water content,w.

Using nitrate ion which is water soluble and negatively charged or using a water soluble switterionic propylviologen sulfonate, it seems reasonable to assume that these quenchers are located inside the water pool. On the other hand, copper lauryl sulfate, $Cu(LS)_2$, is expected to be located at the interface because of the electrostatic interactions between the surfactant head polar groups and copper ions and because of the hydrophobic character of the two alkyl chains [LS]. In the same way, copper ions and methylviologen which are positively charged are attracted to the micellar surface.The locations of these probes are confirmed by a geometrical model (30) described in chapter 3 and tested by small-angle X ray scattering. In all cases, the quenching rate constant of the hydrated electron is determined at various water contents and then at various water pool radii ($r_w = 1.5$ w). By plotting, $Ln k_q$ over Ln w for a given quencher,the observed slope is close to 3 for nitrate ions and propylviologen and close to 2 for copper ions, copper lauryl sulfate and methylviologen (figure 4).

Such differences in the slope can be explained in terms of the location of the quencher in the reverse micelles :It is assumed that in reverse micelles, the average distance between two species, r, is approximately equal to the water pool radius, r_w:

(i) When two species (hydrated electron and quencher) are located inside the droplet,it is experimentally found that k_q varies with w^{-3}, and for a diffusion-controlled reaction, k_q is expected to vary as r^{-3}, where r is the average distance between the two species. Because of the linear relationship between r_w and w, it is concluded that for probes located inside the water pool, the quenching rate constant depends on the volume of the droplet as is observed for nitrate ions and propylviologen sulfonate.

(ii) For two species located in the water pool (hydrated electron) and at the interface (quencher), k_q varies with w^{-2} and is expected to vary with r^{-2} as is observed using copper ions, methylviologen and copper lauryl sulfate. Then for such probes, located at the interface, the quenching rate constant depends on the surface of the droplet.

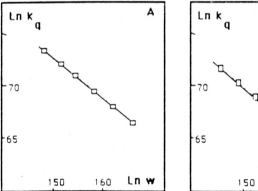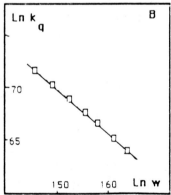

Figure 4: Plot of $Ln\,k_q$ against $Ln\,w$ for various quenchers .A: nitrate ions and B : copper lauryl sulfate

Theoretical studies (32-35) compare the kinetic rate constants of the two reactants. One of the reactants freely diffuses inside the sphere and the other is either fixed at the center or at the surface of the sphere. In the first case it is found that the kinetic rate constant is directly dependent on r^{-3} whereas in the latter case it depends on $r^{-2.5}$. The present results on the change of the variation of hydrated electron quenching rate constant with the water pool radius are in good agreement with those of theoretical calculations.

2.4. Extension of the distribution law to proteins and the change of quenching rate constant with the water content:

By using proteins having a hydrodynamic radius small in comparison to the droplet radius, instead of nitrate ions the results are similar to those previously presented:a linear dependence of LnP_o on protein concentration indicates that the distribution of these proteins in reverse micelles follows a Poisson distribution.At high water content the hydrated electron quenching rate constant, k_q, the water pool concentration and then the water pool radius can be deduced, at different water contents for various proteins. The change of the quenching rate constant of hydrated electron with the water pool radius is determined using various proteins such as chymotrypsin, ribunuclease at pH 4 and 9 and cytochrome c. The slope of $Ln k_q$ over $Ln w$ is close to 3 for chymotrypsin and for ribonuclease at pH 9 and close to 2 for cytochrome c and ribonuclease at pH 4. By analogy to the results obtained with probes, it can be concluded that chymotrypsin and ribonuclease at pH 4 are located inside the droplet and diffuse freely in the water pool whereas cytochrome c (at low pH) and ribonuclease at pH 9 are located inside the water pool. For chymotrypsin using the geometrical model from X-ray scattering measurements similar conclusions are given.

It can be noticed that, for ribonuclease, the quenching rate constant is smaller for protein located at the interface than that obtained for protein located inside the water pool. In such a case, just by changing the location of the probe, the quenching rate constant is changed. These data are very important for using reverse micelles as a microreactor. In chapter 3, from a geometrical model tested by small angle X ray scattering, the average location of the protein can be deduced at high water content. The geometrical model gives macroscopic information on the structural changes brought about by adding probes due to the location of the probes and the present model gives microscopic information from which the location is deduced. Using the two models, the average location of proteins which do not change drastically the micellar structure and have a hydrodynamic radius smaller than that of the water pool, can be deduced.

In this chapter, we have not discussed the cases in which the probes are located in the bulk hydrocarbon phase or form there own aggregates as the following. As has been described previously, the hydrated electron is formed exclusively in aqueous phase. To be observable, a certain number of water molecules in the water pool are required (w > 10). In the case in which the probe is located in the bulk hydrocarbon phase, the hydrated electron is never in contact with its quencher and the probe is not able to react with it. No quenching of hydrated electrons is observed.

For probes creating their own aggregate, not enough water molecules are entrapped in the microphase containing the probe to favour the formation of hydrated electron. Then the hydrated electrons formed are in the empty droplet and no quenching with the probe is observable.

The limitation of this kinetic model is that the structural changes of probes or proteins are negligible and the droplet can be considered as unchanged. This has been observed for chymotrypsin and ribonuclease. For cytochrome c, at high water content, the quenching rate constant does not change with the water pool size. This can be attributed to drastic structural changes (38).

3.CONCLUSIONS

It is shown, from the probability of finding zero quencher per micelle determined independently from the kinetic treatment, that the distribution of quenchers follows a Poisson distribution. This result is extended to enzymes and proteins. From kinetic data, structural data, such as the micellar concentration at various water content and the mass of the water pool and its radius, are deduced. The comparison of the masses determined by pulse radiolysis and by direct structural methods emphasizes the validity of the statistical distribution of probes in monodispersed droplets. The quenching rate constants of hydrated electrons with quenchers such as "small molecules" or enzymes and proteins are determined at various water contents of the water pool. From the change of the quenching rate constant of hydrated electrons with the size of the droplet, the average location of the quenchers in the reverse micelles is given.It is shown that, by changing the pH, the location of the protein can change and its quenching rate constant with hydrated electrons decreases for a protein bonded at the interface.

REFERENCES
-1-M.Wong, J.K.Thomas, and M.Gratzel,, J. Am. Chem.Soc.,98, 2391 , (1976)   -2-V.Calvos-Perez, G.S. Beddard,, and J.H.,Fendler, J. Phys. Chem., 85, 3126   (1981)
-3- M.P.Pileni, B. Hickel and J. Puchault, Chem. PhyLett. 92,308 (1982).
-4- M.P.Pileni, P.Brochette, B. Hickel, and B.Lerebours, J. Colloid Interface Sci., 98, 549 , (1984).
- 5- C.Petit, P.Brochette and M.P.Pileni J.Phys.Chem., 90, 6517, (1986)
-6- M.P.Pileni, J.Mihaud, T.Zemb, P.Brochette and B.Hickel "Surfactant in solu   tion" ed.P.Bothorel and K.Mittal (1987).
-7- M.Wong, J.K.Thomas, T.Nowak J.Am.Chem.Soc, 99, 4730,(1977)
-8- B.Valeur and E.Keh J.Phys.Chem.,83, 3305, (1979)
-9- P.P.Infelta, M. Gratzel and J.K.Thomas J. Phys.Chem.,78,190 (1974).
-10- M.Tachiya Chem. Phys. Lett., 33, 289 (1978).
-11- S.S Atik and L.A. Singer Chem. Phys. Lett., 59, 519 (1978).
-12- A.Yekta, M. Aikawa and N.J. Turro, Chem. Phys.Lett.,63,543 (1979)
-13- J.C.Dederen,M. Van der Auweraer, and F.C. de Schryver, Chem.Phys. Lett., 63, 543 (1979).
-14- G.Rothenberger, P.P.Infelta and M. Gratzel J. Phys. Chem.,83, 1871 (1979).
-15- M.P.Pileni and M.Gratzel J. Phys. Chem., 84, 1822 (1980)
-16- M.P.Pileni and M. Gratzel, M., J. Phys. Chem., 84,1822 (1980)
-17- S.S.Atik and J.K. Thomas J. Am. Chem. Soc., 103, 2550 (1981).
-18- S.S.Atik and J.K. Thomas J. Am. Chem. Soc., 103, 3553 (1981).
-19- M.Tachiya J. Chem. Phys., 76, 340 (1982)
-20- E.Geladé and F.C.de Schryver J.photochem., 18, 223 (1982)
-21- J.M.Furois, P. Brochette and M.P.Pileni J. Colloid Interface Sci., 97, 552(1984).
-22- P.Brochette, J. Milhaud and M.P.Pileni"Surfactant in solution" (1987) ed. P.Bothorel   and K. Mittal
-23- P.Brochette and M.P.Pileni Nouveau J.Chim., 9,551, (1985)
-24- P.Brochette, T.Zemb, P.Mathis and M.P.Pileni J.Phys.Chem.,91, 1444 , (1987).
-25- Robinson, B.H., Steyler, D.C., and Tack, R.D., J. Chem.Soc. Faraday Trans.,75, 481 (1979)
-26- Robinson, B.H., Topracioglu, C., Dore, J., and P.Chieux,J. Chem. Soc., Fara day Trans. 1,80, 13 (1984).
-27- M.Kotlarchyk, S.H. Chen, J.S. Huang and W.Kim Phys. Rev .29, 2054, (1984).
-28- Cabos, C., and Delord, P., J. Phys. Lett., 29, 6 (1969).
-29- Cabos, C., and Delord, P., J. Appl. Cryst., 12, 502 (1979).
-30- C.Petit, P.Brochette and M.P.Pileni Chem.Phys.Lett., 118;414, (1985)
-31- C.Petit, P.Brochette and M.P.Pileni J.Phys.Chem., 90, 6517, (1986)
-32- M.Van der Auwersen, J.C.Derenen, E.Gelade,and F.C.de Schryver J. Chem. Phys., 74, 1140, (1987)
-33- E.Gelade and F.C.de Schryver Reverse micelles P.L.Luisi ed. Plenum

188

Press (1984)

-34- U.Gosele, U.K.A.Klein and M.Hauser Chem.Phys.Lett., 68,291, (1979)

-35- M.Tachiya Chem.Phys.Lett., 68, 291, (1979)

-36- P.Brochette,n C.Petit and M.P.Pileni J.Phys.Chem (in press)

-37- P.L.Luisi private communication

-38- J.P.Huruguen, B.Perly and M.P.Pileni to be published

SYNTHESIS AND SURFACE REACTIONS OF SEMICONDUCTOR CRYSTALLITE CLUSTERS IN REVERSE MICELLE MEDIA

M.L. STEIGERWALD and L.E. BRUS

1. INTRODUCTION

This book illustrates the widespread current interest in reverse micelle solutions. The fact that microscopic regions of hydrophilic and hydrophobic character exist in intimate contact in a simple geometrical structure, in an optically homogeneous liquid medium, implies that these solutions can be novel reaction media. For example, it has been shown that hydrophilic protein enzymes, solubilized in the water pools or at the surfactant interface, retain their biological activity and are able to transform hydrophobic substrates dissolved in the organic phase (1-4).

In this chapter we consider inorganic crystallites of about the same size as these proteins. We describe how one can use the structured nature of the medium, and the differing solvation properties of both reactants and crystallite products, to synthesize and surface derivatize semiconductor crytallites in a general synthetic methodology. While the microscopic reaction mechanisms remain in important aspects obscure, the experimental advantages and possibilities are now becoming clear.

Why would one want to synthesize semiconductor crystallites in the size region of tens of Angstroms (i.e., hundreds to thousands of atoms) ? This size is somewhat below the sizes that can be made in homogeneous colloidal precipitation reactions, and in fact is in the neighborhood of the smallest possible seeds. There has been an active effort in the liquid phase photochemistry of somewhat larger particles, which behave as bulk semiconductors in their electrical properties, for several decades. In 1983 it was realized that ca. 50 Angstrom particles of CdS made by these colloidal methods have a band gap that was larger than that of the bulk material, even though they still had the crystal structure of the bulk material (5). This result implied that a quantum size effect regime had been entered, and that these crystallites were properly termed clusters having properties intermediate between those of molecules and bulk solids. An intriguing aspect was that a crystallite containing thousands of atoms still was not big enough to act as a bulk semiconductor. Systematic efforts then began to find better methods to make, isolate, and study even smaller crystallites of various materials.

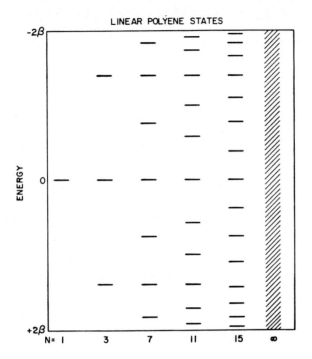

Fig. 1. Diagram showing discrete electronic states as a function of chain length N, for the Huckel molecular orbital model of pi bonding on a linear carbon chain. Note that for N=15, the discrete states near the band edges have a quadratic spacing, characteristic of the simple effective mass, particle-in-a-box model (taken from ref. 6).

2. QUANTUM SIZE EFFECTS IN SEMICONDUCTOR ELECTRONIC SPECTRA

A semiconductor has a filled valence band and an empty conduction band, separated by a band gap E_g. In a bulk crystal these bands have a continuous density of electronic states. Some insight into how discrete molecular states, with increasing size, develop into a continuous band can be obtained from the simple Huckel model for a one dimensional chain of N atoms in figure 1 (Ref. 6).

As chain length increases, new states are created at the band center and push out toward the two band extrema. Near the extrema the discrete states have a quadratic spacing from the band edges. This spacing is an effective mass (quantum particle in a box) type level structure, controlled by an effective mass m_e which is inversely proportional to the nearest neighbor exchange integral. A slow approach of the discrete levels to the band edge with increasing size results from strong chemical bonding (ie, a large exchange integral).

In crystallite clusters the lowest electronic transition occurs from the valence band HOMO to the conduction band LUMO. This transition is shifted to higher energy than E_g by (7)

$$\Delta E = \frac{\hbar^2 \pi^2}{2R^2} \left[\frac{1}{m_e} + \frac{1}{m_n} \right] - \frac{1.8e^2}{\epsilon R} \qquad [1]$$

Here the first term represents the effective mass quantum confinements energies discussed in the preceding paragraph. The second and smaller term is the attractive Coulomb interaction between the electron and hole. Thus the energy of this lowest discrete transition is a convenient measure of the size of crystallite cluster. Actual size dependent optical spectra for ZnSe crystallites appear in figure 2 (Ref. 8). Two discrete transitions appear, albeit with substantial linewidths. Recent temperature dependent, transient hole burning experiments show that approximately 10^3 cm^{-1} of the linewidth is inhomogeneous, due to a range of crystallite sizes and shapes in the sample. Additionally several hundred cm^{-1} is a homogeneous vibrational Franck-Condon contour (Ref. 9). Nevertheless the resulting broad resonance is still a useful measure of cluster size in reacting reverse micelle media, although not a precisely as equation 1 would suggest.

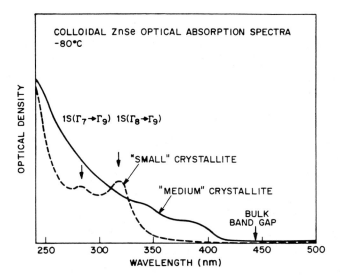

Fig. 2. ZnSe Colloid spectra as a function of particle size (taken from ref. 8). Medium refers to ca. 45 A diameter, and Small refers to ca. 25 A diameter.

As the internal MOs have nodes on the crystallite surfaces, the optical absorption spectra are not sensitive to surface states and/or molecular adsorption. This result has been verified in many colloidal experiments. However, luminescence following optical excitation is quite sensitive to surface conditions.

The electron and hole, once formed, can relax into localized surface states giving a characteristic long lived, red shifted cluster emission.

3. SYNTHESIS IN INVERSE MICELLE MEDIA

Fendler and coworkers in 1984 reported the first synthesis and photochemical study of CdS crystallites in AOT/water/isooctane media, from Cd^{++} in the water pools exposed to gaseous H_2S (10). They noted the crystallites were stabilized against flocculation by enclosure in the water pools. Yet, the crystallites were photochemically active with reagents added either as aqueous or hydrocarbon solutions. Lianos and Thomas reported similar syntheses from inorganic ions at high dilution in the water pools, and recognized that the crystallites formed had blue shifted absorption spectra implying that they were clusters in the quantum size effect regime. They found that smaller crystallites were made and stabilized at low w_0 (smaller water pools), for a fixed concentration of reagents and AOT.

While initially the cluster size was associated with the number of ions in the water pool, later work demonstrated that there had to be growth by exchange of ions (or small aggregates) between water pools (Ref. 11-12). Whitten, McLendon and coworkers demonstrated it was possible to increase CdS cluster luminescence inside the water pool by adsorption of alkylamines from hydrocarbon solution (Ref. 13). They concluded that deep, nonradiative surface states were being eliminated by amine adsorption. Petit and Pileni found the reverse micelle media especially useful in stabilizing smaller and more monodisperse CdS clusters than could be obtained in homogeneous solution (Ref. 14). They also were able to gel the hydrocarbon while preserving the clusters.

This work demonstrated that a significant degree of cluster size control is afforded by the micellar reaction medium, however, by a mechanism that was (and is) far from clear. In this regard, Steigerwald et al conducted the following experiment (Ref. 15). Starting with a sample of micelle stabilized CdSe clusters of small size (ca. 10-20 A in diameter), they added, sequentially, chemical feedstock ; first Cd(as aqueous perchlorate) and then Se (in organometallic form). The goal was to determine whether the nucleation of new particles or the continued growth of the already formed particles would obtain. The optical spectra showed that nucleation predominated at low w_0, while continued growth predominated at higher w_0. Fig. 3 shows the observed spectra at high w_0.

We believe this result indicates that the w_0 size control in the simple arrested-precipitation reactions is due to growth competing with fresh nucleation, there being a fast equilibration of reactants throughout the solution on the time scale of reactant addition.

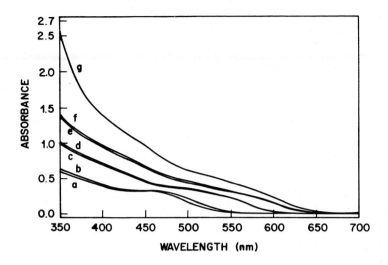

Fig. 3. Optical spectra of CdSe inverse micelle colloids (taken from ref. 15). a-g represent increasing amounts of chemical feedstock added to the initial seed colloid.

The surfactant molecules serve two identifiable functions : they initially disperse the water to give the microemulsion, and they act as a dynamic yet protective coating for the clusters as they form. The "dressed" cluster is analogous to a simple organometallic molecule, with the semiconductor lattice taking the role of the central metal atom, and the surfactant as ligands. A working model is that the organic coating acts as a selective membrane, allowing small inorganic and organic reagents to penetrate and reach the semiconductor surface, but preventing separate clusters from touching. This is one of the principal functions of a ligand in organic and organometallic chemistry.

Once a sample of micelle-encapsulated clusters is prepared, the addition of small amounts of water, increasing w_0, has no effect. A critical point which this experiment proves is that the clusters do not redissolve (disintegrate) to ions on increasing the water, and therefore the size control is not thermodynamic but kinetic in origin.

Addition of a large amount of water results in flocculation. The floc does not redissolve to any degree in any solvent we have tried, and it retains the color of the individual clusters. Sulfonates are only weak ligands for the CdSe surface. A significant component of the AOT-CdSe ligand strength is due to the same hydrophobic/hydrophilic pressure which forms the micelles. In large excess "bulk" water solvates the AOT better than does the CdSe/heptane mixture. When "bulk" water is added the cluster is stripped of some of its ligand, and coordination sites are opened on its surface. Erstwhile separate clusters then touch at these sites, forming aggregates which ultimately flocculate. In an analogous fashion, the opening of the coordination environment in an organometallic molecule typically leads to metal-metal bonds, and finally to metal precipitation. In the semiconductor cluster case, the floc is not redispersible because there are strong chemical bonds at the touching points. Nevertheless, the floc retains the color of the individual clusters because the contact is only pointwise and not crystallographically continuous. There is relatively minor electron delocalization from one cluster to the next.

In AOT stabilized suspension of clusters tends to flocculate spontaneously at high w_0, on a time scale of minutes to hours. However, evaporation of this suspension, effectively to dryness, yields a cluster-embedded surfactant which easily dissolves in organic solvents. This water free solution is stable for many months, showing only modest changes in color. This behavior reflects, as above, the greater effective lability of the AOT ligands at the higher water concentration. Thus the flocculation of large w_0 colloids is not due to large cluster size, but to a changing environment.

4.   SURFACE PASSIVATION AND THE ISOLATION OF PURE CLUSTERS

The experiments described above show that the AOT coordinated surface of CdSe clusters is still reactive. This reactivity has both advantages and disadvantages. One disadvantage is that particles can still fuse in the presence of water ; however, a general advantage is that the reactivity may be exploited by the judicious choice of reagents and reactions. A particularly useful class of surface reactions would covalently terminate the surface. This would preclude irreversible flocculation and enable the isolation of the clusters as chemically distinct species (ie, molecules). Covalent termination implies replacing the labile AOT ligands with tighly bound ones (Fig. 4).

We passivated CdSe surfaces by replacing bis(trimethylsilyl)selenium with phenyl(trimethyl)selenium in the sequential growth reaction described above (ref. 15). The phenyl group terminates the zincblende lattice, making the particle inert to subsequent growth and flocculation, since it ties up the second covalence of the surface selenium atom. When phenylselenide is added to the suspension of CdSe clusters, the microemulsion becomes turbid and a colored solid precipitates. The solid is isolated by either filtration or centifugation, washed to remove residual AOT and solvent, and dried. This gives an intensely colored, free-flowing solid. We believe that the phenyl attachment renders the clusters hydrophobic. The "capped" clusters are not soluble in either water or hydrocarbon, thus they precipitate. The isolated clusters redissolve easily in donor solvents such as pyridine, in contrast to a floc of initially bare clusters.

X-ray, TEM, and Se[77] NMR data all support this analysis of the isolated capped clusters.

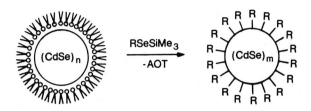

Fig. 4. Schematic representation of CdSe crystallite with AOT ligands reacting to yield a crystallite covalently terminated with R groups (taken from ref. 15).

Capping reactions in reverse micelle media are not limited to reagents which incorporate chalconides. For example, bipyridyl will coordinate to CdSe clusters and lead to precipitation. The dried powder is sparingly soluble in organic solvents. Solubility is markedly improved if both a charged and a neutral capping molecule are utilized : for example, preceding the bipyridyl step with exposure to a reverse micelle solution of potassium benzoate produces a capped powder that is as soluble as the phenyl capped powder. This result appears to imply that adsorption of the charged ligand neutralizes any net charge on the bare cluster. Neutralization enhances solubility in nonpolar solvents. The PhSeTMS capping reaction also neutralizes net positive charge as it bonds phenyl groups to the surface.

These reactions produce size selected semiconductor clusters with surface ligands that can be chosen for various purposes-enhanced temperature stability, alternately, cluster fusion at low temperature, charge transfer relay ability, control of luminescence and excited state lifetimes, specific surface chemical reactivity, specific solubility, etc. This synthetic area will see active development in future years.

5. CRYSTALLITE NUCLEATION AND GROWTH IN BIOLOGICAL SYSTEMS

The water pools in inverse micellar media concentrate the reagents, and the polar head groups of the AOT molecules lower the surface free energy of the smallest clusters. The local amount of water controls the ability of a feedstock ion to grow onto a prexisting cluster. To some degree these parameters are under experimental control in inverse micellar media, and thus degrees of freedom are available that are not present in homogeneous precipitation reactions.

Living systems that use inorganic crystallites for structural, gravitational, and magnetic sensing purposes appear to use similar principles of synthetic control, at a far more advanced (and even more poorly understood!) level of organization (ref. 16-18). Sulfate groups on polysaccharides concentrate calcium ions, and nucleate calcite in skeletal mineralization. Aspartic acid-rich proteins are thought to control the growth of quite specific calcite crystal sizes and habits. In some cases, structural correspondence (ie, epitaxy) between an organic macromolecular surface and a growing crystal surface is thought to control both size and alignment. Increasing insight to the basic molecular biology of these natural processes hopefully will teach us how to more intelligently make use of multicomponent inverse micelle media.

REFERENCES

1. R. Hilhorst, R. Spruijt, C. Laane and C. Veeger, Eur. J. Biochem. 144 (1984), 459.

2. R. Hilhorst, C. Laane and C. Veeger, in E.H. Houwink and R.R. van der Meer (Eds), _Innovations in Biochemistry_, Elsevier, Amsterdam, 1984, p. 81.

3. G.G. Zampieri, H. Jackle and P.L. Luisi, J. Phys. Chem., 90 (1986), 1849.

4. G. Haring, P. Luisi and F. Meussdoerffer, Biochem. Biophys. Res. Comm., 127 (1985), 911.

5. R. Rossetti, S. Nakahara and L.E. Brus, J. Chem. Phys., 79 (1983), 3058.

6   Louis Brus, Nouv. J. de Chimie, 11 (1987), 124.

7   Louis Brus, J. Phys. Chem., 90 (1986), 2555.

8   N. Chestnoy, R. Hull and L.E. Brus, J. Chem. Phys., 85 (1986), 2237.

9   A.P. Alivisatos, A.L. Harris, N.J. Levinos, M.L. Steigerwald and L.E. Brus, . J. Chem. Phys. 89 (1988), 4001.

10   M. Meyer, C. Walberg, K. Kurihara and J.H. Fendler, J. Chem. Soc. Chem. Commun. (1984), 90.

11   P. Lianos and J.K. Thomas, Chem. Phys. Lett., 125 (1986), 299.

12   P. Lianos and J.K. Thomas. J. Coll. Inter. Sci., 117 (1987), 505.

13   T. Dannhauser, M. O'Neil, K. Johansson, D. Whitten and G. McLennon, J. Phys. Chem., 90 (1986), 6074.

14   C. Petit and M.P. Pileni, J. Phys. Chem., 92 (1988), 2282.

15   M.L. Steigerwald, A.P. Alivisatos, J.M. Gibson, T.D. Harris, R. Kortan, A.J. Muller, A.M. Thayer, T.M. Duncan, D.C. Douglas and L.E. Brus, J. Am. Chem. Soc., 110 (1988), 3046.

16   L. Addadi and S. Weiner, Proc. Natl. Acad. Sci. USA, 82 (1985), 4110.

17   L. Addadi, J. Moradian, E. Shay, N.G. Maroudas and S. Weiner, Proc. Natl. Acad. Sci., USA, 84 (1987), 2732.

18   S. Mann, Nature, 337 (1988), 119.

MICROPARTICLE SYNTHESIS AND CHARACTERISATION IN REVERSE MICELLES.

B. H. ROBINSON, A. N. KHAN-LODHI and T. TOWEY

1. INTRODUCTION

A reverse micellar system (often described as a water-in-oil microemulsion) is an optically-transparent, thermodynamically-stable dispersion of two immiscible liquids which are stabilised by a surfactant. Generally small amounts of water are present such that the water-to-surfactant mole ratio is less than ten or so. Fig. 1 shows a schematic triangular phase diagram of the complete three component system.

(The general phase behaviour of surfactant/water/oil systems is considered in more detail in Chapter 2 of this book).

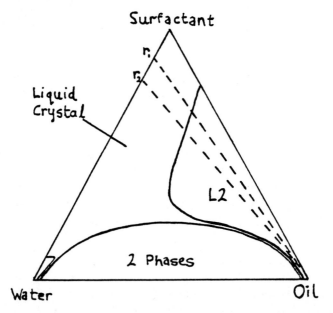

Fig.1. Schematic three component phase diagram biased in favour of an extensive L2 region by the use of a wedge-shaped surfactant such as Aerosol-OT. The dotted lines correspond to a dilution process with oil. Along a line the droplet size is maintained but the concentration of droplets changes. The two dotted lines indicated above refer to different relative concentrations of water to surfactant and have different droplet sizes.

The L2 region represents a reverse micelle (microemulsion) phase of water droplets surrounded by a surfactant skin dispersed in an excess oil phase. The size of the water droplets is determined by the composition of the medium i.e. by w, where w is the molar ratio of water to surfactant. In recent years these water droplets have been used as 'microreactors' for solubilizing guest molecules and macromolecules such as enzymes(1-4) and microparticles(5-7) e.g. Pt, Pd, Rh(5), Au(6), and CdS(7). Both the biological and inorganic systems can exhibit catalytic function. It is our purpose in this review to discuss the preparation and characterisation of the latter systems, with particular reference to metal microparticles.

2. GENERAL PREPARATION PROCEDURES AND SIMPLE KINETIC CONSIDERATIONS FOR THE PREPARATION OF COLLOIDAL METAL PARTICLES

Metal particles of colloidal dimensions are widely used in heterogeneous catalysis (with a typical size of 1-100 nm). There are various methods for their preparation:

(a)   Vapour phase preparation where the metal is vaporised and condensed directly onto a solid support(8).

(b)   Direct generation of colloidal particles on solid supports, where a solid matrix, such as silica or alumina, is impregnated with a metal salt complex (e.g. hexachloroplatinic acid) and then dried yielding microcrystals of the metal salt in the pores of the amorphous solid. Metal particles are then readily obtained by reduction(9).

(c)   A liquid phase synthesis where a metal salt is reduced either in an aqueous solution(10-15) or in a reverse micellar dispersion. In both systems the growth process is constrained; in aqueous solution this is achieved by addition of a chemical stabiliser, in the reverse micelle by the surfactant surrounding the droplet.

It is the potential of method (c) in reverse micelles that is of particular interest to us. The precise mechanistic details of such syntheses are still to be resolved but it would appear that, in order to generate small and monodisperse particles, it is important that the nucleation and particle growth processes occur on a different time scale. To satisfy this condition, it is necessary to produce a large number of metal particle nuclei quickly, which is then followed by a slower particle growth as represented by the scheme below

$$\text{NUCLEATION (Fast) M}^{ox} \xrightarrow[\text{agent}]{\text{reducing}} \text{M}_1^o$$

$$\text{GROWTH (Slow) M}_n^o \, (n{=}1 \rightarrow \frac{m}{2}) + \text{M}_{n'}^o (n'{=}1 \rightarrow \frac{m}{2}) \rightarrow \text{M}^o (2{\rightarrow}m)$$

M^o = particle nuclei which may be metal atoms or small clusters.

m = equilibrium metal particle aggregation number (i.e. the number of metal atoms in a stable cluster.

n,n' = transient aggregation number during growth process.

It is likely that the production of new nuclei during the growth process results in polydispersity within the particulate dispersion. Once the particles, M^o_m , have formed it is necessary that there should be some kind of stabilization of the particles to prevent coagulation. In aqueous solution if there is an electrolyte present which is preferentially adsorbed in the surface region, the particle growth will be arrested by electric double layer coulombic repulsion. It should be noted that adsorption of polymeric materials at the surface can also result in steric stabilisation(16).

3.   CHARACTERISATION OF COLLOIDAL METAL SOLS

Many techniques are available for the characterisation of the size and size distribution of micrometal particles. Transmission electron microscopy (TEM) is the most direct method since it allows visible images of dispersed particles to be obtained and recorded photographically, and polydispersity is readily detected. Stabilized silver colloids in acetonitrile solution formed by reduction by polyethylene-imine have recently been characterised(17,18) by High Resolution Electron Microscopy. There is evidence that the very small particles produced (ca. 40 Å) are produced by growth from nuclei which have decahedral morphology rather than that of the bulk face-centred cubic metal.

Scattering techniques such as small angle X-ray scattering (SAXS) and photon-correlation spectroscopy (PCS) are potentially useful methods for characterising colloidal dispersions since they allow the sampling of a statistically large number of particles in solution, and information on both size and interparticle interactions can be obtained. The characterisation of gold sols by Yudowitsch(19) using SAXS and TEM revealed excellent agreement between the two methods. PCS is now a straightforward and widely used technique for the characterisation of size domains in colloidal dispersions. Since the technique essentially measures the diffusion coefficient of the light scattering species, which can then be related to the particle size (in the case of spheres by the Stokes-Einstein equation), generally only the average hydrodynamic radius may be obtained. This rather complicates the determination of actual particle size since metal colloids are generally stabilised by polymeric materials containing surface-adsorbed ions. Thus the hydrodynamic radius of the metal particle will differ from that of the particle with no surface-adsorbed species. Analysis can be rather complicated when polydisperse systems are studied(20). X-ray diffraction patterns obtained from suspended metal particles (crystallites) also provide information concerning size and size distribution since they exhibit size-dependent line-broadening effects. The line broadening of diffraction signals can be used to calculate the particle diameter using Scherrer's equation(20). Analytical ultra-centrifugation can also be used for measuring the sedimentation coefficients of suspended particles. The sedimentation coefficients are related to the particle mass and hence the size, but the density of the sedimenting entity is often difficult to estimate, as it involves a skin of stabiliser of uncertain density.

Chemical methods have also been developed for the determination of particle sizes of supported metal catalysts(12). This involves the selective chemisorption of gases such as oxygen, hydrogen and carbon monoxide. Then by measuring the volume required for monolayer coverage and evaluating the stoichiometric number of surface atoms interacting with each adsorbed molecule(12), the surface area and hence particle size may be estimated. Turkevich et al have reported good agreement between the chemisorption technique and electron microscopy for platinum particles supported on alumina.

4.   CONSIDERATIONS OF PREPARATION OF COLLOIDAL METAL PARTICLES IN REVERSE MICELLE (MICROEMULSION) MEDIA

The original reason for using w/o (water-in-oil) microemulsions for the 'in situ' preparation of colloidal metal particles was that in many cases they are known to consist of water droplets of a well-defined size. However there are many microemulsions which are not well-defined in terms of structure, but which are still effective for the preparation and stabilisation of microparticles. In the case of droplets, they represent aqueous microcompartments for synthesis and the 'in situ' reduction of metal salts (e.g. of Pt, Rh, Pd) might then be expected to result in the production of particles of a well-defined size and shape. Further these particles would be stabilised by the surfactant shell, which we might suppose would occur when the particles achieved the same size as the droplets from which they were made.

At first sight it might appear that communication of reactants might be difficult, but there is only a small barrier to communication of solubilisates between droplets in a reverse micellar system. Eicke(21) was the first to show that communication was facile (and effectively instantaneous on the timescale of most synthetic reactions). Later in a series of papers, Thomas *et al* (22-24) and Robinson *et al* (25-26) showed that, for well-defined droplet dispersions such as are formed by Aerosol-OT, the rate constant for exchange k_{ex} for the process described below:

$$Ⓐ + ◯ \xrightarrow{\ k_{ex}\ } ◯ + Ⓐ$$

Ⓐ and ◯ are filled and unfilled droplets respectively

was of the order of 10^6 -10^7 dm^3 mol^{-1} s^{-1} ; that is about a factor of 10^3 - 10^4 slower than the limit set by diffusion control of the reaction in an oil solvent of low viscosity. The diffusion controlled limiting rate constant is when exchange occurs at every encounter of droplets in the medium and is calculated by means of the Smolochowski Equation ($k = 8RT/3\eta$, where η is the viscosity of the medium).

The process of exchange is associated with a significant activation enthalpy (the activation energy - RT) which depends on droplet size or w value. Some representative data are given in Table 1 below.

w	$\Delta H^{\dagger}$
10	68 ($\pm$10)
15	83 ($\pm$15)
20	91($\pm$15)

Table 1. Activation enthalpies ($\Delta H^{\dagger}$ /kJ mol^{-1}) for solubilisate exchange in reverse micelle systems based on AOT in n-heptane. Data taken from Ref. (4).

The AOT system seems to represent a lower limit for k_{ex} and thus the activation barriers represented in Table 1 are expected to be among the highest for reverse micellar dispersions. Exchange of hydrophilic species between water droplets or indeed any dispersed phase e.g. glycerol is, however, still a very rapid process and typically the lifetime of a solubilised ion inside an individual droplet is less than 1 ms. Exchange is thought to occur via the formation of a transient dimer species. Therefore on a timescale of greater than 1s, the dispersed droplet phase can be effectively regarded as a pseudo-continuous phase.

The ease of penetration of solubilisates between droplets will also depend on the state of the surfactant film. Recent quasi-elastic neutron scattering experiments(27) have shown that this skin is essentially liquid-like, the lateral diffusion coefficients for AOT around the surface of a water droplet being measured as 6×10^{-10} m^2 s^{-1}. Data is not so far available for the motion of surfactant in the presence of solubilised metal particles.

The initial speculation was that in the case of a well-defined droplet dispersion, the size of the droplets (which is readily controlled) would determine the size of the particles formed. However, it was shown experimentally in an early paper by Stenius(5) and later by other groups(29-32) that the particles produced are generally larger than the initial water droplets. This indicates in part the flexibility of the surfactant shell . It has also been suggested(5) that the surfactant interacts directly with the growing particles, by adsorption on the surface. This could then result in an effectively increased w value for the microemulsion droplets since less surfactant would be available for stabilisation of the oil-water interface; a corresponding larger final particle size would then be obtained. Nevertheless it is still generally true that essentially monodisperse particles can be prepared by initial compartmentalisation of the reactants in the water droplets of the w/o microemulsion .

There have been several reports (Table 2) of metal particles formed by the addition of reducing agents (hydrazine, hydrogen, sodium borohydride) to the metal precursor salt solubilised in the water droplets of the microemulsion. In our own work, both the metal salt and reducing agent are solubilised in the microenvironment of the dispersed polar phase (e.g. water, glycerol, formamide) of well-characterised reverse micelles. However, there are interesting differences in the final product when non-aqueous dispersions are used and these will be discussed presently. Reaction is generally effected in a simple manner by mixing equal volumes of microemulsion solutions containing the metal salt and the reducing agent. AOT was generally used as surfactant, which is, as stated in Section 1, a particularly good surfactant for making reverse micelles of well-defined structure using a variety of dispersed polar phases. In the case of water dispersed in a continuous oil phase, the amount of water present is represented by the value w = [H$_2$O]/[AOT], the molar ratio of water to surfactant and this value for the specific case of AOT is proportional to the size of the droplets present, such that

$$r = \frac{3V_{DP}}{A_s} . w$$

Metal/metal salt	Reverse Micelle system	Ref.
Pd	Water/CTAB/octanol	
Pt	and	5
Rh	Water/PEDGE/hexane (or hexadecane)	
Au	Water/PEDGE/hexane	6
Pt	Water/AOT/n-heptane	
Pd	Glycerol/AOT/n-heptane	33
Au	Formamide/AOT/n-heptane	
Ag		
Ni_2B	water/CTAB/n-hexanol	29
Fe_2B	water/CTAB/n-hexanol	31
Co_2B	water/CTAB/n-hexanol	30
Fe_3O_4	water/AOT/isooctane and	34
	water/NP-6/isooctane	
$CaCO_3$	water/NP-6/cyclohexane	35
$CaCO_3$	water/CaOT/cyclohexane	32
$BaCO_3$	water/BaOT/cyclohexane	36
CdS	water/AOT/heptane	7
	glycerol/AOT/heptane	33

where   CTAB   = cetyltrimethylammonium bromide
         PEDGE = pentaethyleneglycoldodecyl ether
            AOT = sodium 1,2 bis(2-ethyl hexyl)sulphosuccinate
            NP-6 = hexaoxyethylene nonophenyl ether
CaOT, BaOT = calcium and barium salts of AOT

Table 2. Preparation of metal and metal salt particles in reverse micelle media taken from the literature.

where V_{DP} is the molar volume of the dispersed phase species (in this case water), r is the radius of the water droplet and A_S is the molar head group surface area of the surfactant. The same equation holds for other dispersed phases, e.g. glycerol.

Many parameters may be changed, the dispersed phase, the metal, the oil phase, the surfactant. In Table 2 some of these variations are indicated. As a result, it is possible to investigate in a more detailed way the factors which influence particle size.

(a) Preparation of colloidal metal particles in water/AOT/n-heptane reverse micelles.

(i) <u>Platinum</u>

Platinum sols are readily prepared in reverse micellar media; for example by reducing hydrogen hexachloroplatinate (IV) 'in situ' with hydrazine. The correlation between particle size and reactant concentration is shown in Table 3 below, for droplets of initial radius equal to 18Å, as determined by a variety of techniques (37).

The sizes of the Pt microparticles were characterised by transmission electron microscopy, and the standard deviations evaluated by normal procedures.

	$[N_2H_4]$	$[H_2PtCl_6]$	Mean particle radius (r) /Å	Standard Deviation/Å
Pt1	1.12	0.45	84	10
Pt2	2.25	0.45	63	7
Pt3	4.5	0.45	38	5
Pt4	9.0	0.45	24	3
Pt5	4.5	0.112	15	3
Pt6	4.5	0.225	29	6

Table 3. Structural characterisation of Pt sols prepared in a w = 5 water/AOT/n-heptane reverse micelle system at 25°C. Concentration of AOT is 0.1 mol dm^{-3}. All concentrations are overall concentrations (mmol dm^{-3}) in the dispersion.

These results clearly demonstrate (column 4) that particle size is <u>not</u> simply dependent on the finite dimensions of the droplet. It is apparent that the concentrations of both reactants are important. However, this is to be expected from the foregoing discussion (Section 2) since the concentration values will determine the rates of the nucleation and growth processes. (In addition, surfactant adsorption may also be an important factor which may not be dependent on the rate of particle growth). Hence it is clear experimentally that a wide range of particle sizes can be obtained in droplet 'microreactors' which are notionally of the same size. However it should be noted that the primary function of the droplet microreactor system is achieved, in that it provides a compartmentalised medium preventing phase separation of the particles.It is significant by inspection of Table 3 that small particles of similar dimensions to the droplets can indeed be prepared (e.g. Pt 5). On the other hand, altering the relative concentrations results in particle sizes which can be ~6 fold larger.

A kinetic rationalisation for the results is as follows : the contents of the droplets are exchanged rapidly ($t_{1/2}$ exchange $\approx$ 1ms at the concentrations employed). Also as the concentration of hydrazine is increased the average size of the particles produced is decreased. At high hydrazine concentrations the hexachloroplatinate (IV) ion will be reduced rapidly resulting in the generation of a large number of seed nuclei and hence producing smaller particles overall. Furthermore, with lower concentrations of hexachloroplatinate(IV) but at a constant excess concentration of hydrazine the nucleation rate constant is dependent only on the hydrazine concentration. However, there is less growth material and so again smaller particles are produced (Table 3). The particle size can therefore be decreased by increasing the $[N_2H_4]/[PtCl_6^{2-}]$ ratio.

The concentration of droplets can also be readily varied by increasing the AOT concentration at a constant w. This has the effect of reducing the effective reactant concentration which should give slower nucleation and growth rates. Slower nucleation favours large particles and slower growth favours small particles. Hence the effect of changing the total droplet concentration should be relatively small as is indeed observed (Fig.2).The results indicate that a minimum particle size occurs at [AOT]=0.2 mol dm^{-3}, such that [droplets]=1.2 x 10^{-3} mol dm^{-3} and particle radius = 29Å. This perhaps represents the optimum conditions for controlling the relative rates of nucleation and growth, such that small particles are produced for a given concentration ratio of reactants.

The size of the water droplet can be increased by increasing w. As the droplet size is increased, the concentration of droplets and the frequency of encounters for which exchange can occur also decreases. The average number of reactant ions per droplet will increase but the effective concentration of reactants in the dispersed phase decreases.

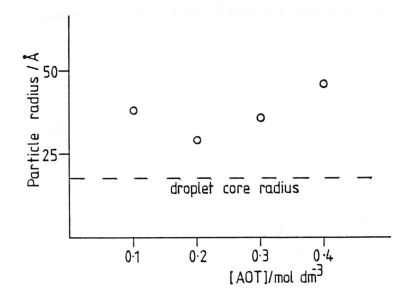

Fig.2. Effect of droplet concentration on particle size in a w=5 water/AOT/heptane microemulsion.Preparation conditions 4.5mmol dm^{-3} hydrazine and 0.45 mol dm^{-3} hexachloroplatinate.

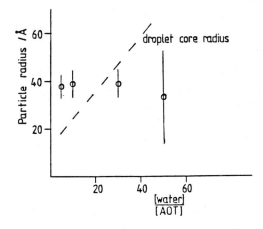

Fig.3. Effect on particle size of variation in water droplet size. Preparation conditions as for Pt 3 (Table 3).

It can be seen (Fig.3) perhaps surprisingly, that the average particle radius is not significantly affected on changing w, suggesting that the relative rates of nucleation and particle growth remain unaltered. However, the particle dispersion is more monodisperse when smaller droplets are used. (The data in Table 3 and figs 2 and 3 are taken from the PhD thesis of Abid Khan-Lodhi).

(ii) Palladium

The preparation of palladium sols is readily carried out by the 'in situ' reduction of potassium tetrachloropalladate(II) with hydrazine by a similar method to that used for platinum. Some typical results are shown in Table 5 for micro-palladium particle formation in AOT-reversed micelles in n-heptane, w = 5, [AOT] = 0.1 mol dm^{-3}. As was the case for platinum, at higher concentrations of reductant, smaller particles are produced as a result of the more rapid nucleation process. It should be noted that, for the same concentrations, the size of Pd particles is much less , e.g. (Pt2 → r = 63 Å; Pd3 → r = 19 Å). It might be supposed from these results that the nucleation rate is much faster for Pd than for Pt.

	$[N_2H_4]$	$[K_2PdCl_4]$	Microparticle radius (r)/Å	Standard Deviation(σ)/Å
Pd1	9	0.45	12	3
Pd2	4.5	0.45	15	4
Pd3	2.25	0.45	19	6

Table 5. Preparation conditions and structural characterisation of palladium sols in a w = 5 water/AOT/n-heptane reverse micelle with 0.1 mol dm^{-3} AOT. All reactant concentrations are overall concentrations (mmol dm^{-3}) in the dispersion.

(b) Preparation of colloidal metal particles in glycerol/AOT/n-heptane reverse micellar systems.

In addition to water, glycerol may also be used as the dispersed phase in a reverse micelle to solubilize hydrophilic species. Nanometre sized droplets are readily formed in the absence of solubilizate(38-39) and several reverse micellar systems based on glycerol as the dispersed phase have now been characterised. The viscosity of the overall microemulsion is of course much lower than that of bulk glycerol, since the viscosity of the dispersion is determined by that of the continuous phase. Such a dispersion is still satisfactory as a reaction medium since exchange of material between droplets is still facile. Typical values for k_{ex} for the exchange process between glycerol droplets are again in the range

$\sim 10^6$-10^8 dm^3 mol^{-1} s^{-1}, which is very similar to the values measured in aqueous microdispersions. Nevertheless the nucleation and growth processes take place exclusively in a glycerol environment. It is found that certain metal colloids coagulate in aqueous microemulsions; however, stable systems can be obtained using glycerol reverse micelles, e.g. colloidal gold.

(i) Colloidal Gold

Gold sols are again easily prepared by the 'in situ' reduction of hydrogen tetrachloroaurate (III) with hydrazine in a w = 1 glycerol/AOT/n-heptane microemulsion. In this case w is defined as [glycerol]/[AOT]. Glycerol is dispersed without difficulty in an AOT solution in the oil, and in this case the reactants are dissolved in the glycerol (by heating to ca. 60°C) before dispersion. Reaction was effected by mixing together equal volumes of the two reactant solutions.

At higher concentrations of hydrazine (see Table 6) smaller particles are again produced so this effect seems to be general for microparticle formation. Changing the droplet concentration has no significant effect, the nucleation rate remaining rapid at all droplet concentrations such that particles of similar size are obtained.

	$[N_2H_4]$	$[HAuCl_4]$	Mean particle radius (r)/Å
Au1	3.65	0.182	25
Au2	1.82	0.182	31
Au3	1.38	0.182	37
Au4	0.91	0.182	42

Table 6. Preparation conditions and characteristation of gold sols in a w = 1 glycerol /AOT/ n-heptane reverse micelle with 0.1 mol dm^{-3} AOT. All concentrations are overall concentrations (mmol dm^{-3}) in the dispersion. Initial droplet radius is 9Å (38-39).

The important point about these 'sols' is that they are apparently stable indefinitely, in marked contrast to gold sols prepared in an aqueous medium.

(ii) Colloidal Platinum

The 'in situ' reduction of hexachloroplatinate (IV) with hydrazine in a w = 1 glycerol/AOT/n-heptane reverse micellar system results in the gradual appearance of a golden brown colour over two weeks in contrast to the usual grey-black colour observed for platinum. It is thought that extremely small particles are generated of approx 5-6Å mean radius , as a result difficulties were encountered in obtaining satisfactory size data by electron microscopy(33).

(c) Preparation of colloidal metal particles in formamide/AOT/n-heptane reverse micelles(33).

Gold particles could also be prepared using formamide as the dispersed phase i.e. using formamide-in-oil reverse micelles. The maximum w value that could be achieved in n-heptane was 1.75. Particles can again be easily generated by the 'in situ' reduction of hydrogen tetrachloroaurate (III) with hydrazine. These particles are in fact found to be even more stable than those prepared with glycerol as the dispersed phase. In glycerol very slight sedimentation was observed with complete sedimentation being immediately induced by the addition of pyridine to the system. However, in formamide-in-oil microemulsions even the addition of pyridine does not induce sedimentation of the gold. This suggests better surface stabilisation of the particle when formamide is used as the dispersed phase. Clearly the dispersed solvent plays an important role in particle stability which is indicative of the operation of the different particle-solvent interactions.

To summarise, for the colloidal metal particles prepared in w/o and g/o AOT microemulsions with n-heptane as the continuous phase it seems to be clear that growth by aggregation is not primarily limited by the initial size of the water droplet. By choosing appropriate reactant concentrations particles of different sizes may be prepared. Since the particle precursor metal salts and the reducing agent hydrazine are both hydrophilic, the reduction, nucleation and growth processes are all expected to take place essentially in the dispersed polar phase. Although the overall concentrations of the reactants are similar to those used for preparation of metal sols in bulk water, the effective concentration of reactants in the dispersed phase is much higher (i.e. for 4.5×10^{-4} mol dm^{-3} $PtCl_6^{2-}$ in a w = 5 water/AOT/n-heptane, the concentration per dm^3 of dispersed water is 0.05 mol dm^{-3}). So reverse micelles not only compartmentalise the reactants but there is a concentration effect making the reduction and nucleation processes more efficient compared to reaction under similar overall concentration conditions in the bulk solvent.

5.   SPECTRAL PROPERTIES OF COLLOIDAL METALS

Colloidal metals are, like bulk metals, good conductors of electrons. The collective oscillations of conduction electrons due to long range Coulombic correlations in the bulk metal are called plasmons(41). In the restricted space of a colloidal particle resonant oscillations can occur and these are called surface plasmons. This effect can be induced by the polarisability of the particle by electromagnetic radiation and the particle then behaves as an electric and magnetic dipole. As a result of these phenomena, a range of temperature stable pigments can be obtained. Mie(41) first showed the relationship between the colours of colloidal metals and bulk optical properties by solving Maxwell's equation for the absorption and scattering of light by spherical particles. In the simple case of very small particles which fall within the Rayleigh scattering regime (diameter <20 nm), the electric field of incident light across the particle will be approximately constant and the particle will essentially behave as an electric dipole. This will lead to the excitation of dipolar surface plasmons. The theory underpinning the optical phenomena observed is complex but a brief description of the essential features is now given.

The scattering (σ_{sca}) and absorption (σ_{abs}) contributions to the extinction cross section per particle for a sphere of radius r can be shown to be given by

$$\sigma_{(sca)} = \frac{128\pi^5 r^6}{3\lambda^4} \left| \frac{\varepsilon - 1}{\varepsilon + 2} \right|^2$$

and

$$\sigma_{(abs)} = \frac{8\pi^2 r^3}{\lambda} \left| I_m \left[\frac{\varepsilon - 1}{\varepsilon + 2} \right] \right|$$

where $\varepsilon = (\varepsilon_1 + i\varepsilon_2)$ is the complex dielectric function of the metal at the optical frequency relative to the surrounding medium and λ is the wavelength of light in that medium. However, for a particle within the Rayleigh regime the scattering contribution to the extinction coefficient is very small and to a reasonable first approximation may be neglected. The absorbance A of a sol may then be written as

$$A = N.\sigma_{abs}. 1$$

where N is the number of particles per unit volume and l is the path length. The absorbance will pass through a maximum when $\varepsilon = -2$; the pronounced colour of metal sols is therefore thought to be due to the resonant excitation of the dipolar surface plasmon.

(i) The U.V. visible absorption spectra of Colloidal Platinum

Typical spectra of platinum sols prepared in a $w = 5$ water/AOT/n-heptane microemulsion are shown in Fig. 4. They are all grey-black in colour and give a smoothly increasing absorbance through the region from $\lambda = 250-850$ nm which agrees well with the spectrum of platinum sols prepared in bulk water . The lack of an absorption maximum suggests that the conditions for the resonant excitation of the dipolar surface plasmons are not satisified. The platinum sols prepared in glycerol /AOT/n-heptane microemulsions do, however, show a pronounced absorption shoulder at $\lambda = 460$ nm and exhibit an unusual golden brown colour. It is suggested that this unusual colour is due to the formation of very small particles or clusters of platinum atoms. In fact for the larger particles, e.g. Pt 4 in Table 3, they display no shoulder and have spectra similar to those expected for larger platinum particles prepared in bulk aqueous and microemulsion media. These small clusters will have electronic properties different from those of larger particles and these will in turn therefore influence the optical properties.

(ii) The UV-visible spectra of colloidal gold

The deep red colour of gold sols is well known(41). The gold sols prepared in glycerol/AOT/n-heptane reverse micelles all displayed an absorption peak at $\lambda_{max} = 515$ nm as seen in Fig. 5. The position of the peak is not affected as the radius changes from 25-42 Å.

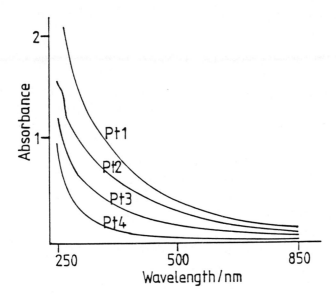

Fig. 4 Absorption spectra of platinum sols in a w = 5 water /AOT/n-heptane microemulsion with 0.1 mol dm^{-3} AOT.

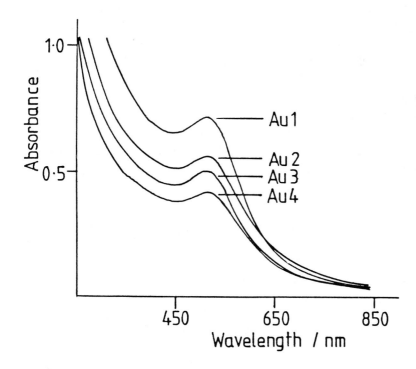

Fig. 5 Absorption spectra of gold sols in w = 1 glycerol/ AOT/n-heptane microemulsion with 0.1 mol dm^{-3} AOT, prepared at 25°C. Preparation conditions are indicated in Table 6.

6.   CATALYSIS USING COLLOIDAL PARTICLES PREPARED IN
      MICROEMULSIONS

Heterogeneous catalysts provide the enabling technology for a number of industrial manufacturing processes(42). Metal catalysts are usually employed in the form of highly divided dispersions which provide a large surface area for reaction. These catalysts not only enhance rates of reaction but they can also facilitate selective reaction pathways which might not otherwise occur(43).

Boutonnet *et al*(44) have carried out a series of experiments comparing hydrogenation reactions of different metal catalysts prepared in microemulsion systems. Platinum particles of radius 30Å were prepared in water/PEDGE/hexadecane microemulsions . The metal precursor salt hydrogen hexachloroplatinate (IV) was initially solubilized in the water droplets of the reversed micelle and hydrazine was then added reducing the salt to platinum. Several different approaches were used to support the catalyst:

(a)   The particles were centrifuged and redispersed in ethanolic solution with the assistance of ultrasound. Aluminium oxide with a high OH content was then added with stirring. The catalyst precipitated onto the support and was dried and washed with ethanol.

(b)   Pumice was added directly to the colloidal metal solution, the pH increased by adding hydrazine and deposition carried out by agitation.

(c)   Neutral aluminium oxide was added to the colloidal platinum and the same procedure as (b) followed but using 1mol dm^{-3} NaOH instead of hydrazine.

(d)   To neutral aluminium oxide the colloidal solution was slowly added, the catalyst was filtered, washed with ethanol and dried.

The kinetics of hydrogenation of 1-hexene were then measured in an ethanolic solution. For comparison the reaction was also carried out with a Pt/SiO_2 commercial catalyst. The activation energies were 21.5 kJ mol^{-1} for the microemulsion-prepared particles and 37.0 kJ mol^{-1} for the commercial catalyst. In the later work(45) monodisperse particles of platinum (20-30Å), rhodium (20-30Å) and palladium (50Å) were prepared in water/PEDGE/hexane or hexadecane microemulsions and then supported on pumice. The deuteration, isomerisation and hydrogen-deuterium exchange of but-1-ene with different catalysts were studied. Interestingly, the catalysts showed no loss of activity which might have been expected due to poisoning by the surfactant in the microemulsion.

The kinetics of catalysis have also been investigated 'in-situ' in reverse micelles (homogeneous catalysis - see Section 8).

7.   OTHER TYPES OF PARTICLE FORMATION

(i) Calcium and barium carbonate particles

Dispersions of these particles are used in engine oils to neutralise corrosive acids formed during the combustion process. They are required to be stable in oil over a wide temperature range. In industrial formulations a variety of surfactants are used to coat the particles(40). Fine colloidal dispersions of these particles have been prepared by Kitahara et al (32,34-36) by bubbling CO_2 in to w/o microemulsions containing $Ca(OT)_2$ and $Ba(OT)_2$ (see Table 2 for surfactants/solvents used). The sizes of the reverse micelles were investigated using PCS and fluorescence depolarization. The particles were sized using electron microscopy. For both the $CaCO_3$ and $BaCO_3$ particles it is found that spherical particles are produced with a diameter (e.g. for $CaCO_3$ 628 Å particles are formed for w = 7.5) which is always much larger than the diameter of the reverse micelle (58Å for w = 7.5) indicating once again that the size of the water core does not primarily determine the size of the particle but nonetheless provides a stabilising medium for the sols. It is also found in this case that the size of the particles increases as the w value increases. It was suggested by Koahara et al that the formation of particles occurs via fusion of reverse micelles containing precursor particles as described previously in this article. The particles are, nevertheless, appreciably smaller than those made in aqueous solutions. Quite a complex procedure is required to obtain a stable system containing particles of < 100 Å which are used commercially(49,50).

(ii) Colloidal iron (II) , cobalt (II) and nickel(II) boride particles

These have been prepared by Nagy and his group(29-31), by reduction of the appropriate ion solubilised in the aqueous phase of the water/CTAB/hexanol systems with $NaBH_4$. The nickel boride catalyst prepared is nearly monodisperse (29) which is quite surprising given that the initial surfactant system does not contain any obvious stable droplet structure (as evidenced by scattering techniques). Nevertheless this does not preclude the existence of less stable structural entities; these are apparently detected by n.m.r. methods. The average size of the particles increases with increasing size of the inner water core (measured by ^{19}F chemical shift variations(48)) but in contrast to the carbonate particles these have about the same size as the original droplet e.g. original droplet at 15.8% water is 18.4Å and depending on the concentration of nickel(II) chloride, r~20-25Å. In the preparation of iron boride, particles sizes of 30-80Å were prepared(31) with sizes again larger than the estimates of the inner water cores of the reverse micelles. The nickel boride catalyst obtained by this procedure is more active than Raney nickel for the hydrogenation of n-1-heptene.

8. KINETICS OF COLLOIDAL CATALYSIS BY MICROMETAL PARTICLES IN REVERSE MICELLES

Over recent years there has been an upsurge of interest in the application of colloidal metal and semi-conductor catalysis of redox reactions in the liquid phase. This has, to some extent, been stimulated by the effort devoted to the development of alternative energy sources e.g. in solar energy conversion. The main reaction investigated in fundamental studies was the oxidation of the methylviologen cation (MV^+):

$$\text{MV}^+ + \text{H}^+ \quad \xrightarrow{\text{catalyst}} \quad \text{MV}^{2+} + {}^1/_2\text{H}_2$$

For such systems, in aqueous media, the dependence of rates on catalyst concentration was generally non-linear, indicating rather complex kinetics. It is therefore of interest to examine the situation in reverse micelles where single redox reactions can be studied and electrochemical theories for microparticle catalysis can be tested. Major advances in our quantitative understanding of such processes have been recently made by Spiro(46,47). A simple reaction which is easily followed spectrophotometrically is the reduction of ferricyanide using thiosulphate:

$$\text{Fe(CN)}_6^{3-} + \text{S}_2\text{O}_3^{2-} \quad \xrightarrow{\hspace{1cm}} \quad \text{Fe(CN)}_6^{4-} + {}^1/_2\text{S}_4\text{O}_6^{2-}$$

This reaction proceeds readily in a reverse micellar system (e.g. AOT/H_2O/heptane) and is straightforward in that the reactants will be located exclusively in the dispersed aqueous phase. In the absence of catalyst, the kinetics are simple, being first order in each reactant (in contrast to the more complex kinetics found in aqueous solution) and $k_2 \sim 5 \times 10^{-1}$ dm^3 mol^{-1}s^{-1}.

In the presence of micrometal particles, the kinetics are more complicated in that according to Spiro, fractional orders are expected for the catalytic rate V_{cat}, as follows:

$$V_{cat} = k_{cat}[\text{Fe(CN)}_6^{3-}]^{r_1P}.[\text{S}_2\text{O}_3^{2-}]^{r_2q}$$

where the parameters r_1 and r_2 are given by

$$r_1 = \frac{(1 - a_1)Z_1}{(1 - a_1)Z_1 + a_2Z_2} \; ; \; r_2 = \frac{a_2Z_2}{(1-a_1)Z_1 + a_2Z_2}$$

The cathodic transfer coefficients are represented by a_1 and Z_1 is the charge-transfer valence of the appropriate couple. The exponents p and q are the electrochemical reaction orders with respect to ferricyanide and thiosulphate respectively, which can be taken as unity.

Then $r_1p + r_2q = 1$, i.e. the <u>sum</u> of the kinetic orders with respect to each component, should be unity. The data shown in Table 7 was obtained for a w = 10 system based on AOT/H_2O/heptane.

Metal Sol	r_1	r_2	$r_1 + r_2$
Pd	0.27($\pm$0.03)	0.76(+0.04)	1.03
Pt	0.33($\pm$0.03)	0.69($\pm$0.04)	1.02

Table 7. Exponents for the reaction between $Fe(CN)_6^{3-}$ and $S_2O_3^{2-}$ in an AOT-stabilised reverse micelle.

Thus it can be seen that good agreement is obtained between theory and experiment.

Plots of initial rate of reaction against metal particle concentration are shown in Fig 6. In both cases the metal particles were prepared 'in situ' and were about 20Å radius. The results show that the reaction is first order in catalyst, the positive intercept reflecting the rate constant in the absence of catalyst. For the data plotted in this way, Pt is seen to be slightly more effective than Pd but the differences are not great. What is clear is that trace amounts of metal present produce a very significant rate enhancement so the catalyst is very effective.

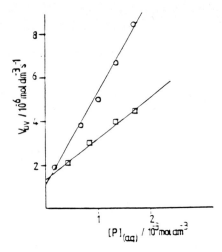

Figure 6. Variation of the overall initial rate V_{ov} with the concentration of palladium and platinum catalysts.Conditions 3.33 x 10^{-3} mol dm $^{-3}$ potassium ferricyanide and 3.33 x 10^{-2} mol dm $^{-3}$ sodium thiosulphate. Temperature = 25°C. W = 10, 0.1 mol dm $^{-3}$ AOT, water/AOT/n-heptane. O = Pt, [ ] = Pd, [Pd]aq = monomer concentration .

This is supported by data for the activation enthalpy $\Delta H^{\ddagger}_{cat}$ in the presence of Pt, which has been measured at 25 kJ mol^{-1}, compared with 65 kJ mol^{-1} in the absence of metal. The reaction is thus effectively catalysed by a system of dispersed microelectrodes which facilitate electron transfer by providing a conducting pathway through the solid metal. Since the reactants are likely to favour an aqueous environment it can be inferred that the metal particles should also have water in its immediate environment. This argues against a dispersion in which the particles are totally coated with surfactant (for stability) since this might then be expected to have a dramatic 'poisoning' effect on the catalyst property.

In conclusion, it is clear that the factors which control the size of the particles generated in the reverse micelle are now beginning to be understood, but much remains to be done to characterise the particle - surfactant and particle dispersed phase interactions.

REFERENCES

1.   K. Martinek, A.V. Levashov, N.L. Klyachko, I.V. Berezin, Biochem. Biophys. Acta 657, (1981), 277.
2.   P.L. Luisi and B. Straub, "Reverse Micelles" , (1984), Plenum Press.
3.   P.L. Luisi and L.J. Magid, Crit. Rev. Biochem. 20, (1986), 409.
4.   P.L. Luisi, M. Giomini, M.P. Pileni and B.H. Robinson, Biochem. Biophys. Acta, 947, (1988), 209.

5. M. Boutonnet, J. Kizling, P. Stenius and G. Maire, Colloid Surf., 5, (1982), 209.

6. K. Kurihara, J. Kizling, P. Stenius and J.H. Fendler, J.A.C.S., 105 (1983), 2574.

7. P. Lianos and J.K. Thomas, Chem. Phys. Lett., 125 (1986), 299.

8. R.C. Baetzold and J.F. Hamilton, Prog. Solid State Chem., 15 (1983), 1.

9. T.A. Dorling, B.W.J. Lynch and R.L. Moss, J. Catal., 22, (1971), 72.

10. D. Silman, A. Lepp and M. Kerker, J. Phys. Chem., 87 (1983), 5319.

11. J. Turkevich, P.C. Stevenson and J. Hillier, Discuss. Faraday Soc. 8, (1981), 348.

12. K. Aika, L.L. Ban, I. Okura, S. Namba and J. Turkevich, J. Res. Inst. Catalysis, Hokkaido Univ. 24, No. 1, (1976), 54.

13. J. Turkevich and G. Kim, Science, 169, (1970), 879.

14. D. Johansen, A. Launokis, J.W. Loder, A.W.H. Man, W.H.F. Sasse, J.D. Swift and D. Wells, Aust. J . Chem., 34 (1981), 981.

15. L.D. Rampino and F.F. Nord, J.A.C.S., 63 (1941), 27.

16. Th.G. Overbeek, Adv. Coll. Int. Sci. 16, (1982), 17.

17. D. Duff, A.C. Curtis, P.P. Edwards, D.A. Jefferson, B.F.G. Johnson and D.E. Logan, J.C.S. Chem. Comm.,(1987), 1264.

18. D. Duff, A.C. Curtis, P.P. Edwards, D.A. Jefferson, B.F.G. Johnson, A.I. Kirkland and D.E. Logan, Angew. Chem. Int. Ed. 20, (1987), 676.

19. A. Guinier and G. Fournet, "Small Angle Scattering of X-rays", John Wiley & Sons, N.Y., (1955), 162.

20. J.P. Jolivet, M. Gzara, J. Mazieres and J. Lefebure, J. Coll. Int. Sci., 107 (1985), 429.

21. H.F. Eicke, J.C.W. Shepherd and A. Steinemann, J. Coll. Int. Sci., 56, (1976), 168.

22. S.S. Atik and J.K. Thomas, J.A.C.S., 103, (1981), 3543.

23. S.S. Atik and J.K. Thomas, J.A.C.S., 103, (1981), 7403.

24. S.S. Atik and J.K. Thomas, J. Phys. Chem., 85, (1981), 3921.

25. P.D.I. Fletcher and B.H. Robinson, Ber. Bunsenges, Phys. Chem., 85, (1981), 863.

26. P.D.I. Fletcher and B.H. Robinson, J. Chem. Soc. Farad. Trans. I, 83, (1987), 985.

27. P.D.I. Fletcher, B.H. Robinson and J. Tabony, J. Chem. Soc. Farad. Trans. I, 82, (1986), 2311.

28. P.D.I. Fletcher, A.M. Howe and B.H. Robinson, J. Chem. Soc. Farad. Trans. I, 83, (1987), 985.

29. J.B. Nagy, A. Gourgue and E.G. Derouane, Stud. Surf. Sci. Catal., 16, (1983), 193.

220

30.   I. Ravet, N.B. Lufimpadio, A. Gourgue and J.B. Nagy, Acta Chim. Hung., 119, (1985), 155.
31.   N.B. Lufimpadio, J.B. Nagy and E.G. Derouane in "Surfactants in Solution", K.L. Mittal, B. Lindman ed. Plenum Press, N.Y.,(1983), 1483.
32.   K. Kandori, K. Kon-No and A. Kitahara, J. Coll. Int. Sci.,122(1), (1988), 78.
33.   A. Khan-Lodhi, Ph.D. thesis, University of Kent, 1988.
34.   M. Grobe, K. Kon-No, K. Kandori and A. Kitahara, J. Coll. Int. Sci. 93, (1983), 293.
35.   K. Kandori, N. Shizuka, K. Kon-No and A. Kitahara, J. Disp. Sci .& Tech. 8 (5&6), 1987, 477-491.
36.   K. Kandori, K. Kon-No and A. Kitahara, J. Disp. Sci. & Tech. 9(1), (1988), 61-73.
37.   A. Howe, Ph.D.thesis, University of Kent, 1984.
38.   P.D.I. Fletcher, M.F. Galal and B.H. Robinson, J. Chem. Soc. Farad. P.D.I. Fletcher, M.F. Galal and B.H. Robinson, J. Chem. Soc. Farad. Trans. I, 81, (1985), 2053.
40.   P. Salino and P. Volpi in "Colloids and Surfactants: Fundamentals and Applications", E. Barni and E. Pelizetti eds., Annali di Chimica, 77, (1987).
41.   G.Mie, Ann. Phys., 25, (1908), 377.
42.   P.Laszlo, Acc. Chem. Res., 19, (1986), 121.
43.   Selectivity in Heterogeneous Catalysis, Farad. Discussion of Chem soc., 72, (1981).
44.   M.Boutonnet, C.Andersson and R.Larsson, Act. Chem. Scan., A34,(1980), 639.
45.   M.Boutonnet, J.Kizling, R.Touroude, G.Maire and P.Stenius, Appl. Catal., 20, (1986), 163.
46.   M.Spiro, J.Chem.Soc.Farad Trans.1, 75, (1979), 1507.
47.   M.Spiro and P.L. Freund, J.Phys.Chem., 89, (1985), 1074.
48.   T. Nguyen and N.N. Ghaffarie, C.R. Acad. Sci. Paris. SERC, 290,(1980),113-115.
49.   J.F. Marsh, Chemistry and Industry, (1987), 470.
50.   I. Markovic, R.H. Ottewill, D.J. Cebula, I. Field and J.F. Marsh, Colloid and Polymer Sci., 262, (1984), 648.

POLYMERIZATION IN AND OF INVERSE MICELLES

G. VOORTMANS AND F.C. DE SCHRYVER

1. INTRODUCTION

Emulsion polymerization is known to be an important process for the production of industrial polymers in the form of colloids or latices, which are the bases of adhesives, paints polishes and other coatings. Most of these emulsion polymerizations are carried out in aqueous medium.

Recently the inverse emulsion polymerization gained a lot of interest for the production of high molecular weight water-soluble polymers as polyacrylamide and derivatives which are widely used for flocculation of colloidal dispersions, sewage treatment and thickener's in enhanced oil recovery (1-4). However, inverse emulsion polymerization can give rise to such problems as instability of the latices resulting in rapid sedimentation of the polymer and a rather broad size distribution of the final particles.

These problems can be overcome with the inverse micro-emulsion polymerization as shown by Leong and Candau (5).

A further advantage of the microemulsion polymerization is the ability to combine the properties of a microemulsion with the rigidity of polymers. An example of this was presented by Arai et al. (6). They polymerized methacrylic acid in dodecylbenzenesulfonic acid inverse micelles. This yielded a triphase catalyst for the hydrolysis of acetates. In this way they avoided the problem of recovery of the catalyst and product which is often encountered when inverse micelles are used for catalysis.

2. STRUCTURE OF MICROEMULSIONS

In contrast to emulsions, which are opaque, unstable, macrodisperse systems of 1-10μm droplet size, microemulsions are thermodynamically stable and transparent oil/water systems stabilized by an interfacial layer of surface-active agents. These structures can exist either alone or in equilibrium with water and (or) oil phase in excess (7). Only the monophasic domain is of importance for the microemulsion polymerization and hence only the structure of the microemulsion in this region will be discussed.

In the monophasic domain, either on the water rich or the oil rich side of the phasediagram, the microemulsion consists of uniform and spherical droplets dispersed in the continuous medium. The particle size is around 0.01 μm resulting in the optical transparancy of the microemulsion. The droplets can solubilize oil and be dispersed in water (direct or o/w microemulsion) or

222

solubilize water and be dispersed in oil (inverse or w/o microemulsion) (figure 1). In the intermediate regions which contain equivalent amounts of water and oil and where phase inversion occurs, the topology of the middle phase is still a matter of discussion. It is commonly described as a bicontinuous structure in which aqueous and oily domains are interconnected (8).

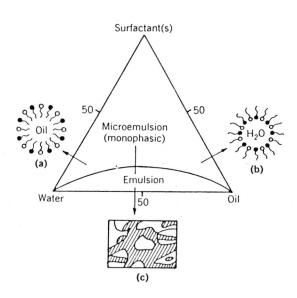

Figure 1: Schematic phase diagram of surfactant-oil-water systems: (a) oil in water droplets; (b) water in oil droplets; and (c) bicontinuous structure

Depending on the chosen purpose, polymerization can be performed on monomers incorporated in the droplets or solubilized in the continuous phase. It is also possible to use a polymerizable surfactant so that the stabilizer itself can be polymerized.

In the following only the polymerization in the oil rich side of the phasediagram will be discussed. First an overview of the polymerization of monomers solubilized in the microemulsion will be given and then the polymerization of the microemulsion itself will be discussed.

3. POLYMERIZATION OF MONOMERS SOLUBILIZED IN THE WATERPOOL
   OF THE INVERSE MICELLES.

Because of the above mentioned advantages of the inverse microemulsion polymerization for the production of high molecular weight water-soluble polymers, this polymerization recently gained attention.

In 1980 Stoffer and Bone reported on the polymerization of methyl acrylate solubilized in a w/o microemulsion (9). The microemulsion consisted of sodium dodecyl sulfate (surfactant), pentanol (cosurfactant), water and methyl acrylate. During polymerization a phase separation occured and no effect of micellization on the molecular weight was found. Moreover the molecular weight was rather low probably caused by the presence of an alcohol which can act as a chain transfer reagent. The phase separation is probably also caused by the alcohol which is a non-solvent for the polymer.

A more detailed investigation of the inverse microemulsion polymerization has been undertaken by the group of Candau. They published several reports on the polymerization of acrylamide in an AOT/water/toluene inverse microemulsion (5,10-13)

The inverse microemulsion polymerization of acrylamide was characterized by high polymerization rates and a high molecular weights. It was found that after polymerization the polymer particles contained one polymer chain on the average which is extremely low compared to the conventional emulsion polymerization where a polymer particle contains about 30000 chains. From this the authors concluded that the inverse emulsion polymerization of acrylamide involves a continuous nucleation mechanism. This means that every radical not involved in termination can nucleate a new particle.

However it was not clear whether this continuous nucleation proceeded by diffusion of monomer through the continuous apolar phase or by collision between the inverse micelles (figure 2).

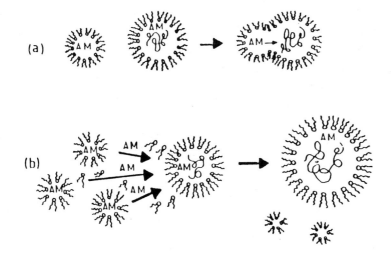

Figure 2: Scheme of the particle growth mechanism: (a) by collisions between particles; (b) by monomer diffusion through the toluenic phase

224

In order to solve this problem they used heptane instead of toluene as the continuous phase. In heptane the solubility of acrylamide is less than in toluene. It was then observed that the final particle size was less which suggests a diffusion of the monomer through the continuous phase. However this difference might also be caused by a difference in the dynamics of the initial microemulsions.

A study of the copolymerization of acrylamide with sodium acrylate gave an answer to this problem (14,15). This study showed that the composition of the copolymer always equalled the composition of the feed and the sequence distribution was close to a Bernouillian behavior. This behavior can only result from a collision mechanism since only in this case the local monomer concentration ratio's at the reaction site would be maintained.

It was furthermore interesting to compare the obtained reactivity ratio's in the inverse microemulsion polymerization of the two monomers with the reactivity ratio's ($r_A=r_M=1$) of the monomers in solution and inverse emulsion polymerization ($r_A=0.3$ $r_M=0.95$).

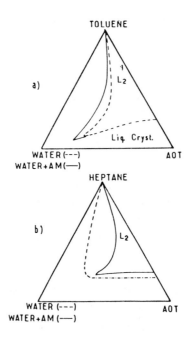

Figure 3: Schematic phase diagram (w/w) of the systems AOT/oil/water (or water + acrylamide in a weight ratio of 1) with toluene (a) or heptane (b) as the oil. The full line delimits the homogeneous L_2 area. The dashed line delimits the L_2 region in the absence of acrylamide.

According to the authors the difference between the solution polymerization and the microemulsion polymerization can be explained by an enhanced screening of the carboxylate group of the acrylate by the sodium ions due to a higher local concentration of the monomers in the microemulsion, leading to less repulsion between two acrylate units. This causes the probability of AA sequences to increase.

This screening cannot explain the difference between the inverse emulsion and microemulsion polymerization since the local concentration of the monomers is comparable in both cases. The difference is probably caused by a different nucleation mechanism. In inverse emulsion polymerization the monomer diffuse through the continuous phase while in the inverse microemulsion a collision mechanism takes place. Since the two monomers have a different solubility in the apolar phase this would cause an apparent difference in the reactivity ratio's of the two monomers in case of the inverse emulsion polymerization.

In addition to the polymerization mechanism the authors also compared the initial microemulsion with the latex after polymerization (10-13). It was shown that addition of acrylamide to a Toluene/AOT/water microemulsion extended the microemulsion domain. In heptane the reverse was observed (figure 3).

This is caused by the fact that acrylamide dissolves better in toluene than in heptane resulting in the location of acrylamide in the interface in case toluene is used as the continuous phase. This location leads to a strong interaction between the particles before polymerization. During polymerization the acrylamide is transferred from the interface to the core of the particles which results in a decrease of the attraction between the particles. This was clearly shown by viscosimetry and by dynamic light scattering data. Sedimentation experiments further showed the existence of empty inverse micelles together with particles of much larger size containing polymer. The empty inverse micelles form because of a decrease of the total interfacial area partly caused by the transfer of acrylamide to the core of the particles. Finally, viscosity data suggested a high internal viscosity of the droplets much larger than the continuous phase.

4. POLYMERIZATION OF THE SURFACTANT IN A W/O MICROEMULSION.

Nagai et al. studied the effect of micellization of higher alkyl salts of dimethylaminoethyl methacrylate (figure 4) (16,17).

Polymerization of these monomers in water showed an increase in the degree of polymerization with the length of the alkyl chain. Since the aggregation number of the different alkyl derivatives in aqueous micelles increases with increasing chain length, this leads to a better orientation of the surfactant and thus resulting in the observed increase of the polymerization degree. The authors also performed the polymerization of the dodecyl derivative in benzene.

It was found that the degree of polymerization increased dramatically which can be explained by the fact that the polymerizable groups are packed together in the center of the inverse micelles formed by the surfactant in benzene.

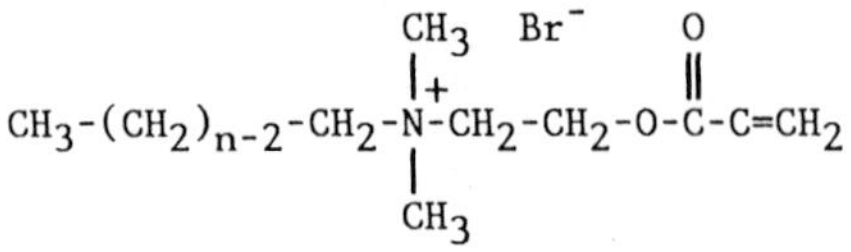

$$CH_3-(CH_2)_{n-2}-CH_2-\overset{\overset{\displaystyle CH_3}{|}}{\underset{\underset{\displaystyle CH_3}{|}}{\overset{+}{N}}}-CH_2-CH_2-O-\overset{\overset{\displaystyle O}{\|}}{C}-C=CH_2 \qquad Br^-$$

Figure 4: Structural formula of alkyl ammonium salts of dimethylaminoethyl methacrylate (n=4-16).

Together with the high degree of polymerization a high polymerization rate was observed. In solvents were no organized structureş are formed, the rate of polymerization was found to be much less in addition to a lower polymerization rate. These data clearly show the influence of the formation organized structures on the polymerization.

Voortmans et al. studied the stabilization of an inverse micellar system by means of polymerization (18,19). For this purpose a polymerizable ammonium surfactant was prepared (figure 5).

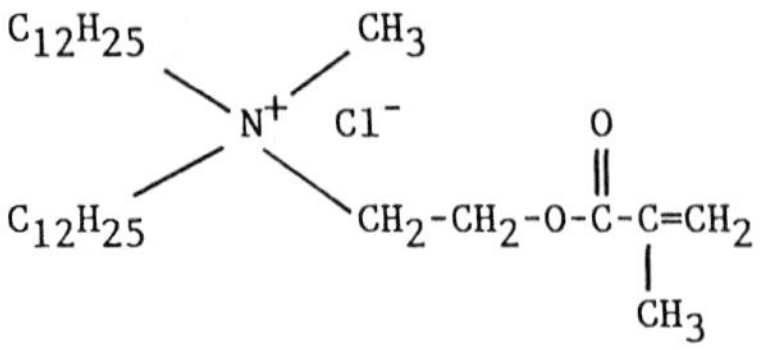

$$\begin{array}{c} C_{12}H_{25} \qquad\qquad CH_3 \\ \diagdown \qquad \diagup \\ N^+ \quad Cl^- \qquad \overset{\overset{\displaystyle O}{\|}}{} \\ \diagup \qquad \diagdown \\ C_{12}H_{25} \qquad\qquad CH_2-CH_2-O-C-C=CH_2 \\ | \\ CH_3 \end{array}$$

Figure 5: Structural formula of the polymerizable surfactant

Polymerization of the above monomer leads to molecular weights several times larger than the molecular weight of the inverse micelle. This result shows that the polymerization does not proceed topologicaly as was found to be the case for the polymerization of vesicles (20,21). The dynamics of the inverse micellar system probably causes this behavior since the rate constant of exchange via dimer formation was of the order of 10^8 where as a typical value for the rate constant of propagation of a methacrylate is about 1000 and as a consequence the propagating radical will exchange several times before the addition of the next monomer. The same behavior has also been found by Hammid and Sherigthon for the polymerization of aqueous micelles were the dynamics of the microemulsion results from the continuous build up and break down of a micelle (22,23).

To investigate whether polymerization leads to stabilization of the inverse microemulsion a viscosity study was performed. It has been shown that the Huggins' hydrodynamic interaction parameter (k_H) in equation [1] is related to the dynamics of the system.

$$n_{sp}/c = [n] + k_H [n]^2 c \qquad\qquad [1]$$

Where [n] presents the intrinsic viscosity of the solution. The Huggins' hydrodynamic interaction parameter was determined for the monomeric and polymeric system. From this comparison it could be concluded that the dynamics of the initial inverse microemulsion is lowered upon polymerization. It was furthermore found that this decrease in dynamic character of the microemulsion increased when the molecular weight of the polymer increased and was function of the history of the system. In addition if the polymeric aggregates were separated and than redissolved with the same amounts of water an increase in the interaction between the polymeric particles was observed. This last result indicates that the organization in the initial inverse micellar system determines the stability of the final polymeric microemulsion. Since this high organization cannot be reached starting from the polymer chains, an increased interaction between the particles is observed.

The observed molecular weight dependence was explained by the fact that the swelling of the polymer decreases if its molecular weight decreases. As a result the particles will have a more rigid interphase which causes more repulsion between the particles.

Dynamic light scattering further showed that the size of the polymer particles ranges from 150 to 260 Å depending on the initial amount of water used. From these data and the intrinsic viscosity obtained from equation [1] for the polymeric systems the corresponding aggregation numbers can be estimated according to equation [2]

$$N_{agg} = \frac{2.5 \ N_A \ V_H}{[n] \ (M_0 + R \ M_w)} \qquad\qquad [2]$$

Where N_A presents the Avogadro number, M_0 the molecular weight of the monomer, M_w the molecular weight of water and R the number of water molecules per surfactant monomer.

Comparing the resulting values with the degree of polymerization it could be shown that polymeric aggregates must consist of more than one polymer chain.

Using the fluorescence probing technique the aggregation behavior of the monomer and polymer were investigated. For the monomer a CMC was found at 10^{-4}M. The polymeric aggregates were stable against dilution when the molecular weight of the polymer exceeded the molecular weight of the inverse micel several times. When the polymerization degree approached the aggregation number of the inverse micellar system, again a CMC at 10^{-4}M was found. This behavior can be explained by the fact that the polymeric aggregates consist of more than one polymer chain which was shown by viscosimetry and dynamic light scattering data. This aggregation is unfavourable for the polymer since it is in a good solvent. The aggregation of the polymer chain is probably caused by the fact that micellization within one chain would be impossible because of the swelling of a polymer in a good solvent while the existence of a long separate chain in solution would be equivalent to an open aggregate of the same length formed by the monomer which is known to be unfavourable. As a consequence two competing processes are determining the aggregation behavior of the polymer. For the lower molecular weights the polymer characteristics will take over on the micellization characteristics.

5 POLYMERIZATION OF ALL THE COMPONENTS IN THE W/O MICROEMULSION EXCEPT WATER

Chew and Gan used the inverse microemulsion for the production of water containing polymers. In a first approach they studied the polymerization of styrene in a pentanol/water/sodium dodecyl sulfate microemulsion (24). However it was that polystyrene reduces the stability of the microemulsion and causes turbidity. This is caused by a reducing of the number of conformations of the polymer chains by the microemulsion droplets. Furthermore the interaction of the aromatic nucleus with the polar groups reduces water solubility in the inverse microemulsion.

Their attention was therefore drawn to the polymerization of an inverse microemulsion of methylmethacrylate (continuous phase), acrylic acid (co-surfactant), sodium acrylamidostearate (surfactant) and water. The polymerization of such a microemulsion resulted in transparent polymers which could contain up to 16% water. Increasing the amount of surfactant leads to a "softer" polymer. A water sorption study on this terpolymer showed that after drying the polymers were able to take up water from a water-vapor chamber. This was also the case for the undried polymers, but this leads to turbidity which reverted back to transparancy upon drying.

REFERENCES

1. V. Glukhikh, C. Graillat and C. Pichot, J. Polym. Sc.: Polym. Chem. Ed. 25 (1987) 1121
2. C. Graillat, C. Pichot, A. Guyot and M.S. El-Aasser, J. Polym. Sc.: Polym. Chem. Ed. 24 (1986) 427-449
3. C. Pichot, C. Graillat, V. Glukhikh, Makromol. Chem. Suppl. 10/11 (1985) 199-214

4. J.W. Vanderhoff, D.L. Visioli and M.S. El-Aasser, Polym. Mater. Sc. Eng. 54 (1986) 375-380
5. Y.S. Leong and F. Candau, J. Phys. Chem. 86 (1982) 2269-2271
6. K. Arai, Y. Masiki and Y. Ogiwara, Makromol. Chem. Rapid Commun. 7 (1986) 655-659
7. L.M. Prince, "Microemulsions : Theory and Practice", Academic Press New York (1977)
8. S.J. Chen, D.F. Evans, B.W. Winham, D.J. Mitchell, F.D. Blum and S. Pickup, J. Phys. Chem. 90 (1986) 842-847
9. J.O. Stoffer and T. Bone, J. Polym. Sc.: Polym. Chem. Ed. 18 (1980) 2641
10. Y.S. Leong, S.J. Candau and F. Candau, "Surfactants in solution", Ed. K.L. Mittal and B. Lindman, Plenum Press, N.Y. (1984) 1897
11. F. Candau and Holtzscherer , Journal de Chimie Physique 82 (1985) 691-694
12. F. Candau, Y.S. Leong, G. Pouyet and S. Candau, J. Colloid and Interface Sci. 101 (1981) 167-172
13. F. Candau, Y.S. Leong and R.M. Fitch, J. Polym. Sc.: Polym. Chem. Ed. 23 (1985) 193-214
14. F. Candau, Z. Zekhnini and F. Heatley, Macromolecules 19 (1986) 1895-1902
15. F. Candau, Z. Zekhnini, F. Heatley and E. Franta, Colloid and Polymer Sc. 264 (1986) 676-682
16. K. Nagai, Y. Ohishi, H. Inaba and S. Kudo, J. Polym. Sc.: Polym. Chem. Ed. 23 (1985) 1221-1230
17. K. Nagai and Y. Ohishi, J. Polym. Sc.: Polym. Chem. Ed. 25 (1987) 1-14
18. G. Voortmans, A. Verbeeck, C. Jackers and F.C. De Schryver, Macromolecules in press.
19. G. Voortmans, C. Jackers and F.C. De Schryver, British Polymer Journal in press.
20. R. Büschl, T. Folda and H. Ringsdorf, Makromol. Chem. Suppl. 6 (1984) 245-258
21. W. Reed, Macromolecules 18 (1985) 2402-2409
22. S.M. Hamid and D.C. Sherrington, Polymer 28 (1987) 325-331
23. S.M. Hamid and D.C. Sherrington, Polymer 28 (1987) 332-339
24. L.M. Gan, C.H. Chew, S.E. Friberg and T. Higashimura, J. Polym. Sc.: Polym. Chem. Ed. 14 (1981) 1585

ENZYMATIC CATALYSIS IN REVERSED MICELLES

Yu. L. KHMELNITSKY, A. V. KABANOV, N. L. KLYACHKO,
A. V. LEVASHOV and K. MARTINEK

1. INTRODUCTION

It is well known that the main structural pattern of biological membranes is the flat bilayer of lipid molecules. However, the notion of the lipid bilayer as the only possible way of organization of membrane lipids, which represents the essence of the widely accepted fluid mosaic model (1) of biological membranes, does not agree with established facts of structural rearrangements of lipids, for example, from the bilayer to the hexagonal phase (see, for example, (2) and references therein). Further investigations of the structure of lipid membranes, mainly by the Dutch group, for recent reviews see (3-5), resulted in the discovery of other types of non-bilayer lipid stuctures, in particular, so-called lipidic particles, representing reversed lipid micelles sandwiched between monolayers of the lipid bilayer. The concept of non-bilayer structures in lipid membranes made a basis for a new "metamorphic mosaic" model of biomembranes (6) which explains elegantly many processes occurring in the living cell, such as fusion and compartmentalization of membranes, exo- and endocytosis, lipid flip-flop, etc.

From the point of view of our discussion, the ability of certain proteins to induce the formation of non-bilayer structures upon incorporation into model and biological membranes is of particular significance. Evidence of this kind has been obtained for cytochrome c (5), methaemoglobin (7), cytochrome P450 (8), proteins of the erythrocyte membrane (9), and hydrophobic polypeptides (10), the latter being regarded as a model of anchoring fragments of integral membrane proteins. It is quite possible that the formation of non-bilayer structures represents a general mechanism for the protein incorporation into membranes (11-13) and for the regulation of their activity. This assumption is supported by the fact that ATPase (14) and mannosyl transferase (15) show a maximal catalytic activity exactly under conditions when the formation of intramembraneous lipidic particles occurs. The same is true also in the case of cell lipases (16,17), for which non-bilayer lipids have shown to be the best susbstrates.

Taken together, the above data allow one to conclude that model studies of enzymes and enzymatic catalysis in systems of reversed surfactant micelles in organic solvents are of great importance in understanding of the enzyme functioning in natural lipid systems. By the present time many reversed micellar systems containing dozens of different enzymes have been studied (for reviews, see (18-25)). In this chapter some important new developments in the field of micellar enzymology will be discussed.

2. METHODS FOR THE INCORPORATION OF ENZYMES INTO REVERSED MICELLES

Homogeneous (optically transparent) solutions of enzymes in organic solvents may be obtained by one of the three following procedures : (a) a complete solubilization of an aqueous protein solution in a surfactant-organic solvent system, i.e. a complete dissolution of the introduced aqueous component ; (b) a complete or partial solubilization of a dry (lyophilized) protein in a solution of surfactant in organic solvent containing a predetermined amount of water, and (c) a partial capture of aqueous protein solution by surfactant solution in organic solvent.

The first procedure was proposed by ourselves (26) and is nowadays most widely used. A small amount (of the order of several volume per cent) of the aqueous protein solution is introduced into a surfactant solution in an organic solvent (the actual ratio of the volumes of the aqueous and organic solutions is determined by the object of the experiment, the required degree of hydration of the surfactant, etc.). The resulting mixture is vigorously shaken (for several tens of seconds) until the formation of an optically transparent solution. Exposure of the mixture to ultrasound can also be used instead of mechanical shaking (27,28). It must be emphasized that the choice of the method of stirring is important in the study of enzymes and particularly enzyme kinetics in reversed micellar systems, because the time required to attain equilibrium in the colloidal system depends on its efficiency : in some cases the equilibration time in reversed micellar systems may be measured in hours and tens of hours (29). One must bear in mind, however, that the action of ultrasound can inactivate the enzyme (30).

The second procedure, proposed in Menger's laboratory (31), consists in the initial introduction of the required amount of water into the surfactant solution in an organic solvent (in order to attain the required degree of hydration of the surfactant), after which a dry (lyophilized) protein is dissolved in the resulting transparent solution with vigorous shaking. The time required for the dissolution of the dry protein is usually greater than in the procedure employing aqueous solutions but it diminishes with the increase in the degree of hydration of the surfactant. The complete dissolution of the dry protein takes place only at moderate protein concentrations, comparable to those obtained by the first method (18).

In order to obtain highly concentrated protein solutions, an excess of the dry protein is introduced into the surfactant solution in an organic solvent and its undissolved part (after several hours) is separated by centrifugation (18). The repetition of the procedure for the dissolution of the protein in the resulting supernatant increases the protein content in the micellar solution. When this method is used to obtain concentrated protein solutions, it is necessary to monitor the surfactant and water contents in the supernatant after separating the insoluble component, because the latter can entrain both the surfactant and water.

The third solubilization method, proposed by Luisi and co-workers (32,33), involves the spontaneous transfer of the protein in a two-phase system consisting of approximately equal volumes of the aqueous protein solution and the organic solvent containing the surfactant. After the completion of the solubilization, carried out with slight stirring, the organic phase containing the dissolved protein is separated. A disavadvantage of this method is the long time required for the solubilization process. In addition, the problem of determining the amount of water solubilized by micelles (the degree of hydration of the surfactant) arises.

We draw attention to the fact that, for very low values of $w \leq 1–2$, proteins are almost insoluble in reversed micellar systems. The solubilization of the protein begins only at a certain value of w, beyond which an increase in the degree of hydration of the surfactant is accompanied by an almost linear increase in the solubility of the protein up to concentrations comparable to those attainable in aqueous solutions. The actual solubilization limit depends on the temperature, the pH of the aqueous solution, the nature of the surfactant, the concentrations of water and the surfactant in the system, and also on the nature of the protein.

3. STRUCTURE OF PROTEIN-CONTAINING REVERSED MICELLES

The process of the enzyme solubilization in reversed micellar systems results in the formation of hydrated protein-containing reversed micelles, no matter which of the above listed solubilization procedures has been used. Enzyme molecules are located (entrapped) in the inner water cavities of these micelles, and are surrounded by a water layer and a surfactant shell which protect the enzyme against the inactivation by the bulk organic phase. Usually the protein-containing micelle carries a single enzyme molecule, although under certain conditions two or more enzyme molecules may be present inside the same micelle (34). In its main features, the above view of the protein-containing reversed micelle is accepted by all authors working in the field (34-42). However, a considerable controversy has existed as to their fine structural organization. At the present time, two main opposing models of protein-containing reversed micelles are being discussed in the literature. According to the water-shell model (Fig. 1a), suggested by Luisi and coworkers (21,35,36), the entrapment of the protein molecule is <u>invariably</u> accompanied by an increase in the micelle size and the filled micelle contains more surfactant and water molecules than the initial

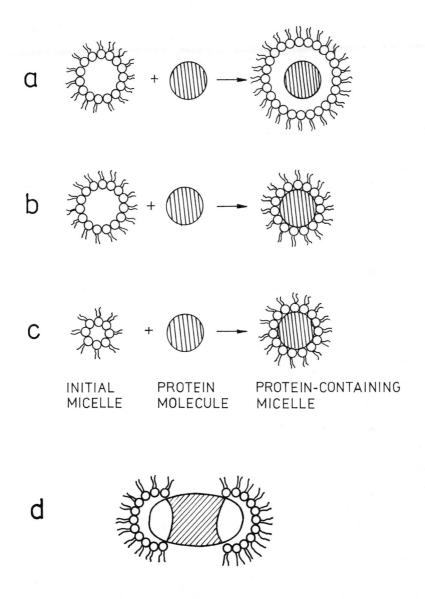

INITIAL   PROTEIN   PROTEIN-CONTAINING
MICELLE   MOLECULE   MICELLE

Fig.1. Models for solubilization of protein molecules (radius r_p) in reversed micelles (outer radius R_m, inner cavity radius r_m) leading to the formation of protein-containing micelles (outer radius R_{mp}, inner cavity radius r_{mp}) :
a) water shell model : $R_{mp} > R_m$ and $r_{mp} > r_p$, independent of the r_p/r_m ratio ;
b) induced fit model : $R_{mp} > R_m$ and $r_{mp} \approx r_p$, when $r_p > r_m$;
c) fixed size model : $R_m \approx R_{mp}$, when $r_p \leq r_m$;
d) reversed micelles containing hydrophobic proteins, such as bacteriorhodopsin (52) and porin (53)

empty micelle (as the result of redistribution of these micellar components between filled and unfilled micelles). An alternative model, suggested by ourselves (34), assumes that such increase in size occurs only when the inner cavity of the initial empty micelle is smaller than the protein molecule (the induced fit model, see Fig. 1b) ; in this case the entrapment of the protein can, in principle, result in the increase in the aggregation number as well as the hydration degree of the surfactant. If, on the other hand, the size of the initial water cavity exceeds (or is approximately equal to) that of the protein molecule, then, in contrast to the water-shell model, the protein entrapment may not lead to any susbstantial increase in the size of the reversed micelle (the fixed size model, see Fig. 1c). Ultra-centrifugation measurements (34) confirmed the validity of the fixed size model for a number of water-soluble (hydrophilic) proteins, such as trypsin, chymotrypsin, lysozyme, egg white albumin, horse liver alcohol dehydrogenase and γ-globulin. It was found (34) that the increase in volume, if any, of the reversed micelle as a result of the protein entrapment did not exceed 10% of its initial value, even when the protein molecule and the water cavity of the initial micelle were of the same size. The protein-containing micelle contains practically the same number of both surfactant and water molecules as the unfilled one ; in other words, the observed mass of the new aggregate is equal to the sum of masses of the protein and the empty micelle.

Recent experiments with the use of direct physical methods, such as quasi-elastic neutron scattering (37,38) and photon correlation spectroscopy (38) have unambiguously confirmed the validity of the fixed size model. Nevertheless, the model has been subjected to criticism (21,36) on the grounds that it contradicts "the common sense", because it implies the "disappearance" of water from the micelle upon the entrapment of the enzyme molecule without any appreciable loss in mass. In order to resolve this discrepancy, we have undertaken ^{13}C-NMR investigations (43) aimed at the elucidation of structural rearrangements occurring in the micelle upon the insertion of the protein molecule, which could explain the surprising phenomenon of the apparent disappearance of water.

As a model system, we used α-chymotrypsin dissolved in the reversed micellar solution of Aerosol OT (AOT) in octane, because it is this system that has been shown to be best described by the fixed size model by several methods (34,37,38). The dimensions of α-chymotrypsin molecule (44) are 40x40x50 Å. It may be approximated by a sphere of the same volume (about 41000 Å^3) with a radius of 21.5 Å. The size of inner cavities of reversed AOT micelles in octane depends (45) on the hydration degree of the surfactant expressed as $w = [H_2O]/[AOT]$. In our experiments we used $w = 12$ when the water cavity of the reversed micelle has a radius of 22.1 Å (45). This means that the volume of the inner cavity of the initial empty micelle (45200 Å^3) slightly exceeded that of the enzyme molecule, i.e. the conditions required for the realization of the fixed size model (Fig. 1c) were satisfied. According to this model (34), the amount of water and the aggregation number of AOT are the same both in protein-containing and protein-free micelles. Simple calculations show that in this case water molecules in the protein-containing micelle should form a layer of about 6 Å thickness around the entrapped enzyme (at the above indicated volume ratio of the water cavity to the α-chymotrypsin molecule). In order to keep overall dimensions of the protein-containing micelle unchanged, this water layer should be expelled from the inner cavity of the micelle, which is now almost completely occupied by the enzyme, and penetrate into the surfactant shell towards the organic phase. The situation is depicted schematically in Fig. 2).

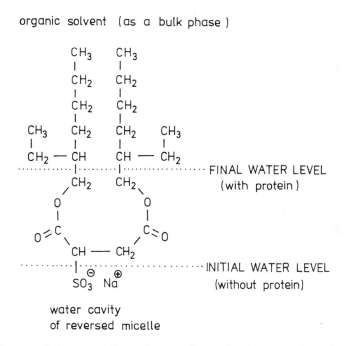

Fig. 2. Scheme of the expulsion of water from the inner cavity of a reversed AOT micelle on entrapment of a protein molecule (the fixed size model in Fig.1).

^{13}C-NMR measurements indeed confirmed the validity of the above conclusions. It was found that the addition of α–chymotrypsin into the solution of AOT in octane at $w = 12$ caused dramatic changes in the spin-lattice relaxation time T_1 for carbon atoms situated inside the range of 5-7 Å from the inner micellar interface (between the dotted lines in Fig. 2), whereas T_1 values for carbons lying further along AOT aliphatic chains remained essentially unchanged ; experimental details will be published elsewhere (44). The observed shifts in T_1 values were ascribed (4) to the change in the microenvironment of corresponding carbon atoms caused by their immersion into the expelled water layer. Quite remarkably, the thickness of the expelled water layer determined from the NMR measurements (5-7 Å) excellently coincides with the value predicted by the fixed-size model (about 6 Å). Inspection of molecular models, constructed on the basis of available data on the conformation of AOT molecules in reversed micelles (46-49), confirms that the suggested expulsion of water from the micellar core is indeed possible : in the areas where AOT polar heads are bonded to the hydrocarbon tails (Fig. 2) there are cavities quite capable of hosting the expelled water molecules.

Our description of the structural rearrangement of water and surfactant molecules in AOT reversed micelles is in excellent agreement with recent results (50) of studies on the hydration of lipid bilayers in biological membranes. It has been shown (50) that under certain conditions the apparent partial molar volume of hydration water in biological membranes can reach zero value due to the inclusion of water molecules into large voids in the head-group region of the lipid bilayer, which means, in fact, apparent "disappearance" of water exactly in the same way as in our model micellar system in the presence of protein.

It should be stressed that the above considerations are strictly valid only for hydrophilic proteins which do not possess the ability to interact strongly with interfaces and do not contain large hydrophobic surface areas. In the latter cases deviations from the model may be found because of possible distorsions of the micellar membrane. For example, AOT reversed micelles containing solubilized myelin basic protein (MBP) (41,42), which is known (51) to interact strongly with lipid membranes, do not completely conform to the ideal fixed size model, although under certain conditions the validity of the model has been confirmed for this protein as well (42). Another example is given by integral hydrophobic proteins bacteriorhodopsin (52) and porin (53). These proteins, when solubilized in reversed micellar solution, induce the formation of dumb-bell-like micellar aggregates in which polar parts of protein molecules are incorporated into "local" reversed micelles, whereas the central hydrophobic part of the protein freely contacts with the organic solvent. This type of protein-containing reversed micellar aggregates is shown in Fig. 1d.

4.   REGULATION OF CATALYTIC ACTIVITY OF ENZYMES IN
      REVERSED MICELLES
      Variation of the hydration degree (w = [H_2O]/[Surf.])
      One of the most striking effects observed in the study of enzymes in reversed micellar systems is the dependence of the catalytic activity of solubilized enzymes on the degree of hydration, w, of reversed micelles. The dependence of the catalytic activity on w has been observed for all enzymes studied in reversed

micellar systems (see reviews (18,19) and references therein). As a rule, this dependence is bell-shaped, i.e. there is an optimal value of **w** for which the catalytic activity of the solubilized enzyme is maximal. The actual value of the optimal **w** depends on the nature of the enzyme and surfactant . Several characteristic examples are given in Fig. 3.

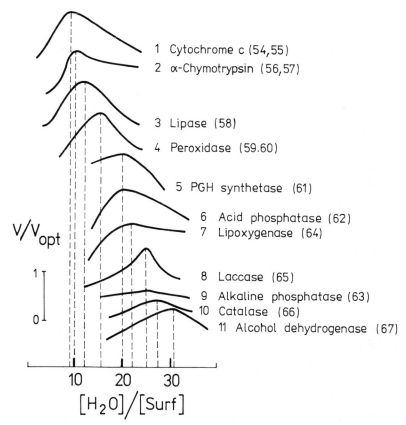

Fig. 3. Regulation of the relative catalytic activity, V/V_{opt}, of different enzymes solubilized in AOT reversed micelles, by variation of the surfactant hydration degree. Numbers in brackets are the references from which corresponding data were taken.

In order to explain the observed bell-shaped dependence of the catalytic activity on **w**, one must recall that **w** is the parameter which determines the size of the aqueous core of hydrated reversed micelles : the higher **w**, the greater the size of the core (45). Analysis of the data presented in Fig. 3 shows that the maximal catalytic activity of solubilized enzymes is observed at **w** values which correspond to reversed micelles with the size of the inner cavity equal to that of the enzyme molecule. This conclusion is illustrated by Fig. 4 which is based on the data from Fig. 3 and shows the correlation between radii of different enzymes

238

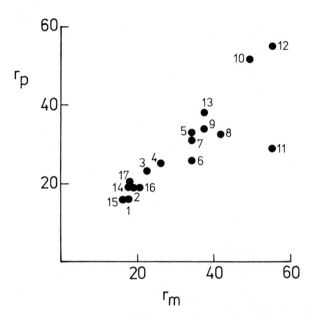

Fig. 4. Correlation between effective radii of entrapped enzyme (r_p) and corresponding optimal aqueous micellar cavities (r_m). Point numbers : 1 - 11, same as in Fig. 3 ; 12 - alcohol dehydrogenase (69) (r_p = the longest semiaxis of the enzyme molecule) : 13 - lactate dehydrogenase (70) : 14 - trypsin (71,72) ; 15 - lysozyme (73) : 16 - ribonuclease (74) ; 17 - pepsin (105). Dimensions of enzyme molecules were taken from (68,75,76).

(r_p) and corresponding optimal aqueous micellar cavities (r_m). According to our concept of protein-containing reversed micelles, as discussed in the preceding section, under optimal conditions the fixed size model is operative (Fig. 1c, $r_p \cong r_m$). This means that the highest catalytic activity is observed when the entrapped enzyme molecule is forced to a close contact with the inner micellar interface, or, to put it differently, when it is tightly fit into a tailor-made reversed micelle. Such a fit probably helps the enzyme to retain its most active conformation by making it more rigid, or "frozen".

It should be noted that r_p in Fig. 4 represents an effective radius, i.e. the radius of the sphere of the same volume as that of the enzyme molecule. It is clear, therefore, that in the case of highly asymmetric enzymes deviations from the correlation between r_p and r_m may be found. Such a deviation was indeed observed, for example, for horse liver alcohol dehydrogenase (point n°11 in Fig. 4) which represents an ellipsoid with the axial ratio 1:2.5 (45x55x110 Å) (68). This enzyme reveals the highest catalytic activity in reversed micelles with the diameter of the aqueous cavity equal to the longest dimension of the enzyme molecule (point n°12 in Fig. 4) (67,69).

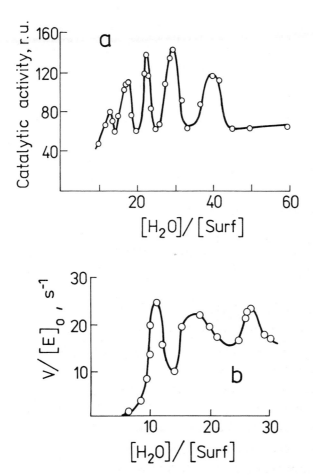

Fig. 5. Regulation of catalytic activity of oligomeric enzymes solubilized in AOT reversed micelles in octane by variation of the surfactant hydration degree. (a) Lactate dehydrogenase (70) ; (b) γ-glytamyl transferase (77).

The case of oligomeric enzymes

The dependence of catalytic activity of oligomeric enzymes on **w** reveals several maxima, as shown in Fig. 5 for lactate dehydrogenase (70) and γ-glutamyl transferase (77) solubilized in AOT reversed micelles in octane. The molecule of lactate dehydrogenase is composed of four identical subunits (total molecular weight 140 kD) and represents an ellipsoid with dimensions 74x74x84 Å (68). Following the above formulated concept of the geometric fit of the enzyme molecule to the aqueous cavity of reversed micelles under conditions of the optimal catalytic activity, and using the data for the dimensions of differently packed subunits of lactate dehydrogenase (78), one may conclude that the maxima at w = 14, 18, 22, 31, and 40 on the profile of catalytic activity in Fig. 5a correspond, respectively, to the monomer, dimer, trimer and/or tetramer, and octamer of subunits of lactate dehydrogenase. Following the same line of reasoning, the maxima on the catalytic activity profile for γ-glutamyl transferase (Fig. 5b), which is composed of two subunits with molecular weights 20 and 53 kD (79,80), were assigned to the light (w = 11) and heavy (w = 17) subunits and to their dimer (w = 26).

Ultracentrifugation experiments (77) confirmed the presence in both systems of subunit aggregates predicted on the basis of the catalytic activity studies (Fig. 5). Morover, in the case of γ-glutamyl transferase we succeeded (77) in the separation of the light and heavy subunits by means of preparative ultracentrifugation of the solution of the enzyme in AOT/octane reversed micellar system under conditions when no dimers were present in solution. After the removal of the heavy fraction the catalytic activity profile measured in the supernatant revealed a single maximum corresponding to the light subunit of γ-glutamyl transferase (as the maximum at w = 11 in Fig. 5b). On the other hand, the heavy fraction pellet, being redissolved in octane, also gave a single maximum, but corresponding to the heavy subunit of the enzyme (as the maximum at w = 17 in Fig. 5b). Finally, the solution obtained by mixing back the light and heavy fractions produced the profile characteristic for the native enzyme (Fig. 5b). From the results we conclude that reversed micellar systems provide convenient means for the regulation of the composition of protein complexes, their artificial build-up and disintegration.

Substitution of water inside reversed micelles by water-miscibel organic solvents

Another approach (81) to the regulation of catalytic activity of enzymes dissolved in reversed micellar systems consists in a (partial) dehydration of inner cavities of reversed micelles by susbtitution of water by a mixture of water and water-miscible organic solvents.*

As a example, Fig. 6a shows dependences of catalytic activity of α-chymotrypsin on w in systems AOT-octane-water-glycerol with different concentrations of glycerol (up to 94 vol. %, referred to the volume of the water-glycerol mixture used to solvate reversed micelles). Similar profiles have been obtained for peroxidase and laccase (87). Butanediol and dimethyl sulfoxide may also be used instead of glycerol (88).

---

* Fragmentary results on the influence of water-miscible organic solvents on enzyme catalysis in reversed micellar systems have been published also by others (82-86).

Characteristic features of enzymatic catalysis in reversed micelles solvated by water-miscible organic solvents are (a) the shift of the profile of the catalytic activity on **w** towards lower values of **w**, and (b) the increase in the catalytic

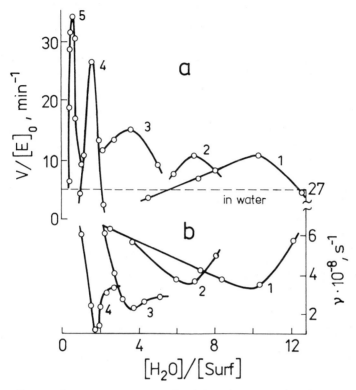

Fig. 6. Dependence on w of (a) the maximal velocity of α-chymotrypsin-catalyzed hydrolysis of N-benzoyl-L-tyrosine p-nitro-anilide, and (b) the rotational frequency of the spin label 4-(2-iodoacetamide)-(2,2,6,6-tetramethylpiperidine-N-oxyl) covalently attached to the active site of α-chymotrypsin in the system AOT/water/glycerol/octane. Water/glycerol volume ratios : 1, 100/0 ; 2, 80/20 ; 3, 50/50 ; 4, 20/80 ; 5, 6/94. The dashed line shows the values of V/E_0 and ν in aqueous solution. From (81,88).

activity observed under optimal conditions with increasing concentration of the water-miscible organic solvent (Fig. 6a). The first trend can be explained in the framework of the above-formulated concept of the geometrical fit of the enzyme molecule to the inner cavity of the reversed micelle under optimal conditions. We suggest that the size of inner cavities of glycerol-containing micelles existing under optimal conditions is approximately the same for all catalytic activity profiles in Fig. 6a and corresponds to the size of the enzyme molecule (it should be stressed that the **w** value in Fig. 6a is calculated as [H_2O]/[AOT] and does not include glycerol present in the micelle). This conclusion is based on the literature data (89) on the size of glycerol-containing AOT reversed micelles and is further supported by our results (88,90) of EPR measurements of the rotational frequency, v, of a spin probe covalently attached to the active site of α-chymotrypsin. Dependences of v on $\mathbf{w}$ obtained for spin-labeled α-chymotrypsin dissolved in AOT reversed micellar systems containing different concentrations of glycerol, corresponding to those used in Fig. 6a, are shown in Fig. 6b. From comparison of Fig. 6a and Fig. 6b one can see that in a given system the highest catalytic activity is attained under conditions of the lowest conformational mobility of the enzyme, described in terms of v. This observation is in full agreement with the notion that under optimal conditions the enzyme molecule is tightly squeezed, or "frozen", by a tailor-made reversed micelle in a highly active conformation.

The EPR data in Fig. 6b also enable one to explain the increase in the optimal catalytic activity with increasing content of glycerol in the system (Fig. 6a). It is clearly seen from Fig. 6b that the increase in the concentration of glycerol results in a considerable decrease in the rotational frequency measured under optimal conditions. In other words, the enzyme becomes more and more stabilized in the most active conformation with concomittant enhancement of the catalytic activity. This effect becomes even more pronounced when 2,3-butanediol is used instead of glycerol. In this case (88) the optimal catalytic activity of solubilized α-chymotrypsin is 20 times higher than in aqueous solution, as compared to the maximal 7-fold increase evidenced by Fig. 6a for glycerol-containing systems. The observed influence of water-miscible organic solvents on the catalytic activity can be explained in terms of high microviscosity of the inner cavity (91) and increased rigidity of the surfactant shell (38) of reversed micelles containing solubilized viscous solvents such as glycerol. Both these factors would tend to decrease the conformational mobility of the entrapped enzyme molecule.

The ability of reversed micelles solvated by water-miscible organic solvents to damp conformational fluctuations in the enzyme molecule leads also to another interesting phenomenon, namely the alteration of the observed susbstrate specificity of the enzyme. The relation between the catalytic constant of α-chymotrypsin-catalyzed hydrolysis of various substrates, determined under optimal conditions, and the conformational mobility of the enzyme, expressed in terms of the rotational frequency, v, of the spin label in the active site of the enzyme, is shown in Fig. 7 for reversed-micellar systems of different compositions. It is evident that with decreasing conformational mobility of the enzyme (caused by the increase in the concentration of organic cosolvent inside reversed micelles) the difference in k_{cat} between "good" and "bad" susbstrates strongly diminishes, so that α-chymotrypsin becomes almost equally highly effective towards any substrate introduced into the system. The enzyme acquires the ability to hydrolyze even p-nitroacetanilide, which is completely stable against chymotryptic cleavage in aqueous solutions. A possible reason for such behaviour is that in the "frozen" state the active site of the enzyme molecule is forced into

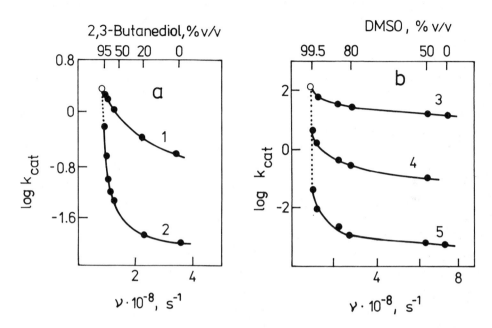

Fig. 7. Relation between the catalytic constant, k_{cat}, of α-chymotrypsin-catalyzed hydrolyses of various substrates, determined under optimal conditions, and the conformational mobility of the enzyme, expressed in terms of the rotational frequency, ν, of the spin label in the active site of the enzyme. (a) AOT/water - butanediol/octane, (b) CTAB/water - dimethyl sulfoxide/octane-chloroform. 1, N-benzoyl-L-tyrosine p-nitroanilide ; 2, N-succinyl-L-phenylalanine p-nitroanilide ; 3, N-benzyloxycarbonyl-L-tyrosine p-nitro-phenolate ; 4, p-nitrophenylcaprylate ; 5, p-nitrophenyltrimethylacetate. Concentrations of cosolvents (upper axes) are referred to the volume of the aqueous phase. Unfilled circles are extrapolated values. From (92).

the most favourable configuration, which otherwise can be created only as a result of the binding of a "good" substrate.

Variation of the surfactant concentration

A change in the concentration of surfactant in reversed micellar systems carried out at a fixed w value results in the alteration of the concentration of identical micelles without influencing their size and other properties (93,94). Hence, the catalytic activity of solubilized enzymes should not depend on the surfactant concentration, if the value of w is kept contant. The validity of this conclusion has been demonstration for a number of enzymes, such as α-chymotrypsin (57,95), trypsin (95), alkaline phosphatase (63) and lipoxygenase

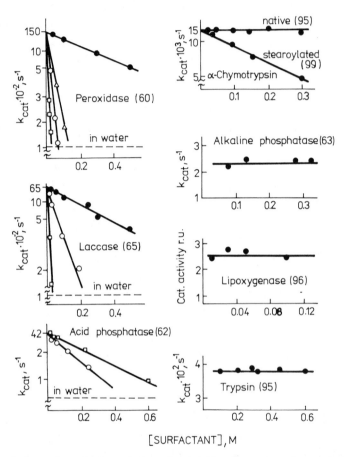

244

(96) (see Fig. 8, right column). However, there is a group of enzymes, such as peroxidase (60), acid phosphatase (62), laccase (65), and prostaglandin synthetase (61), whose activity strongly depends on the surfactant concentration. The change

[SURFACTANT], M

Fig. 8. Regulation of catalytic activity of solubilized enzymes by variation of the surfactant concentration at a constant [H2O] / [Surfactant] in systems : •, AOT/water/octane ; Δ, dodecylammonium propionate/water/diethyl ether-benzene ; x, Brij 95 (or 56)/water/cyclohexane ; □, lecithin/water/methanol-pentanol/octane. Dashed line shows levels of corresponding catalytic activities in aqueous solution. Numbers in brackets are the references from which corresponding data were taken.

in catalytic activity can be as high as 1 or 2 orders of magnitude, as illustrated by Fig. 8 (left column). All enzymes from this group, in contrast to enzymes from the first group are characterized by the presence in their molecules of anchoring groups of different nature (97,98). We suggest that it is the presence of such groups capable of interacting with micellar membrane, that causes the dependence of catalytic activity on the surfactant concentration. The validity of this suggestion was confirmed experimentally (99) in comparative studies of catalytic behaviour of α-chymotrypsin in the native state and covalently modified with stearoyl residues. When solubilized in AOT reversed micelles in octane, the native α-chymotrypsin, being a member of the first group of enzymes, did not show any dependence of catalytic activity on the surfactant concentration at a fixed value of **w**. On the other hand, in the case of the hydrophobized α-chymotrypsin a profound dependence of catalytic activity on the surfactant concentration was observed. These results are shown in Fig. 8 on the panel corresponding to α-chymotrypsin. Thus, by introducing a hydrophobic anchoring group it is possible to convert artificially an enzyme from the first group to an enzyme belonging to the second group. An important corollary of these findings is that the occurrence of the dependence of catalytic activity of an enzyme dissolved in a reversed micellar system on the surfactant concentration can serve as a convenient test for the ability of the enzyme to interact with micellar (and probably also with biological) membranes.

Possible molecular mechanisms explaining the dependence of catalytic activity of solubilized enzymes on the surfactant concentration are discussed in the following section. Here, we draw attention to the fact that the limiting high level of catalytic activity reached at low surfactant concentrations (Fig. 8, left column) does not depend on the nature of the reversed micellar system and hence reflects a fundamental property of the enzyme itself. We suggest that this limiting level represents a true catalytic potential of the enzyme, which is not attainable in aqueous solutions (cf. dashed line levels in Fig. 8, left column).

5. KINETIC THEORY OF ENZYMATIC REACTIONS IN REVERSED MICELLES

Substrate partioning effects

The kinetics of chemical reactions catalyzed by enzymes entrapped in reversed micelles obey, as a rule, the classic Michaelis-Menten equation. However, the kinetic theory of enzymatic reactions proceeding in such a microheterogeneous medium should take into account the partition of substrates between the pseudophase of micelles and the bulk phase of the organic solvent (18,95,100). Because of the substrate partitioning the apparent value of the Michaelis constant becomes dependent on the volume ratio of the bulk (organic solvent) and the micellar (aqueous) phases.

Consider a reaction between enzyme E and substrate S that obeys Michaelis kinetics :

$$E + S \; \rightleftharpoons \; ES \longrightarrow E + products \tag{1}$$

and proceeds in areversed micellar system. Using the assumption of the substrate partitioning, one obtains :

$$(S) \text{ in the bulk phase } \overset{P_S}{\underset{}{\rightleftharpoons}} (S) \text{ in micelles} \qquad [2]$$

where the partition coefficient P_S is given by :

$$P_S = \frac{[S]_{mic}}{[S]_b} \qquad [3]$$

Hereafter the subscripts "mic" and "b" refer to the micellar and bulk phrases, respectively. We shall not take into account the distribution in hydrophobic solvents. Therefore, we shall assume that all catalytics activity is confined to the micellar phase. The initial velocity of the enzymatic reaction, related to the volume of the whole system, is then given by :

$$v = \frac{k_{cat,mic} [E]_{o,mic} [S]_{O,mic}}{K_{m,mic} + [S]_{O,mic}} \cdot \Theta \qquad [4]$$

where Θ is the volume fraction of the micellar phase and the subscript "O" denotes initial concentrations. We further assume that the exchange of susbstrate molecules between the phases is sufficiently fast, i.e. the enzymatic reaction does not disturb the partitioning equilibrium [Eq. 2]. Then concentrations of reagents can be found from Eq. 3 and material balance equations :

$$[S]_{O,t} = [S]_{O,mic} \cdot \Theta + [S]_{O,b} (1 - \Theta) \qquad [5]$$

$$[E]_{O,t} = [E]_{O,mic} \cdot \Theta$$

Substitution of Eqs.3 and 5 into Eq. 4 gives :

$$v = \frac{k_{cat,app} [E]_{O,t} [S]_{O,t}}{K_{m,app} + [S]_{O,t}} \qquad [6]$$

where

$$k_{cat,app} = k_{cat,mic} \tag{7}$$

and

$$K_{m,app} = K_{m,mic} \frac{1 + \Theta (P_s - 1)}{P_s} \tag{8}$$

In the case of a charged substrate, Eq. 8 can be simplified if one assumes that molecules of the substrate are confined exclusively to the micellar phase, so that only $P_S \gg 1$, but also $P_S\Theta \gg 1$.

Then

$$K_{m,app} = K_{m,mic} \, \Theta \tag{9}$$

On the other hand, in the case of a hydrophobic susbstrate which is localized mainly in the bulk phase, one can assume that $P_S \ll 1$, so that

$$K_{m,app} = K_{m,mic} \frac{1 - \Theta}{P_S} \tag{10}$$

The validity of the kinetic theory based on the pseudophase approach has been confirmed experimentally for different enzymes and reversed micellar systems (57,95,101,102). As an example, Fig. 9 shows the kinetic data (95,102) obtained for the hydrolysis of the water-soluble susbstrate N-benzoyl-D, L-arginine p-nitroanilide catalyzed by trypsin solubilized in reversed micelles of cetyltrimethylammonium bromide (CTAB) in an octane-chloroform mixture. In a full agreement with Eqs. 7 and 9, the value of $K_{m,app}$ linearly depends on the volume fraction of the micellar phase, Θ, whereas $k_{cat,app}$ is fairly constant and represents the true reactivity of the enzyme in the micellar phase = $k_{cat,mic}$). (It is to be stressed that Θ should be varied in such a way that $\mathbf{w}$ remains constant so as to prevent the alteration of the properties of the micellar phase proper). Similar results were obtained (57) for α-chymotrypsin-catalyzed hydrolysis of N-trans-cinnamoyl imidazole in the systeme of AOT reversed micelles in heptane. As to the case of hydrophobic substrates, the validity of Eq. 10 was checked (101) by the example of hydrolysis of p-nitrophenyl alkanoate esters, catalyzed by lipase in AOT reversed micelles in heptane.

The simplest version of the pseudophase model discussed above assumes a uniform distribution of the substrate over the entire volume of the hydrated reversed micelle. In a general case, however, the model should take into account

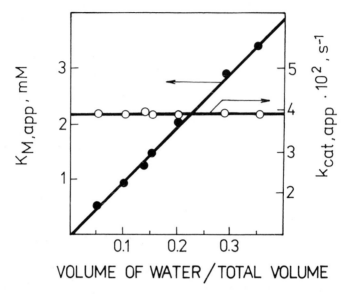

Fig. 9. Hydrolysis of N-benzoyl-D, L-arginine p-nitroanilide catalyzed by trypsin in the reversed micellar system CTAB/water/chloroform-octane (1:1). The Michaelis parameters $K_{m,app}$ and $k_{cat,app}$ are plotted as a function of the volume fraction of water in the system (w = 25, [CTAB] = 0.1 - 0.8M) from (95,102).

the structural heterogeneity of reversed micelles and allow for the partition of the substrate between the aqueous cavity and the surfactant shell of the reversed micelle. The development of such a model is now being completed in our laboratory (103).

It should be stressed that the above theory is strictly valid only for enzymes that do not interact with the micellar membrane (see Fig. 8, right column). Otherwise, the value of k_{cat} becomes dependent on the surfactant concentration at constant **w**, i.e. on the volume fraction of the micellar phase, Θ (Fig. 8, left column), so that Eq. 7 is not valid any more. An extension of the kinetic theory explaining the dependence of enzymatic activity on the surfactant concentration is given in the following section. The extended version of the theory also accounts for the observed bell-shaped dependence of catalytic activity on **w** (Fig. 3).

Influence of **w** and surfactant concentration on catalytic activity of solubilized enzymes : a theoretical consideration

In this section, we shall deduce a general equation relating the catalytic activity of solubilized enzymes to concentration parameters of micellar systems, such as the hydration degree, **w**, and the concentration of the surfactant (104).

First of all, we must take into account the fact that reversed micelles are usually characterized by a rather broad size distribution. Even in the case of AOT reversed micelles, which are generally regarded as an example of a fairly monodisperse system, the size distribution cannot be totally neglected (37,106-108). All micelles present in the system can be subdivided into the following three categories.

a) "Optimal" micelles (M_{opt}), with the radius of the inner cavity lying within the range between $r_{opt} - \Delta r$ and $r_{opt} + \Delta r$. When entrapped into optimal micelles, the enzyme exhibits the highest catalytic activity (cf. Fig. 3).

b) "Small" micelles (M_s), with the radius of the inner cavity smaller than $r_{opt} - \Delta r$.

c) "Large" micelles (M_1), with the radius of the inner cavity exceeding $r_{opt} + \Delta r$.

According to the concept that the catalytic activity of enzymes is correlated with the size of reversed micelles (see above), the distribution of micellar sizes should inevitably lead to the variation of enzymatic activity inside the same system. We suggest that catalytic constants of enzymes molecules entrapped into optimal, large, and small micelles are different and equal to $k^{(opt)}_{cat}$, $k^{(1)}_{cat}$, and $k^{(s)}_{cat}$, respectively. For the sake of simplicity we further assume that in small micelles the enzyme is inactive, i.e. $k^{(s)}_{cat} = 0$ (cf. fig. 3). Enzyme molecules can be exchanged between micelles of different sizes according to the scheme :

$$\overline{M}_{opt} + M_s \overset{K_1}{\underset{}{\rightleftarrows}} M_{opt} + \overline{M}_s \qquad [11]$$

$$\overline{M}_{opt} + M_1 \overset{K_2}{\underset{}{\rightleftarrows}} M_{opt} + \overline{M}_1 \qquad [12]$$

where $\overline{M}_{opt}$, $\overline{M}_1$ and $\overline{M}_s$ are enzyme-containing micelles ; M_{opt}, M_1 and M_s are empty micelles ; K_1 and K_2 are corresponding equilibrium constants equal to :

$$K_1 = \frac{[M_{opt}]\,[\overline{M}_s]}{[M_s]\,[\overline{M}_{opt}]} \qquad \text{and} \qquad K_2 = \frac{[M_{opt}]\,[\overline{M}_1]}{[M_1]\,[\overline{M}_{opt}]}$$

Let us now consider the dependence of catalytic activity on the surfactant concentration at a constant **w**, observed for membrane-active enzymes (Fig. 8, the left column). This dependence has been accounted for (104) by a scheme according to which interactions (collisions) between enzyme-containing and empty micelles lead to a reversible formation of new surfactant aggregates, similar to deformed or enlarged micelles (109) where the solubilized enzyme shows very low (if any) catalytic activity. Obviously, the fraction of such micelles increases with the concentration of the surfactant (the number of empty micelles), which naturally decreases the observed value of the catalytic activity. The intermicellar interaction can be described by the following set of equations :

$$\bar{M}_s + M \overset{K_3}{\underset{}{\rightleftharpoons}} \bar{M}^*_s \qquad\qquad [13]$$

$$\bar{M}_{opt} + M \overset{K_4}{\underset{}{\rightleftharpoons}} \bar{M}^*_{opt} \qquad\qquad [14]$$

$$\bar{M} + M \overset{K_5}{\underset{}{\rightleftharpoons}} \bar{M}^*_1 \qquad\qquad [15]$$

Here M denotes empty micelles of all three types, $[M] = [M_s] + [M_{opt}] + [M_1]$; $\bar{M}^*_s$, $\bar{M}^*_{opt}$ and $\bar{M}^*_1$ are "deformed" micelles, containing inactive enzyme ; and K_3, K_4 and K_5 are the equilibrium constants :

$$K_3 = \frac{[\bar{M}_s]\,[M]}{[\bar{M}_s^*]} \;,\quad K_4 = \frac{[\bar{M}_{opt}]\,[M]}{[\bar{M}_{opt}^*]} \quad \text{and} \quad K_5 = \frac{[\bar{M}_1]\,[M]}{[\bar{M}_1^*]}$$

The observed catalytic constant (k_{cat}) can be expressed via the maximal reaction rate (V) and the overall enzyme concentration (C_E) as follows :

$$k_{cat} = \frac{V}{C_E} = \frac{1}{C_E}\,(k_{cat}^{(opt)}\,[\bar{M}_{opt}] + k_{cat}^{(1)}\,[\bar{M}_1] \qquad\qquad [16]$$

The equilibrium concentrations $[\bar{M}_{opt}]$ and $[\bar{M}_1]$ in Eq. 16 can be obtained from the expressions for equilibrium constants [Eqs. 11-15] the material balance equation.

$$C_E = [\bar{M}_{opt}] + [\bar{M}_s] + [\bar{M}_1] + [\bar{M}*_{opt}] + [\bar{M}*_1] + [\bar{M}*_s]$$

and the distribution function of micellar sizes at a given **w**, $\rho(r)$. In the first approximation, the latter can be described (106) by the normal law :

$$\rho(r) = \frac{1}{<r>\sqrt{2\pi\sigma}} \exp\left\{-\frac{1}{2\sigma^2}\left(\frac{r}{<r>} - 1\right)^2\right\} \tag{17}$$

where $<r>$ is the mean radius of the inner cavity at the given hydration degree and σ is the standard deviation. The value of σ characterizes the width of the size distribution : the larger σ, the broader the distribution. Since the radius of the inner cavity is linearly related to the degree of the surfactant hydration (45), one can easily replace r in Eq. 17 by **w**. Thus, after all necessary substitutions, as outlined above, Eq. 16 becomes (104) :

$$k_{cat} = \frac{k_{cat}^{(opt)} + k_{cat}^{(1)} K_2 \beta}{1 + K_1\alpha + K_2\beta + \gamma\left(\dfrac{1}{K_4} + \dfrac{1}{K_3}\alpha + \dfrac{K_2}{K_5}\beta\right)[SURF]} \tag{18}$$

where [SURF] is the concentration of the surfactant, and

$$\alpha = \frac{\displaystyle\int_{0}^{w_{opt}-\Delta w} \exp\left\{-\frac{1}{2\sigma^2}\left(\frac{w(r)}{w} - 1\right)^2\right\} dw(r)}{\displaystyle\int_{w_{opt}-\Delta w}^{w_{opt}-\Delta w} \exp\left\{-\frac{1}{2\sigma^2}\left(\frac{w(r)}{w} - 1\right)^2\right\} dw(r)} \tag{19}$$

$$\beta = \frac{\displaystyle\int_{w_{opt+\Delta w}}^{\infty} \exp\left\{-\frac{1}{2\sigma^2} \left(\frac{w(r)}{w}\right) - 1\right\}^2 dw(r)}{\displaystyle\int_{w_{opt+\Delta w}}^{w_{opt+\Delta w}} \exp\left\{-\frac{1}{2\sigma^2} \left(\frac{w(r)}{w}\right) - 1\right\}^2 dw(r)} \qquad [20]$$

$$\gamma = \frac{1}{(\sigma^2 + 1)\, w^2} \qquad [21]$$

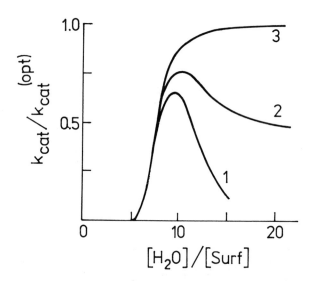

Fig. 10. The relative catalytic constant, $k_{cat}/k_{cat}^{(opt)}$, of an enzymatic reaction in the system of reversed micelles as a function of the hydration degree. Calculated by equation [18] for the conditions : $K_1 = K_2 = 1$, $K^{-1}_4 = K_1/K_3 = K_2/K_5 = 10^{-3}$; $w_{opt} = 10$; $\Delta w = 2$; $\sigma = 0.2$.
(1) $k_{cat}^{(1)}/k_{cat}^{(opt)} = 0$.
(2) $k_{cat}^{(1)}/k_{cat}^{(opt)} = 0.5$;
(3) $k_{cat}^{(1)}/k_{cat}^{(opt)} = 1$.

Here, w is the overall hydration degree of the surfactant, w(r) is the hydration degree of the population of micelles with the radius of the inner cavity r, w_{opt} is the hydration degree of optimal micelles with the radius of the inner cavity equal to r_{opt}, and Δw is the variation in the hydration degree causing the variation Δr in the radius of the inner cavity.

Using Eq. 18, it is possible to explain qualitatively the observed dependences of the catalytic activity of solubilized enzymes on the hydration degree and concentration of the surfactant.

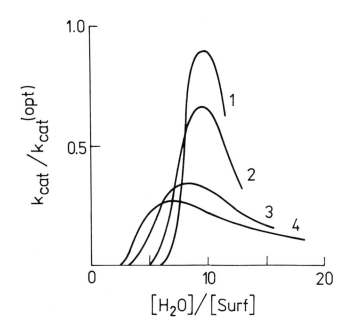

Fig. 11. The relative catalytic constant, $k_{cat}/k_{cat}^{(opt)}$, of an enzymatic reaction in the system of reversed micelles as a function of the hydration degree at various widths of the micellar size distribution. Calculated by equation [18] for the conditions given in Fig. 10 : $k_{cat}^{(1)}/k_{cat}^{(opt)} = 0$; (1) $\sigma = 0.1$; (2) $\sigma = 0.2$; $\sigma = 0.5$; (4) $\sigma = 0.8$.

Catalytic activity of solubilized enzymes as a function of the hydration degree. The role of the polydispersity of reversed micelles.

Theoretical dependences of the relative catalytic constant, $k_{cat}/k^{(opt)}_{cat}$, on the hydration degree, calculated using Eq. 18, are shown in Fig. 10. In calculations, we assumed that the parameters k^{-1}_4, K_1/K_3 and K_2/K_5 in Eq. 18

are sufficiently small, so that the catalytic activity is independent of the surfactant concentration. Curve 1 is the case when the enzyme is totally inactivated in large micelles, $k^{(1)}_{cat}/k^{(opt)}_{cat} = 0$. The "$k_{cat}$ vs.w" dependence of this kind was observed experimentally, for instance, for acid phosphatase, prostaglandin synthetase, and cytochrome c (see Fig. 3). Curve 3 corresponds to another extreme case when catalytic activities of the enzyme entrapped in optimal and large micelles are equal, $k^{(1)}_{cat}/k^{(opt)}_{cat} = 1$. This dependence (curve with "saturation") was found for phospholipase A_2 (110). Finally, curve 2 (incomplete inactivation of enzyme in large micelles) is typical of α-chymotrypsin (see Fig.3).

Fig. 11 illustrates how the shape of the "k_{cat} vs.w" dependence can be deeply changed by the micelles polydispersity. As one could expect, the increase in the standard deviation (σ), making the size distribution of micelles broader and thus reducing the fraction of optimal micelles in the system, decreases the observed catalytic activity of solubilized enzyme and, hence, flattens the corresponding curve.

Another point of interest is that with increasing width of the size distribution the optimum of the catalytic activity shifts towards low hydration degrees, i.e. lower mean micellar radii (Fig. 11). This shift is caused by the asymmetry of the size distribution function (for details, see (104)). An important

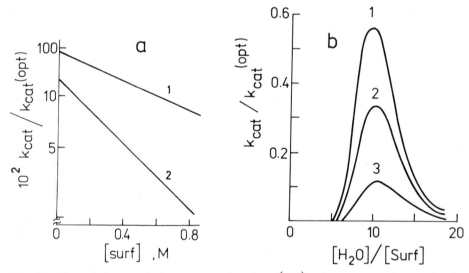

Fig. 12. The relative catalytic constant, $k_{cat}/k_{cat}^{(opt)}$, of an enzymatic reaction in the system of reversed micelles as a function of the surfactant concentration (a) and the hydration degree (b) calculated by equation [18] for the conditions : $k_{cat}^{(1)}/k_{cat}^{(opt)} = 0$; $K_1 = K_2 = 1$; $K_4^{-1} = K_1/K_3 = K_2/K_5 = 10^3$; $w_{opt} = 10$; $\Delta w = 2$; $\tilde{\sigma} = 0.2$. Curves in Fig. (a) correspond to [SURF] values : (1) 0.02 M ; (2) 0.1 M ; (3) 0.5 M.

consequence of such a shift is that in reversed micellar systems with different size distributions the optimal catalytic activity of the same enzyme can be observed at different values of the mean micellar radius. This means that the coincidence of the size of the entrapped enzyme molecule with the radius of the aqueous cavity of the reversed micelle under optimal conditions, observed in AOT micellar systems (Fig. 4), may well represent an exception, rather than a general rule.

Catalytic activity of solubilized enzymes as a function of the surfactant concentration

Fig. 12 shows dependences of the catalytic activity on the surfactant concentration and hydration degree calculated using Eq.18 for the most general case when all the equilibria described by Eqs. 11-15 are taken into account. The shape of these curves somewhat differs from those observed experimentally. First, for peroxidase (59,60), the limiting values of the catalytic constant, obtained by extrapolation of the surfactant concentration to zero, are the same at various w values. Second, the limiting value of k_{cat} is independent of the nature of the micelle-forming surfactant (Fig. 8, left column). Contrary to these experimental data, the straight lines in Fig. 12a do not converge on Y-axis to one point.

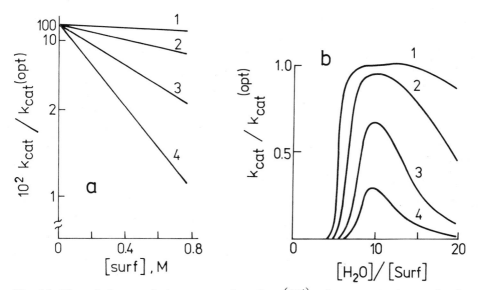

Fig. 13. The relative catalytic constant, $k_{cat}/k_{cat}^{(opt)}$, of an enzymatic reaction in the system of reversed micelles as a function of surfactant concentration (a) and the hydration degree (b). Calculated by equation [18] for the conditions $k_{cat}^{(1)}/k_{cat}^{(opt)} = 0$; $K_1 = K_2 = K_4^{-1} = 10^{-3}$; $K_1/K_3 = 10^3$; $w_{opt} = 10$; $\Delta w = 2$. Curves in Fig. (a) correspond to w values : (1) and (2) 10 ; (3) 7 ; (4) 20 : σ : (1), (3) and (4) 0.2 ; (2) 1. Curves in Fig. (b) correspond to [SURF] values : (1) 0.001 M ; (2) 0.01 M ; (3) 0.1 M ; (4) 0.5 M ; $\sigma = 0.2$.

To satisfy the experimental data in terms of the theoretical model under consideration, it should be admitted that the constants K_1, K_2 and K^{-1}_4 are sufficiently low (Fig.13). This implies that in the optimal micelle the solubilized enzyme forms a stable complex with surfactant molecules.

CONCLUSION

By the present time, more than three dozen enzymes have been studied in reversed micellar systems owing to efforts of about 20 laboratories all over the world. Evaluating the rapid development of micellar enzymology, it is to be noted that a simple idea of placing the enzyme molecule into a microreactor formed by the reversed micelles had led to consequences that a decade ago, when studies in this field started, were difficult to predict. In fact, together with new fundamental approaches to modeling the enzyme-membrane organization and to the mechanistic study of enzymatic catalysis, micellar enzymology has also given principal results opening new prospects in practical chemistry, fine organic synthesis, clinical and chemical analysis, bioconversion of energy and other applied areas (20,23,25).

In our opinion, such a successful progress is a result of two following main advantages of reversed micellar systems. First, these microheterogeneous systems provide means for easy dissolution of both hydrophylic and hydrophobic substances under standard conditions, thus giving extremely wide possibilities to vary the nature of molecular objects used in research. It is this property of reversed micellar systems that has brought about considerable achievements in such achievements in such areas as enzymatic transformation (111-113) and analysis (114) of hydrophobic compounds and selective modification of proteins with water-insoluble reagents (115). Second, reversed micelles ensure a strict nanocompartmentalization of solubilized macromolecules, thus permitting a controlled buildup of subunit structures (see, for example Fig. 5 and accompanying discussion) and synthesis of conjugates of natural and synthetic macromolecules of a predetermined composition (116).

As an example of the use of the nanocompartmentalization approach, let us consider our recent results (117) on homogeneous enzyme immunoassay in reversed micelles (Fig. 14). When thyroxine-specific antibodies are added to a peroxidase-thyroxine conjugate solubilized in reversed micelles, the immunocomplex is formed, which significantly differs in size from the initial enzyme-antigene conjugate. Since the catalytic activity of solubilized biocatalysts is influenced by the ratio of sizes of the enzyme molecule and the inner cavity of the reversed micelle, as discussed above in the chapter, the formation of the immunocomplex leads to an alteration of the enzymatic activity. Addition of free thyroxine into the system causes the dissociation of the complex with concomittant restoration of the catalytic activity to an extent dependent on the concentration of added free antigen (Fig. 14). The sensitivity of the assay procedure can be optimized by adjusting the size of reversed micelles (i.e. the value of **w**).

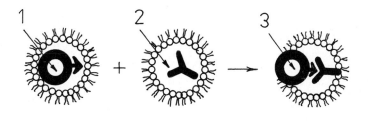

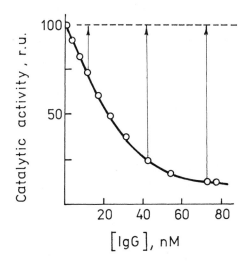

Fig. 14. Homogeneous enzyme immunoassay in the reversed micellar system of AOT in octane. 1, peroxidase-thyroxine conjugate 2, antibody ; 3, immunocomplex. Vertical arrows show the change in catalytic activity produced by the addition of 10 nM thyroxine. Experimental conditions : [AOT] = 0.1 M, w = 38, [peroxidase] = 0.1 nM. Substrates : pyrogallol 10 nM, hydrogen peroxide 0.2 nM. From (117).

REFERENCES

1 Singer S.J., Nicholson G.L., Science **175**, 720-731 (1972).
2 Borovjagin V.L., Vergara J.A., McIntosh T.J., J. Membr. Biol. **69**, 199-212 (1982).
3 De Kruijff B., Cullis P.R., Verkleij A.J., Hope M.J., Van Echteld C.J.A., Tarashi T.F., in : The enzymes of Biological Membranes (2nd ed.) A.N. Martonosi (Ed.), vol. 1, Membrane Structure and Dynamics, Plen, N.Y. 1985 pp. 131-204.
4 Cullis P.R., Hope M.J., Talcock C.P.S., Chem. Phys. Lipids **40**, 127-144 (1986).
5 Cullis P.K., Kruijff B., Hope M.J., Verkleij A.J., Nayar R., Farren S.B., Tilcock C., Madden T.D., Bally B., in : R.C. Aloia (Ed.), Membrane fluidity in biology, Academic Press, New York, 1983 pp. 39-81.
6 Cullis P.R., De Kruijff B., Hope M.J., Nayar R., Schmid S.L., Can. J. Biochem. **58**, 1091-1100 (1980).
7 Chupin V.V., Ushakova I.P., Bondarenko S.V., Vasilenko I.A., Serebrennikova G.A., Evtigneeva R.P., Rosenberg G.Ya., Kol'tsova G.N., Bioorgan. Khim. (Russ.) **8**, 1275-1280 (1982).
8 Stier A., Finch S.A.E., Börsterling B., FEBS Lett. **91**, 109-112 (1978).
9 Hah J.-S., Hui S.W., Jung C.Y., Biochemistry **22**, 4763-4769 (1983).
10 Zvonkova E.N., Habarova E.I., Vasilenko I.A., Evstigneeva R.P., Bioorgan. Khim. (Russ.) **10**, 1401-1408 (1984).
11 Van Echteld C.J.A., De Kruijff B., Verkleij A.J., Leunissen-Bijvelt J., De Gier, J. Biochim. Biophys. Acta **692**, 126-138 (1982).
12 Murphy D.J., FEBS Lett. **150**, 19-26 (1982).
13 Stier A., Finch S.A.E., Greinert R., Hohne M., in : Kitani K. (Ed.), Liver and aging, Elsevier Biomedical Press, Amsterdam, 1982 pp. 3-14.
14 Hui S.W., Stewart T.P., Yeagle P.K., Albert A.D., Arch. Biochem. Biophys. **207**, 227-240 (1981).
15 Jensen J.W., Schitzbach J.S., Biochemistry **23**, 1115-1119 (1984).
16 Dawson R.M.C. J. Am. Oil Chem. Soc. **59**, 401-406 (1982).
17 Dawson R.M.C., Irvine R.F., Brag J., Quinn P.J., Biochem. Biophys. Res. Commun. **125**, 836-842 (1984).
18 Khmelnitsky Yu.L., Levashov A.V., Klyachko N.L., Martinek K. Rus. Chem. Revs. **53**, 319-331 (1984).
19 Martinek K., Levashov A.V., Khmelnitsky Yu.L., Klyachko N.L., Berezin I.V., Eur. J. Biochem. **155**, 453-468 (1986).
20 Martinek K., Berezin I.V., Khmelnitsky Yu.L., Klyachko N.L., Levashov A.V. Coll. Czech. Chem. Commun. **52**, 2589-2602 (1987).
21 Luisi P.L., Magid L.J., CRC Crit. Rev. Biochem. **20**(4), 409-474 (1986).
22 Luisi P.L., Steimann-Hofmann B., Methods Enzymol. **136**, 188-216 (1987).
23 Laane C., Hilhorst R., Veeger C., Methods Enzymol. **136**, 216-229 (1987).
24 Waks M., Proteins (Struct. Function Genet.) **1**, 4-15 (1986).
25 Shield J.W., Ferguson H.D., Bommarlus A.S., Hatton T.A., Ind. Eng. Chem. Fundam. **25**, 603-612 (1986).
26 Martinek K., Levashov A.V., Klyachko N.L., Berezin I.V., Dokl. Akad. Nauk SSSR (Russ.) **236**, 920-923 (1977).
27 Balny C., Hui Bon Hou G., Douzou P., Jerusalem Symp. Quantum Chem. Biochem. **12** (Catal. Chem. Biochem. Theory Exp.) 37-50 (1979).
28 Balny C., Douzou P., Biochimie **61**, 445-452 (1979).
29 Kabanov A.V., Namyotkin S.N., Matveeva E.G., Klyachko N.L., Martinek K., Levashov A.V., Mol. Biol. (Russ.) **22**, 473-484 (1988).
30 Joly M., A physicochemical approach to the denaturation of proteins, Academic Press, London, New York, 1965.
31 Menger F.M., Yamada K. J., Amer. Chem. Soc. **101**, 6731-6734 (1979).
32 Luisi P.L., Henninger F., Joppich M., Biochem. Biophys. Res. Commun. **74**, 1384-1389 (1977).

33   Luisi P.L., Bonner F.J., Pellegrini A., Wiget P., Wolf R., Helv. Chim. Acta **62**, 740-753 (1979).

34   Levashov K. J., Khmelnitsky Yu.L., Klyachko N.L., Chernyak. V.Ya, Martinek. K., Colloid Interface Sci. **88**, 444-457 (1982).

35   Bonner E.J., Wolf R., Luisi P.L., J. Solid-Phase Biochem. **5**, 255-268 (1980).

36   Zampieri G.G., Jaeckle H., Luisi P.L., J. Phys. Chem. **90**, 1849-1853 (1986).

37   Fletcher P.D.I., Howe A.M., Perrins N.M., Robinson B.H., Toprakcioglu C., Dore J.C., in : K.L. Mittal, B. Lindman (Eds.), Surfactants in Solution, Plenum, New York, London, 1984, pp. 1745-1758.

38   Fletcher P.D.I., Robinson B.H., Tabony J., J. Chem. Soc., Faraday Trans. I **82**, 2311-2321 (1986).

39   Petit C., Brochette P., Pileni M.P., J. Phys. Chem., **90**, 6517-6521 (1986).

40   Shen E., Goklen K.E., Hatton T.A., Chen S.H. Biotechnolol. Progr. **2**, 175-186 (1986).

41   Chatenay D., Urbach W., Cazabat A.M., Vacher M., Waks M., Biophys. J. **48**, 893-898 (1985).

42   Chatenay D., Urbach W., Nicot C., Vacher M., Waks M., J. Phys. Chem. **91**, 2198-2201 (1987).

43   Shapiro Yu.E., Budanov N.A., Levashov A.V., Klyachko N.L., Khmelnitsky Yu.L., Martinek K., Coll. Czech. Chem. Commun. (1988), in press.

44   Birktoft J.J., Blow D.M., Henderson R., Steitz T.A. Phil. Trans. Royal. Soc. **B257**, 67-80 (1970).

45   Eicke H.F., Rehak J. Helv. Chim. Acta **59**, 2883-2891 (1976).

46   Ueno M;, Kishimoto H., Kyogoku Y., J. Colloid Interface Sci. **63**, 113-119 (1978).

47   Martin C.A., Magid L.J., J. Phys. Chem. **85**, 3938-3944 (1981).

48   Maitra A.N., Eicke H.F., J. Phys. Chm. **85**, 2687-2691 (1981).

49   Maitra A., J. Phys. Chem. **88**, 5122-5125 (1984).

50   Scherer J.R., Proc. Natl. acad. Sci. USA 84, 7938-7942 (1987).

51   Riccio P., Masotti L., Cavatorta P., De Santis A., Juretic D., Bobba A., Pasquali-Ronchetti I., Quagliariello E., Biochem. Biophys. Res. Commun. **134**, 313-319 (1986).

52   Montal M., in : P.L. Luisi, B.E. Straub (Eds.), Reverse Micelles, Plenum, New York, 1984 pp. 221-229.

53   Wirz J., Rosenbuch J.P., in : P.L. Luisi, B.E. Straub, (Eds.), Reverse Micelles, Plenum, New York, 1984 pp. 231-238.

54   Davydov R.M., Belovolova L.V., Genkin M.V., Mol. Biol. (Russ.) **19**, 1338-1349 (1985).

55   Lysko A.I., Surkov S.A., Arutyunyan A.M., Khmelnitsky Yu.L., Klyachko N.L., Levashov A.V., Martinek K. Biofizika (Russ.) **31**, 231-236 (1986).

56   Levashov A.V., Klyachko N.L., Martinek K., Bioorgan. Khim. (Russ.) **7**, 670-679 (1981).

57   Fletcher P.D.I., Rees G.D., Robinson B.H., Freedman R.B., Bioch. Biophys. Acta **832**, 204-214 (1985).

58   Malakhova E.A., Kurganov B.I., Levashov A.V., Berezin I.V., Martinek K., Dokl. Akad. SSSR (Russ.) **270**, 474-477 (1983).

59   Martinek K., Klyachko N.L., Levashov A.V., Berezin I.V., Dokl. Akad. Nauk SSSR (Russ.) **269**, 491-493 (1983).

60   Klyachko N.L., Levashov A.V., Martinek K., Mol. Biol. (Russ.) **18**, 1019-1031 (1984).

61   Mevkh A.T., Sud'ina G.F., Lagutina I.O., Levashov A.V., Biokhimiya (Russ.) **50**, 1719-1723 (1985).

62   Levashov A.V., Klyachko N.L., Pshezhetskii A.V., Berezin I.V., Kotrikadze N.G., Lomsadze N.G., Lomsadze B.A., Martinek K., Dokl. Akad. Nauk SSSR (Russ.) **289**, 1271-1273 (1986).

63   Klyachko N.L., Levashov A.V., Pshezhetsky A.V., Bogdanova N.G., Berezin I.V., Martinek K., Eur. J. Biochem. **161**, 149-154 (1986).

64   Kurganov B.I., Tsetlin L.G., Malakhova E.A., Chebotareva N.A., Lankin
        V.Z., Levashov A.L., Glebova G.D., Berezovskii V.., Martinek K., Berezin
        I.V., Dokl. Akad. Nauk SSSR (Russ.) **282**, 1263-1267 (1985).
65   Pshezhetskii A.V., Klyacho N.L., Levashov A.V., Bushueva M.V.,
        Martinek K., Biokhimiya (Russ.) **53**, N 6 (1988).
66   Klyachko N.L., unpublished results.
67   Klyachko N.L., Pshezhetskii A.V., Vakula S.V., Levashov A.V., in
        preparation.
68   Squire P.G., Himmel M.E., Arch. Biochem. Biophys. **196**, 165-177
        (1979).
69   Pshezhetskii A.V., Ph. D. Thesis, Moscow University (1988).
70   Klyachko N.L., Merker S., Vakula S.V., Ivanov M.V., Berezin I.V.,
        Martinek K., Levashov A.V., Dokl. Akad. Nauk SSSR (Russ.) **298**,
        1479-1481 (1988).
71   Douzou P., Keh E., Balny C., Proc. Natl.Acad. Sci. USA **76**, 681-684
        (1979).
72   Walde P., Peng Q., Fadnavis N.W., Battistel E., Luisi P.L., Eur. J.
        Biochem., **173**, 401-409 (1988).
73   Grandi C., Smith R.E., Luisi P.L., J. Biol. Chem. 256, 837-844 (1981).
74   Wolf R., Luisi P.L., Biochem. Biophys. Res. Commun. **89**, 209-217
        (1979).
75   Righetti P.G., Caravaggio T., J. Chromatogr. **127**, 1-28 (1976).
76   Horiike K., Tojo H., Yamano T., Nazaki M., J. Biochem., **93**, 99-106
        (1983).
77   Namyotkin S.N., Kabanov A.V., Evtushenko G.N., Chernov N.N., Berezov
        T.T., Shegolev A.A., Ryzhova V.V., Klyachko N.L., Martinek K.,
        Levashov A.V., Bioorgan. Khimiya (Russ.) (1988), in press.
78   Friedrich P., Supramolecular enzyme organization, Pergamon Press,
        Oxford, 1984.
79   Tate S.S., Meister, A. Mol. Cell. Biochem. **39**, 357-368 (1981).
80   Loginov V.A., Chernov N.N., Berezov T.T., Bull. Exp. Biol. (Rus.) 7,
        58-60 (1980).
81   Klyachko N.L., Bogdanova N.G., Levashov A.V., Kabanov A.V.,
        Pshezhetskii A.V., Khmelnitsky Yu.L., Martinek K., Berezin I.V., Dokl.
        Akad. Nauk SSSR (Russ.) **297**, 483-487 (1987).
82   Eremin A.N., Metelitsa D.I., Biokhimiya (Rus.) **50**, 102-108 (1985).
83   Eremin A.N., Kazilunene B.M., Vaitkiavicius R.K., Metelitsa D.I.,
        Biokhimiya (Russ.) **51**, 856-863 (1986).
84   Han D., Rhee J.S., Biotechnol. Lett. **7**, 651-656 (1985).
85   Han D., Rhee J.S., Biotechnicol. Bioeng. **28**, 1250-1256 (1986).
86   Fletcher P.D.I., Freedman R.B., Robinson B.H., Rees G.D., Schomäcker
        R., Biochim. Biophys. Acta **12**, 278-282 (1987).
87   Klyachko N.L., Bogdanova N.G., Martinek K., Levashov A.V., Bioorgan.
        Khim. (Russ.) (1989), in press.
88   Klyachko N.L., Bogdanova N.G., Koltover V.K., Martinek K., Levashov
        A.V., Biokhimiya (Russ.) (1988), in press.
89   Fletcher P.D.I., Galal M.F., Robinson B.H., J. Chem. Soc., Faraday Trans.
        I, **80**, 3307-3314 (1984).
90   Belonogova O.V., Likhtenstein G.I., Levashov A.V., Khmelnitsky Yu.L.,
        Klyachko N.L., Martinek K., Biokhimiya (Ruus.) **48**, 379-386 (1983).
91   Visser A.J.W.G., Vos K., van Hoek A., Santema J.S., J. Phys.Chem., **92**,
        759-765 (1988).
92   Levashov A.V., Klyachko N.L., Bogdanova N.G., Martinek K., Dokl.
        Akad. Nauk SSSR (Russ.) (1989), in press.
93   Bedwell B., Gulari E., in : K.L. Mittal, E.J. Fendler (Eds.), Solution
        Behavior of surfactants, Plenum, New York, 1982, pp. 833-846.
94   Kotlarchyk M., Chen S.H., Huang J.S., Kim M.W., Phys. Rev. **A 29**,
        2054-2069 (1984).
95   Martinek K., Levashov A.V., Klyachko N.L., Pantin V.I., Berezin I.V.,
        Biochim. Biophys. Acta **657**, 277-294 (1981).

96   Shevchuk T.V., unpublished results.
97   Welinder K.G., Smillie L.B., Can. J. Biochem. **50**, 63-90 (1972).
98   Reinhammer B., in : B. Brown (Ed.), Copper Proteins and Copper Enzymes, Academic Press, New York, 1979, pp. 1-35.
99   Kabanov A.V., Levashov A.V., Martinek K., Vestnik MGU (Series 2, Chemistry (Russ.) **27**, 591-594 (1986).
100   Levashov A.V., Khmelnitsky Yu.L., Klyachko N.L., Martinek K., in : K.L. Mittal, B. Lindman (Eds.), Surfactants in Solution, Plenum, New York, 1984 pp. 1069-1091.
101   Fletcher P.D.I., Robinson B.H., Freedman R.B., Oldfield C., J. Chem. Soc., Faraday Trans I, **81**, 2667-2679 (1985).
102   Levashov A.V., Pantin V.I., Martinek K., Berezin I.V., Dokl. Akad. Nauk SSSR (Russ.) **252**, 133-136 (1980).
103   Khmelnitsky Yu.L., Polyakov V., Grinberg V.Ya., Neverova I.N., Levashov A.V., Martinek K., in preparation.
104   Kabanov A.V., Klyachko N.L., Pshezhetskii A.V., Namyotkin S.N., Martinek K., Levashov A.V., Mol. Biol. (Rus.) **21**, 275-286 (1987).
105   Zima J., Klyachko N.L., Martinek K., in preparation.
106   Zulauf M., Eicke H.F., J. Phys. Chem., **83**, 480-486 (1979).
107   Robinson B.H., Toprakcioglu C., Dore J.C., Chieux P., J. Chem. Soc., Faraday Trans. I, **80**, 13-27 (1984).
108   Eicke H.F., Kvita P., in : P.L. Luisi, B.E. Straub (Eds.), Reverse Micelles, Plenum Press, New York, 1984 pp. 21-35.
109   Eicke H.F., Shepherd J.C., Steinemann A., J. Colloid Interface Sci. **56**, 168-176 (1976).
110   Rakhimov M.M., Tuichibaev M.U., Gorbataya O.N., Kabanov A.V., Levashov A.V., Martinek K., Biol. Membr. (Russ.) **3**, 1030-1036 (1986).
111   Hilhorst R., Spruijt R., Laane C., Veeger C., Eur. J. Biochem. **144**, 459-466 (1984).
112   Morita S., Narita H., Matoba T., Kito M., J. Amer. Oil Chem. Soc. **61**, 1571-1574 (1984).
113   Hilhorst R., Laane C., Veeger C., FEBS Lett. **159**, 255-258 (1983).
114   Kurganov B.I., Tsetlin L.G., Malakhova E.A., Chebotareva N.A., Lankin V.Z., Glebova G.D., Berezovsky V.M., Levashov A.V., Martinek K., J. Biochem. Biophys. Methods **11**, 177-184 (1985).
115   Kabanov A.V., Levashov A.V., Martinek K., Annals New York Acad. Sci. **501**, 63-66 (1987).
116   Kabanov A.V., Alakhov V.Y., Klinskii E.Y., Khrutskaya M.M., Rakhnanskaya A.A., Polinskii A.S., Yaroslavov A.A., Severin E.S., Levashov A.V., Kabanov V.A., Dokl. Akad. Nauk SSSR (Russ.) (1988), in press.
117   Kabanov A.V., Khrutskaya M.M., Budavari M., Eremin S.A., Klyachko N.L., Levashov A.V., Dokl. Akad. Nauk. SSSR (Russ.) (1988)., in press.

SOME GENERAL CONSIDERATIONS ABOUT THE FIELD OF PROTEIN-CONTAINING REVERSE MICELLES

P.L. LUISI

1.   INTRODUCTION

The first uv-spectrum of a hydrophylic enzyme (α–chymotrypsin) solubilized in cyclohexane with the help of methyltrioctyl ammonium chloride was reported over ten years ago[1]. This was followed by the observation that in the ternary system AOT/hydrocarbon/water (AOT stands for the anionic surfactant bis 2-ethylhexyl sodium sulfosuccinate) enzymes could retain their activity[2]. Since then, the field of enzymes in reverse micelles has enjoyed a lot of attention, and nowadays over a dozen groups all around the word are active in the area. I like to mention a few of these : Martinek in Prague (formerly in Moscow) ; Laane, Van Riet and Hilhorst in Wageningen (NL) ; Robinson in Norwich, and Fletcher in Kent, both in U.K. ; in France, Pileni and, independently, Waks in Paris, Biellmann in Strasbourg and Clausse in Compiègne ; in the States, Menger, who was actually one of the pioneers in the field, L. Magid in Knoxville, (TN) and Hatton at the M.I.T. in Boston ; in Italy, Maestro and Caselli in Bari, Strambini in Pisa, Palmieri in Bologna, Rialdi in Genoa ; Garcia-Carmona in Murcia, Spain ; Fadnavis in Hyderabad, India, Rhee and Han in Seoul ; - and this list is certainly not complete.

A large number of review articles have appeared in the last few years [3-9] covering exaustively all aspects of the problem - actually one can say that there are already too many reviews. For this reason I refrain to write here yet another one. I like instead to present some critical remarks about the field in general.

2.   THE MAIN QUESTIONS IN FUNDAMENTAL RESEARCH

The type of questions in the field of enzyme in reverse micelles are indeed challenging. Restricting the analysis for the moment to the basic science questions (we will consider biotechnology later on), I propose the following ones :

1) What is the mechanism of uptake of proteins into the micelles ? What are the driving forces, and what are the kinetically relevant steps affecting the uptake ?

2) What is the structure of the macromolecular host-guest complex (protein in the micellar aggregate) ? Where is the protein located in the micellar system ? How is the structure (conformation) of the protein modified by the novel environment and in particular by the w value ? (w is the molar ratio water to surfactant, i.e. $w = [H_2O] / [SURF]$).

3) How and why is the stability of proteins changed, due to the incorporation into the reverse micelles ?

4) How and why is the enzyme mechanism of action modified by the incorporation of an enzyme in the micellar system ? Is for example the enzyme specificity altered ?

5) Why are enzymes in reverse micelles so active at very low water content ? (in particular in the w range 3-15, and this also for water soluble hydrolases).

6) What are the physical properties of water which interact with the enzymes under those conditions-e.g. the pH, or the water activity, or its dielectric constant ?

7) Why do some enzyme show "superactivity", i.e. a k_{cat} which is larger in reverse micelles than in water ?

8) How does the structure of the surfactant affect all the above mentioned enzymatic properties ? (most of the data until now are obtained with AOT, i.e. in hydrocarbons ; or with cetyl trimethyl ammonium bromide (CTAB) in chloroform and chloroform/hydrocarbon mixtures).

9) How can water-soluble enzymes, which with all likehood are compartimented in the water pool of reverse micelles, act upon water-insoluble substrates which are localized in the bulk organic phase ?

All these different aspects, and others, are partly discussed in the reviews cited above (3-9), and one could ask whether and to what extent the answers given until now are satisfactory and definitive.

Of course, the answer is bound to be subjective. I personally think that we have no satisfactory answer (i.e. accepted by the large majority of workers in the field, and intellectually satisfactory as well) to questions 5, 6, 7 ; and that we can have for the other questions only intelligent guesses. This may then not appear a very brilliant situation. However, as mentioned before, the field is still relatively new, and both the concepts underlying the above questions as well as the experimental instrumentation have still to be sharpened and tailored to the particular issues. Also, theoreticians have still to get into the field- or they barely begin to do so.

The involvement of theoreticians is important, as it is really necessary that the many sparse data be incorporated into a thermodynamic unitarian view. This would also help the experimentalists to choose the more correct approach to a given problem.

3.   THE THERMODYNAMICS : THEORY AND EXPERIMENTS

A first attempt to handle the thermodynamics of the uptake of proteins into reverse micelles has been made by Maestro (10,11). The starting parameters are the initial concentrations of all components, and the radii of the protein and of the initial unfilled (i.e. without protein) micelles. The parameters at equilibrium (after uptake of the protein) are then calculated with a procedure which essentially is a minimization of the free energy of the system. The energy is calculated as the sum of four contributions, of which the most important ones are an electrostatic contribution and an entropic one (the uptake of the protein into the micelles brings about an expulsion of small ions from the micelle itself). In turn, these calculations are carried out on a model which depicts the protein-containing micelle as a double condensator.

Based on first calculations, it appears that the theoretical model gives a good agreement with experimental data (calculations are however rather cumbersome and are still partly in progress).

Thermodynamically speaking the micellar system is seen as an equilibrium state, characterized by a set of parameters (aggregation number,w, radius of the micelle, number of micelles, and so on) - which after the perturbation (uptake of the protein) reaches a new thermodynamic equilibrium state, which is characterized by a completly new set of parameters. There is no preferential set of parameters for the protein-contained micelle in absolute : the set depends upon the boundary conditions, and in particular upon the initial concentration of all species. However the free energy difference for the uptake of the protein has one minimum under conditions where the radius of the protein and the radius of the initial unfilled micelles are close to each other.

One obvious consequence arising from this thermodynamic treatment is the following : that the percentage of micelles which are filled by protein molecules is not known a priori in the final state. This is so even if one assumes that there is only one protein molecule per micelle. Therefore, one is bound to deal with a population of micelles which is binodial : "filled" and "unfilled" micelles, i.e. containing and non-containing protein, respectively.

Once this view is accepted, one should choose an experimental method which is best suited to the situation. In this specific case, it should be an experimental technique which is able to discriminate between filled and unfilled micelles and possibly simultaneously record a signal from the filled and from the unfilled micelles in the same solution sample. It should be only too evident that methods which provide a number average value between filled and unfilled micelles (like most osmometric or scattering techniques) are not so good.

In fact, one method which permits the simultaneous recording of the signals of filled and unfilled micelles has been described, and consists in the "double dye" ultracentrifugation (12) : accordingly, different samples at different concentrations are recorded in the same run, and the UV-scanner records directly the concentration of water and surfactant simultaneoustly in one sample after another. The amount of water and/or surfactant sedimenting with filled and unfilled micelles is thus directly measured by uv-spectroscopy.

It is quite obvious that, at least until now, this is the method par excellence for the determination of the structural change following the uptake of the protein in the micellar solution. In our group, we have tried to approach the problem also by small angle neutron scattering and light scattering (also in collaboration with Robinson and Magid). In fact, one can try to interpret scattering data (a single curve) in terms of the contributions (related to filled and unfilled micelles) - but everybody who has worked in the field knows how hard and how finally unreliable this is going to be. Of course scattering experiments should be pursued and eventually they will bring important contributions, as it happened in other fields. But, in the restricted topic of protein-containing micelles, the light scattering technique is not yet one which can be of general validity.

There are problems with the analytical uv-scanning ultracentrifugation as well. One is of practical nature. It is a costly instrument, not available in all labs involved with micellar work. Furthermore, the Beckman company has disconnected production of the old Model E, and the new instrument, (Beckmann L 8 - 60 M) disgracefully enough, has no serious monochromator -only optical filters- which is totally insufficient for the double dye ultracentrifugation technique (as one has to record simultaneously at two wavelengths rather close to each other). We are presently working for modyfing the new Beckman instrument, incorporating in it a suitable monochromator. Another problem with the double dye method may be seen in the psychological difficulty, that some groups still have, to use a device found by somebody else.

The picture which was obtained from the few already published ultracen-trifugation data (12) is however rather clear : each time a protein is uptaken, there is a re-shuffling of the material (water, surfactant) to reach new values of w, micellar diameter, aggregation number, and so on. This is what one would expect on the basis of thermodynamics and dynamic behaviour of these systems, and what common sense would also predict. It is perhaps important to stress again that there is nothing like a definite (specific, constant) molecular weight (or dimension) for the protein-containing micelle : this size depends on the initial conditions. In other words a protein -containing AOT micelle is not characterized by its own molecular properties- those change each time you change the initial concentrations.

Of course the enlargement of the original micelle can be small depending on the initial experimental conditions (size of the protein, initial w, concentration, etc.). That it could be so was actually proposed in 1980, in the earlier paper dealing with the water-shell model for proteins in reverse micelles (13) and latter on re-enforced by other authors (14,15). In particular from Martinek's group comes the notion that the micelle does not change its size upon uptake of the protein (15). This model has been criticized before (5). As the thermodynamic treatment and the hard ultracentrifugation data show, this can be true under some particular sets of conditions- but one should not confuse these particular conditions with a general law and push forward for antithermodynamic models.

4.   IS IT A SUCCESSFUL FIELD ?

A first simple way to evaluate the "success" of a field is to measure its popularity in terms of publications and growth. Judged in this way, the field of protein-containing reverse micelle is certainly a successful one.

Another criterion of success can be evaluated on the basis of the question, whether and to what extent the "local" knowledge obtained in a given field can be "exported" into more general areas. In our case one can for example ask the question, whether the field of enzymes in reverse micelles has brought something qualitatively new into enzymology, biochemistry ou biotechnology.

For example the investigation of the first two questions (of our previous list) may in principle shed some light on the field of proteins in membranes. It is actually a point which is often done, and actually the idiom "biomimetic membrane chemistry" is often used in this sense.

I personally have doubt that this is so- or at least, I would use much precaution into this kind of "exporting". In fact, most of the available data on proteins in reverse micelles are collected using AOT, and it is becoming more and more clear that this surfactant has a series of quite peculiar properties which put it aside from all others- one cannot certainly generalize from AOT to the whole class of surfactants, and even less from the AOT data to membranes in vivo.

A better kind of mimetic membrane chemistry can perhaps be achieved if phospholipids are used as the surfactants for protein-containing reverse micelles. The replacement of AOT and other synthetic surfactant with phosholipids is not an easy matter, but it is one of the new research directions (3,17).

The other area where the research on enzymes in reverse micelles has been considered useful for biochemistry at large is summarized in the questions nr. 5 and 6, which pivot on the role of the minimal amount of water on enzyme activity. Indeed, the study of enzymes in reverse micelles has shown that even hydrolytic enzymes are able to display their maximal catalytic activity under conditions of quasi water starvation, i.e. when there is only enough water to build 2-3 layers around the protein surface. This feature is not limited to AOT, but shared by different surfactant systems.

I believe that clarifying the question of the enzyme activity in reverse micelles under conditions of water starvation can shed some light on enzymology at large.

It would be interesting to look in this way to all questions listed above, namely under the perspective : is this question going to bring something to enzymology at large ? Due to self-imposed space limitations, I restrain to do so here. I rather consider briefly the question of biotechnology.

5.   THE QUESTION OF BIOTECHNOLOGY

Several papers in the field begin with the statement, that enzymes in reverse micelles are very important biotechnologically because they permit enzymatic reactions in essentially apolar solvents. This is certainly true in principle. As a matter of fact, however, this biotechnological exploitation seems to be slow to come.

There is only one criterion in order to establish whether a certain system is or is not of biotechnological importance : whether it is used or not for industrial processes. And up to now, to the best of my knowledge, no reverse micellar system containing enzymes is being used industrially.

On the other hand, it is fair to say that the other well known method to utilize enzymes in organic solvents, the two phases system has been around for over a quarter of century, and it was never used industrially. It may well be that the only way to use enzymes industrially is by immobilization.

Of course, all this may change, and it is right to keep pointing out the potential importance of reverse micelles for enzymatic conversions. Recently, it has been shown that also cells can be solubilized in organic solvents by the use of reverse micellar solutions (18) - which is again of potential biotechnological importance (microbiology in organic solvents), but it has not been used yet in any practical way.

Industry has rather taken an interest in another aspect of the field : the use of reverse micellar systems to extract, isolate and purify proteins. I know of at least three companies which are pursuing this research, based on the observations that extraction of proteins with reverse micelles can be a specific process (and can even be useful for water insoluble proteins) (16,19-20). In Wageningen, as well as the MIT in the group of Hatton, the possibility of large scale reactors based on reverse micellar has been under investigation for the last few years.

The pharmaceutical industry has also taken an interest in the field. In one example I know, EGF (epidermial growth factor) has been solubilized in Span/Tween/squalene/water, and the system successfully tested in rabbits as an eye-drop formulation. The rationale is the following : water solutions are not very efficient as a vehicle for EGF, as a large part is lost from the eye during administration. An oily viscous solution would be better, but the protein EGF is not soluble in oil. The solubilization of EGF in a reverse micellar solution combines the water solubility with an oily viscous bulk medium.

It is possible that more examples of this kind will come. The problem here lies in the biocompatibility of the micellar systems. Clearly, AOT in gasoline is not the best pharmaceutical formulation. But there are several natural surfactants and plenty of biocompatible organic solvents.

Finally, among the new things appeared in the last couple of years, I like to mention the so-called "microemulsion gels" : under certain conditions, water-in- oil microemulsions and reverse micellar systems can be converted into gels (a family of organogels, in which mainly the main component is the organic solvent hosting the micellar system). Following the original observations of the formation of gelatine gels (21) and lecithin gels (22), several papers have already appeared on the topic (23-25) and there is a growing interest on the field for pharmaceutical application (transdermal transport in particular as well as for more technical applications (as membranes or films).

On the basis of this short presentation we can make a list of question/problems in biotechnology :

1) the use of enzymes in reverse micelles for the catalytic conversion of water-insoluble susbstrates.

2) the use of bacteria - and cell - containing reverse micellar systems as a way of extending microbiology to organic solvents.

3) the use of reverse micellar systems for protein extraction and purification

4) applications to pharmacology (controlled drug release and vehiculation) of both solutions and gels.

5) applications in technical chemistry (films, membranes, separation techniques).

The activity is high in all of these five sections, and the years to come will tell us whether we can talk about "success" in biotechnology. It is however important to remark, by comparing the two list of questions advanced here, that the field is indeed one, where the interplay between basic and applied science is fulfilled.

REFERENCES

1. P.L. Luisi, F. Henninger, M. Joppich, A. Dossena, G. Casnati, Biochem. Biophys. Res. Com. 74, 1384-1389 (1977).

2. K. Martinek, A.V. Levashov, N.L. Klyachko, L.V. Berezin, Dokl. Akad. Nauk. SSSR 236, 951 (1978).

3. K. Martinek, Eur. J. Biochem. 155, 453 (1986).

4. M. Waks, Proteins 1, 14 (1986).

5. P.L. Luisi, L. Magid, Crit. Rev. Biochem. 20, 409 (1986).

6. A review (Luisi, Giomini et al.).

7. P.L. Luisi, Angewandte Chemie 24, 439 (1985).

8. P.L. Luisi, C. Laane, Trends Biotech. 4, 153 (1986).

9. P.L. Luisi, B. Straub (eds.), Reverse Micelles, Plenum Press, New York (1984).

10. M. Maestro, P.L. Luisi, J. Phys. Chem., in press.

11. M. Maestro, P.L. Luisi, Symposium on Surfactants (K. Mittal, ed.), Plenum Press, New York (1988), in press.

12. G.G. Zampieri, H. Jäckle, P.L. Luisi, Phys. Chem. 90, 1849 (1986).

13. F.J. Bonner, R. Wolf, P.L. Luisi, J. Solid Phase Biochem., 5, 255 (1980).

14. D. Chatenay, W. Urbach, A.M. Cazabat, M.Vacher and M. Waks, in press.

15. A.V. Levashov, Y.L. Khmelnitsky, N.I. Klyachko, V.Y. Chernyak, K. Martinek, Colloid Interface Sci. 88, 444 (1982).

16. K. Van't Riet, Proceedings of the 3rd European Conference on Biotechnology, 541 (1984).

17. L. Magid, P. Walde, G. Zampieri, E. Battistel, Q. Peng, E. Trotta, M. Maestro, P.L. Luisi, Colloids and Surfaces, 30, 193-207 (1988).

18. G. Haering, P.L. Luisi, F. Meussdörffer, Biochem. Biophys. Res. Commun., 127, 911 (1985).

19. M.E. Leser, G. Wei, P.L. Luisi, M. Maestro, Biochem. Biophys. Res. Commun., 135, 629-635 (1986).

20. K.E. Göklen, T.A. Hatton, Biotech. Progr. 1, 69 (1985).

21. G. Haering, P.L. Luisi, F. Meussdörffer, Biochem. Biophys. Res. Commun. 127, 911 (1985).

22. R. Scartazzini, P.L. Luisi, J. Phys. Chem. 92, 829-833 (1988).

23. C. Gitler, M. Montal, FEBS Lett. 28, 329 (1972).

24. L. Magid, P. Walde, G. Zampieri, E. Battistel, Q. Peng, E. Trotta, M. Maestro, P.L. Luisi, Colloids and Surfaces, 30, 193-207 (1988).

25. D. Capitani, A.L. Segre, G. Haering, P.L. Luisi, J. Phys. Chem. 92, 3500-3504 (1988).

INTERPHASE TRANSFER FOR SELECTIVE SOLUBILIZATION OF IONS, AMINO ACIDS AND PROTEINS IN REVERSED MICELLES

E. B. LEODIDIS and T. A. HATTON

1. INTRODUCTION

Among the rich variety of structures observed in surfactant-cosurfactant-salt-oil-water systems (1-3), reversed micelles have attracted particular attention over the past twenty years. This is attested to by the growing number of survey papers on surfactant aggregation and micellization in nonpolar solvents (4,5), the application of various analytical methods in the study of reversed micelles (6-8), the study of solubilization, catalysis and enzymatic reactions in reversed micelles (8-12), and the application of reversed micelles for protein separations using liquid-liquid extraction (13,14). Many of these topics have been considered elsewhere in this book.

In this chapter, we focus on the important problem of solubilization in reversed micelles. The term "solubilization" was originally coined for aqueous micellar solutions but is easily generalized to include nonpolar solvents and reversed micelles. It refers to the increased solubility of substances normally insoluble or only slightly soluble in the continuous solvent phase, when surfactants at concentrations greater than the critical micelle concentration (CMC) are added to that phase.

Much of the initial work on solubilization in reversed micellar solutions concentrated on the understanding of the solubilization of water, and justly so (11,15,16), since surfactant-mediated water solubilization in nonpolar solvents is of primary importance in enhanced oil recovery (17). The "secondary" solubilization of electrolytes was also studied at the time (18), because of the important effects electrolytes can have on the maximum water solubilization in reversed micellar solutions, and of the salt effect on the phase behavior of the solutions obtained. More recently, it has been recognized that the extraction of metal ions from solution can in many cases proceed via solubilization in reversed micelles (19-22), and this may be the dominant mechanism in some liquid ion-exchange processes.

Hydrophilic enzymes can also be solubilized in reversed micellar solutions, without loss of enzymatic activity (7,12), with potential applications in the important areas of biocatalytic synthesis and protein separations. Consequently, this area has received considerable attention over the past decade. In parallel with biological macromolecules, many small polar organic molecules have been solubilized in reversed micellar solutions, and their properties have been examined using a variety of analytical methods (11,12).

Irrespective of the size or the specific properties of the solubilized (guest) molecule very little is known about the thermodynamics or the kinetics of the solubilization process. In their recent review, Luisi et al. (12) identified a number of important questions that must be addressed for our understanding of the solubilization process to be complete, among them being identification of the driving forces responsible for solubilization, the localization of the solubilizate in the reversed micelles, and the size or shape perturbations induced in the reversed micelles upon solute uptake.

There are two simple methods of solubilizing hydrophilic molecules in reversed micelles. In the injection method, employed in the majority of studies, an aqueous solution is injected into an organic solvent containing surfactant (and cosurfactant, if necessary). The association of the solute with the interface can be checked using techniques capable of yielding detailed microscopic information on the molecular level (e.g., NMR, ESR, fluorescence, hydrated electrons), and quantitative information on solute binding to the interface can be obtained from these results, limited of course by the assumptions inherent in their analysis (12). Details on the size perturbations induced by the solute can be obtained through scattering (23-28) and other (29-32) methods.

In the phase transfer method shown in Figure 1, an aqueous electrolyte solution is contacted with an organic solvent containing the oil-soluble surfactant. After the mixture is vigorously shaken and interphase equilibrium is attained, the phases are separated by centrifugation and analyzed for bulk solute concentrations. Similar phase equilibrium methods have been used in the past for the determination of the association behavior of various organic molecules (33), but with a different emphasis than that given here.

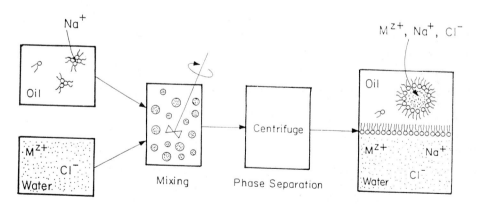

Fig.1. Schematic of the phase transfer process for solubilization in reversed micelles.

There are two distinct advantages of the phase transfer technique over the injection method. First, any separation method utilizing reversed micellar solutions must rely on interphase transfer, and the required solute partitioning between the phases is studied directly using this experimental approach. Second, material balance constraints for all species in the system provide a means by which significant information on the distribution and localization of solutes can be obtained. This information can be used to infer details of the intermolecular interactions between the solubilizate and the surfactant layer without the need for sophisticated spectroscopic methods.

Specific examples illustrating the use of the phase transfer method in analyzing selective solubilization of ions, amino acids and proteins in reversed micellar systems are given in the sections that follow. The solubilization of cations is discussed in section 2, and that of small polar organic molecules, with emphasis on amino acids, in section 3. Protein solubilization is covered in section 4, drawing on the insights gained from our earlier consideration of the solubilization of ions and small organic molecules. Most of the results presented here were obtained with the AOT/isooctane system, which is favored for its simplicity, although reference will occasionally be made to other surfactant/organic solvent systems. We do not draw any distinctions between reversed micellar solutions and water-in-oil microemulsions, preferring to use the term reversed micelles even for droplets with well developed water pools.

2. SOLUBILIZATION OF IONS IN REVERSED MICELLES

Electrolytes are important components in the two-phase systems under consideration. It has long been established that a high salt concentration is a prerequisite for the formation of a "Winsor Type II" system (34,35), defined as a water-in-oil (W/O) microemulsion in equilibrium with an excess aqueous phase. Thus a background electrolyte must be present when the solubilization of small organic molecules or biopolymers is investigated. An understanding of electrolyte behaviour, coupled with a knowledge of the ion distributions in the two-phase system, is desirable if the solubilization characteristics of these other, organic solutes are to be interpretted correctly.

The effects of electrolytes on reversed micelles have been examined from various perspectives in the past. Kunieda and Shinoda investigated the effects of temperature and sodium chloride concentration on the maximum water uptake by AOT reversed micellar phases before an excess phase is formed (36-38), while Gosh and Miller (39) focused on the influence of salinity on the phase behaviour of AOT/water/oil systems. Of more direct relevance to this paper is the work of Tosch et al.(40), Barthe et al.(41), Fletcher (42), and Leodidis and Hatton (21), all of whom investigated the partitioning of sodium salts between W/O microemulsions formed with anionic surfactants and an excess aqueous phase. In

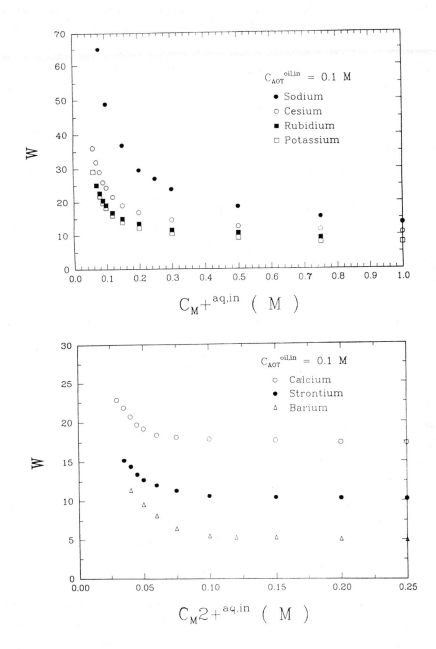

Fig. 2. Equilibrium water uptake by an AOT/isooctane solution as a function of initial

aqueous phase salt concentration for (a) Monovalent cations , and (b) Divalent cations.

addition, Leodidis and Hatton (21) examined the effect of cations other than the surfactant counterion on the two phase equilibrium. In all cases it was noted that water uptake by the W/O microemulsion increased dramatically at low external electrolyte concentration (a manifestation of the fact that inverted micelles are not thermodynamically stable under conditions of low external salt concentration), and dropped dramatically as the external salt concentration was increased, approaching an asymptotic value at very high external salt concentrations. In general, the anion was excluded from the water pools of the reversed micelles, its "partition coefficient" towards the micellar phase being substantially less than unity. This differs significantly from the injection method, where electrolyte is forced into the reversed micelles.

In Figure 2 the water uptake by a 0.1M AOT/isooctane solution in contact with an excess electrolyte solution is shown as a function of the concentration and type of the cation introduced in the excess aqueous phase (21). It is evident that the minimum cation concentration necessary for the stability of the two-phase system was different for each cation studied, as was the asymptotic value of w, the molar ratio of solubilized water to surfactant in the organic phase, obtained at high salt concentrations. It should be noted that Winsor II systems were not obtained with Li^+, Be^{2+} and Mg^{2+}, at least over the concentration ranges studied. Significant differences in water uptake were observed with different ions, even with those of equal charge and almost equal hydrated size (e.g., K-Cs, Ca-Sr). The nature of the anion did not influence water uptake significantly, although divalent anions seemed to lead to slightly higher water uptake values. The anions F^-, Cl^-, Br^-, I^-, SCN^-, SO_4^{2-}, CO_3^{2-}, PO_4^{3-} and CrO_4^{2-} were investigated.

Tosch et al. (40) predicted that the reversed micelle radius should increase almost linearly with water uptake, or w. This was verified by Aveyard et al. (43), who also showed that in the case of AOT most of the surfactant was situated at the micellar interfaces, and negligible amounts of AOT were solubilized in the bulk aqueous phase. Leodidis and Hatton (21) used these observations together with the correlation of Eicke (44) for AOT surface head area variations with w to calculate the radii of reversed micelles formed as a function of salt type and concentration. The results are shown in Figure 3.

Fletcher has reported that water uptake depends only on external salt concentration and not on AOT concentration; however, the salt used in his phase equilibrium experiments was sodium chloride (42). Leodidis and Hatton (21) found that when a different cation (e.g., potassium) was introduced externally the water uptake w was an almost linear function of AOT concentration at constant external salt concentration (Figure 4). This was related to the redistribution of cations between the reversed micellar water pools and the bulk aqueous phase as discussed below. The concentration ratio of sodium (from the AOT) to externally-introduced cation in the final equilibrium aqueous phase was found to depend only on the

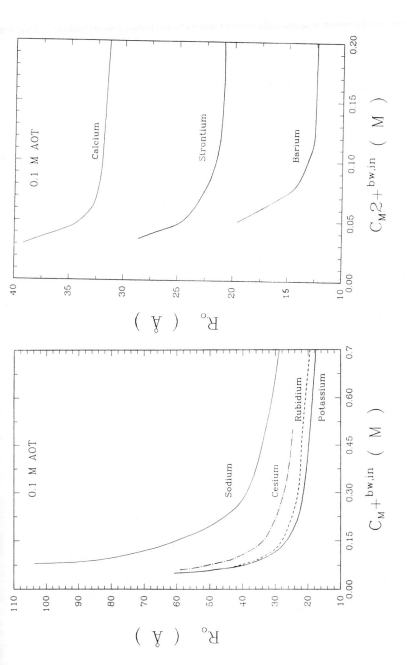

Fig. 3. Equilibrium size of reversed micelles formed as a function of initial aqueous phase salt concentration for (a) Monovalent cations , and (b) Divalent cations.

ratio of total moles of AOT (N^o_{AOT}) in the system to total moles of externally introduced electrolyte (N^o_{MClz}), irrespective of experimental procedure, suggesting that cation distribution was affected only by specific ion parameters such as charge, size and polarizability, and not by micellar size, surfactant concentration or phase volume ratio. Results for cation redistribution are plotted in Figure 5. For clarity, the experimental data for only one of each of the monovalent and divalent ions are shown on these figures; the results for the other ions are well represented by the continuous curves predicted by the theory of Leodidis and Hatton (21), as discussed later. It is evident that in all cases sodium was rejected by the organic phase, there being a significant selectivity of the micellar pools towards K^+, Rb^+ and Cs^+; this must be attributed mostly to the size difference between the larger hydrated Na^+ and the three other ions. Divalent cations were more strongly drawn towards the water pools because of the stronger electrostatic interaction with the anionic surfactant. There was a significant difference in selectivity between the three ions Ca^{2+}, Sr^{2+} and Ba^{2+}, which cannot be understood in terms of hydrated sizes since they are almost equal.

A number of significant theories have appeared in the last decade directed at the thermodynamics of reversed micellar solutions (45-51), many of them concentrating on Winsor II systems. The dependence of water uptake on electrolyte concentration in the

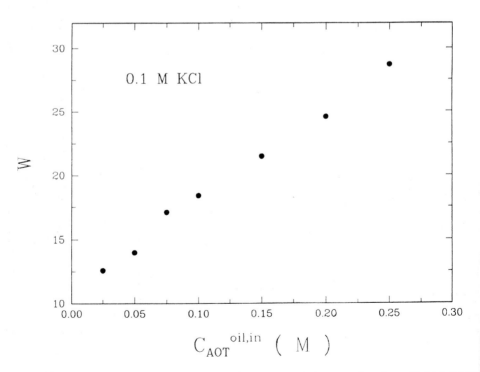

Fig. 4. Equilibrium water uptake by an AOT/isooctane solution as a function of initial AOT concentration.

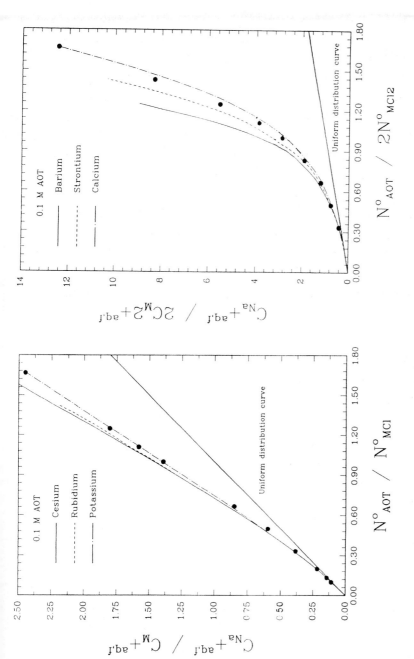

Fig. 5. Comparison of experimental and theoretical results for cation redistribution between

the two phases of the Winsor II system for (a) Monovalent cations , and (b) Divalent

cations.

external aqueous phase is now understood to be due mostly to changes in the electrostatic part of the free energy of the double layers formed inside the reversed micellar water pools. Modification of the electrostatic free energy is probably the reason for the observed differences in equilibrium reversed micellar radii obtained with different electrolytes.

Leodidis and Hatton (21) have recently developed a phenomenological model capable of predicting the experimentally observed selectivity of the water pools for the different cations in order to elucidate which properties determine cation behaviour in the AOT system. Ion distributions within the reversed micelle were calculated using the equation

$$C_i^{mwp}(r) = C_i^{bw} \left(\frac{1-f(r)^{mwp}}{1-f^{bw}} \right)^{\tau_i} \exp\left[-\frac{ze\psi(r)}{kT} - A_i \left(\frac{1}{\varepsilon^{bw}} - \frac{1}{\varepsilon(r)^{mwp}} \right) \right]$$

(1)

where the electrostatic potential ψ was obtained from the solution to Poisson's equation

$$\frac{1}{r^2}\frac{d}{dr}\left(r^2 \varepsilon(r)\frac{d\psi}{dr} \right) = -\frac{\rho(r)}{\varepsilon_o}$$

(2)

and $\varepsilon(r)$, the dielectric constant of the water at a radial position r within the water pool, was assumed to depend only on the local electric field $E(r) = -d\psi/dr$, according to an expression suggested by Booth (52). The local charge density $\rho(r)$ was determined by the local ion concentrations and charges in the reversed micelle.

Equation (1) is the heart of this model, and relates the concentration of any ion within the micellar water pools to its bulk phase value. The pre-exponential term is an entropic correction factor originally proposed by Ruckenstein and Schiby (53) to account for volume exclusion effects owing to finite hydrated ion sizes. Here f is the fractional solution volume occupied by the ions, the superscripts bw and mwp referring to the bulk water and the micellar water pool, respectively, while the parameter τ_i is the ratio of the hydrated volume of ion i to the volume of a water molecule. These τ_i values, which were assumed to scale as the hydrated volumes of the respective ions, were calculated using the hydrated radii tabulated by Conway (54) and an empirically-determined value of τ_{Na+} obtained by fitting the model to a single experimental point.

The first term in the exponent in Equation (1) is simply the electrostatic contribution to the electrochemical potential of the ions in solution, while the second term accounts explicitly for the effect of the local water polarization on the ion-solvent interactions (55). The constants A_i are the experimentally-determined values of the electrostatic free energy of hydration tabulated by Noyes (56).

Figure 5 shows that the model can predict the ion redistribution between the two phases for both monovalent and divalent cations. For clarity, the experimental data for only one of each of the monovalent and divalent ions are shown on these curves. Figure 6 shows the ion concentration profiles for a particular case with KCl as the external electrolyte. The maximum and the subsequent sharp decrease of concentrations close to the micellar "wall" are due not only to hydrated volume exclusion and dielectric exclusion effects, but also to the fact that the AOT heads were assumed to be distributed over a 5 -thick layer, creating an additional volume exclusion effect. Figure 6 shows that the anion is largely excluded from the water pools and that the cations concentrate close to the micellar wall. Cations with smaller hydrated radii and smaller electrostatic free energies of hydration show a higher affinity for the micellar interface. This is illustrated by the relative heights of the concentration peaks between sodium and potassium.

From this work it is suggested that ionic charge, hydrated size and electronic properties are all important in determining the selectivity of the water pools, and that larger ions with

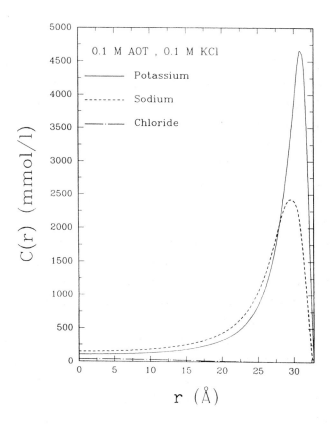

Fig. 6. Cation concentration profiles inside the water pool of a reversed micelle.

greater electrostatic free energies of hydration can be exchanged in favor of the smaller ions introduced with the excess aqueous phase. Moreover, the facility of ion exchange between the phases and the satisfactory explanation of the ion redistribution results by the present model imply that the concept of binding (57,58) of counterions to the surfactant headgroups is not necessary for interpreting system behaviour.

Numerical calculations indicated that for micelles of the same size the electrostatic potential inside the water pools increases in absolute value as the hydrated size and/or electrostatic free energy of hydration of the externally introduced cation increase. Consequently, the total electrostatic free energy for a given micellar water pool radius is higher for cations with larger $r_{i,h}$ or $\Delta G^h_{i,el}$, and these cations should therefore lead to the formation of larger reversed micelles having lower electrostatic free energies. Experimental results agree qualitatively with this simple argument (see Figures 2 and 3), although cesium does induce the formation of larger micelles than does potassium, which implies that the simple picture that we present is still not complete. Specific hydration effects may well be of importance. For the cations Li^+, Be^{2+}, Mg^{2+}, which do not form stable Winsor II systems, it appears that the electrostatic free energy for spherical micelles is so high that other geometries having lower free energy are favored.

In summary, it is evident that electrolytes are important system components in the phase transfer method. For anionic surfactants the use of cations with larger hydrated sizes or electrostatic free energies of hydration leads to the formation of larger reversed micelles in the equilibrium W/O microemulsion. Lower salt concentrations also favor the formation of larger reversed micelles, although a lower limit of salt concentration exists, beyond which a Winsor II system is no longer stable. Surfactant concentration can have a significant effect on the equilibrium radius of the micelles formed if a cation other than sodium is introduced in the aqueous phase, as sodium, which is normally displaced from the reversed micelles by practically all other ions, is rejected less from the reversed micelles because of its increasing abundance. The selectivity of various cations towards the water pools can be explained in terms of their hydrated sizes and electronic properties, suggesting that the feasibility of using W/O microemulsions in a heavy metal ion removal process can be assessed by looking at the size and electronic properties of the ions. The phenomenological model outlined here has been successful in predicting cation behaviour in Winsor II systems involving AOT as the surfactant.

3. SOLUBILIZATION OF SMALL ORGANIC MOLECULES IN REVERSED MICELLES

The solubilization of small guest molecules in reversed micelles is by no means as well-studied as in normal micellar systems (59), although it has attracted much attention over the past twenty years. Solubilization of specific probe molecules that interact with the reversed micellar interface can yield a wealth of information on the interface under consideration, eventually leading to a full characterization of the interface with respect to its flexibility, hydrogen-bonding capability, possibility for chemical interactions, hydration properties, etc. Similarly, solubilization of guest molecules with specific fluorescent properties, or of spin labelled or radioactive compounds, can give important information on the properties of water in reversed micelles, including the microviscosity, dielectric constant and mobility of water in the water pools. Information on the size of the reversed micelles can be obtained from scattering, fluorescence or NMR experiments, although these methods do require a number of assumptions regarding the localization of the solute probe and its interactions with the solvent environment.

Fluorescence techniques for the analysis of reversed micellar solutions have been reviewed recently by Verbeek and DeSchryrer (60), while other spectroscopic methods were also included in the discussion by Vos et al. (61). Kitahara (11) described the use of gas chromatography, vapor pressure osmometry and NMR in obtaining solubilization isotherms for various polar solutes in reversed micelles. Significant advances in the applications of NMR and ESR to reversed micelles have been made, much of the NMR work before 1986 being summarized in the article by Chachaty (62). Small-angle neutron scattering (24) and small-angle X-ray scattering (23) have been used recently to measure the perturbations induced in reversed micellar sizes by various solutes. A number of other analytical methods (e.g., pulse radiolysis, electron transfer, IR spectroscopy) have also been used in the past decade to provide valuable information on reversed micellar solubilization although all have been in conjunction with the the injection method. The recent review by Luisi et al. (12) provides a comprehensive overview of this topic.

The phase transfer method provides another approach for the determination of the level of interaction of the solubilizate with the surfactant headgroup layer, since partition coefficients or binding constants of the solutes with the reversed micellar interface can be determined directly, permitting the investigation of the effect of specific hydrophobic, electron donor or electron acceptor groups on interfacial interactions. The solubilization of amino acids by reversed micelles is particularly interesting for many reasons, not least of which is that there is a wide range of specific interactions with the interface that can be probed using different side chains of various natural and synthetic amino acids. Since these zwitterionic compounds have significant water solubilities, but negligible solubility in the organic solvent itself, there is no ambiguity as to their being located within the reversed micelles. Furthermore, amino acids are fairly simple probe molecules, unlike many of the solutes that have been used with fluorescence, ESR, etc. methods.

Fendler et al. (63) carried out a careful experimental study of partitioning of twelve amino acids in reversed micelles of dodecylammonium propionate in hexane. They concluded that electrostatic and "hydrophobic" interactions determine the solubility of amino acids in reversed micelles and that the free energy of transfer from bulk water to the micelles correlates with existing hydrophobicity scales (64,65). Boiceli et al. (66) solubilized various amino acids in phospholecithin inverted micelles, utilizing IR and NMR to study the effect of the solutes in the water pools. No decisive information was obtained apart from the observation that water appeared to be significantly perturbed by the solute presence. Rodgers and Lee (67) studied the fluorescence quenching of tryptophan by singlet molecular oxygen in reversed micelles and concluded that tryptophan may be largely located in the reversed micellar interface.

Luisi's group reported the first attempts to solubilize tryptophan in reversed micelles via the phase transfer method (68,69). They found that very little phase transfer was obtained when the net charge on the tryptophan molecule was either zero or of sign equal to that of the surfactant. Lee et al. (70,71) investigated the solubilization of many amino acids as a function of pH in the AOT/isooctane system, showing that a favourable electrostatic interaction leads to increased solubilization, that a salting-out of amino acids occurs at high salt concentrations and that at low pHs the solubilization seems to be limited by the packing of the amino acids in the micellar core. The slope of the solubility curve as a function of net solute charge was found to correlate well with existing hydrophobicity scales, implying that the hydrophobicity of the side chain could play an important role in amino acid solubilization. A complicating factor associated with these experiments was that because of the pH variation it was impossible to decouple the electrostatic and hydrophobic effects, and therefore to obtain information about the association of the amino acids with the reversed micellar interface.

A more detailed analysis of amino acid solubilization in AOT reversed micelles has been undertaken by Leodidis and Hatton (72) who have shown that the phase transfer experiment provides two easy approaches for discriminating between amino acids that associate strongly with the reversed micellar interface, and those that do not. They exploited the fact that when a dilute electrolyte solution of solute i is contacted with a reversed micellar organic phase, the partitioning of the solute to this phase will depend on the strength of the interactions with the interface, and on the hydrophobic effect on the amino

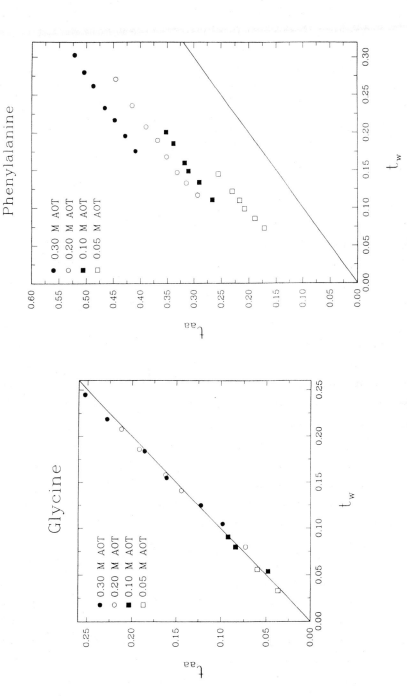

Fig. 7. Fractional amino acid transfer to the reversed micellar phase as a function of the fractional water transfer for (a) Glycine , and (b) Phenylalanine.

acid side chain. The final equilibrium bulk aqueous phase concentration can be used as shown below to determine the interfacial partition coefficient.

We use the mole fraction-based partition coefficient between the micellar interface and the water pools

$$K_{x,i}^{int/w} = \frac{x_i^{int}}{x_i^w} = \frac{n_i^{int}/(n_i^{int} + n_{AOT})}{C_i^w/55.5} \simeq \frac{n_i^{int} 55.5}{n_{AOT} C_i^w} \tag{3}$$

where we have assumed that $n_{int}^i \ll n_{AOT}$. This definition has been useful in the past in quantifying the transfer of small organic solutes from water to normal micelles and phospholipid vesicles and liposomes (73-77). If the solute is not charged and the reversed micelles are sufficiently large ($w \geq 20$ for AOT), so that the water pools are well developed and their properties are close to those of bulk water, then it is reasonable to assume that the average concentration of solute in the water pools themselves is equal to its concentration in the excess aqueous phase. Then, together with the overall material balance constraint on the solubilizate we obtain

$$K_{x,i}^{int/w} = \frac{55.5}{n_{AOT}} \left[V^w \left[\frac{C_i^{w,in} - C_i^{w,f}}{C_i^{w,f}} \right] - K_i^{o/w} V^{oil} \right] \tag{4}$$

which shows that a simple measurement of the final concentration of solute i in the excess aqueous phase can be used to determine the association constant of the solute with the interface. Additional assumptions inherent in this development are that the presence of the reversed micelles does not affect the solute solubility in the bulk oil phase, that the surfactant is all located at the micellar interfaces, and that there is no dimerization or other association of the solute in the organic phase.

A plot of fractional amino acid transfer (t_{aa}) vs fractional water transfer (t_w) to the reversed micellar phase for an amino acid not associating with the micellar interface ($K_x^{int/w} \ll 1$) should result in a single curve, independent of surfactant concentration, with a slope of unity, provided the assumption of equal solute concentrations in the micellar water pools and the excess aqueous phase is valid. This is indeed the case, as seen from Figure 7(a) for glycine. Figure 7(b) shows a similar plot for phenylalanine, an amino acid that associates strongly with the interface. In contrast to the glycine case, the phenylalanine curves depend on AOT concentration, and the separation between the curves is an indication of the strength of the association of the amino acid with the interface, i.e., of $K_x^{int/w}$. It has also been noted that the interfacial partition coefficient $K_x^{int/w}$ is independent of amino acid concentration for all amino acids studied.

Another qualitative indication of the strength of interfacial association is given in Figure 8, where water uptake t_w is plotted vs amino acid concentration in the initial aqueous phases, at a specific surfactant and salt concentration. For glycine, which does not associate with the interface, t_w is independent of amino acid concentration, while for the interfacially-active phenylalanine and leucine t_w is a strong function of the amino acid concentration. These latter amino acids act as cosurfactants, and larger slopes of t_w vs $c_a^{w,in}$ curves imply stronger cosurfactant effects.

Fletcher (42) found that salt concentration significantly affected the value of $K_x^{int/w}$ for the solutes with which he worked, but could not determine whether this was due to salting out from the aqueous phase at high salt concentrations or due to the increased curvature of the reversed micellar interface at the lower w values obtained with higher salt concentrations. We investigated this effect with para-nitro-phenylalanine and tryptophan. Figure 9(a) confirms that $K_x^{int/w}$ decreased with increasing salt concentration. This cannot be attributed to tryptophan salting out, as tryptophan's solubility in water is quite high under these conditions. A plot of ln $K_x^{int/w}$ vs. w^{-1} shows a very high degree of linearity for both tryptophan and para-nitro-phenylalanine (Figure 9(b)). w^{-1} is proportional to the micellar interface curvature, which fact strongly suggests that the salt effect on $K_x^{int/w}$ is an indirect one. The surfactant interfacial density close to the micelle-water interface becomes higher

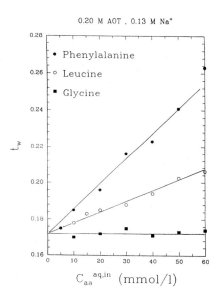

Fig. 8. Illustration of the amino acid cosurfactant effect. Water uptake by the reversed micelar solution as a function of amino acid concentration in the initial aqueous phase.

286

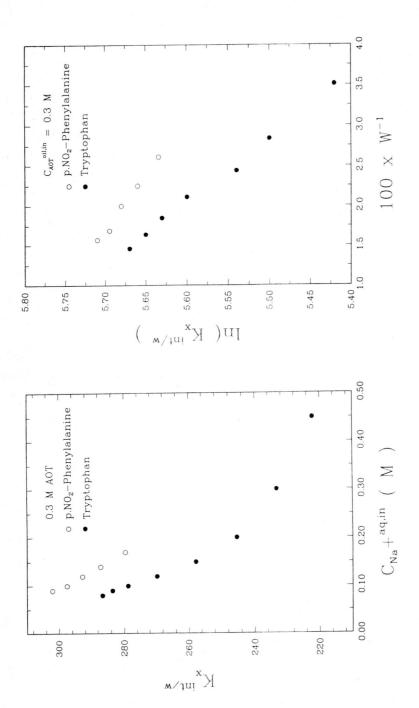

Fig. 9. (a) Interfacial partition coefficients of two amino acids as a function of external sodium concentration. (b) Correlation of the logarithm of the partition coefficient with the interfacial curvature of the reversed micelles.

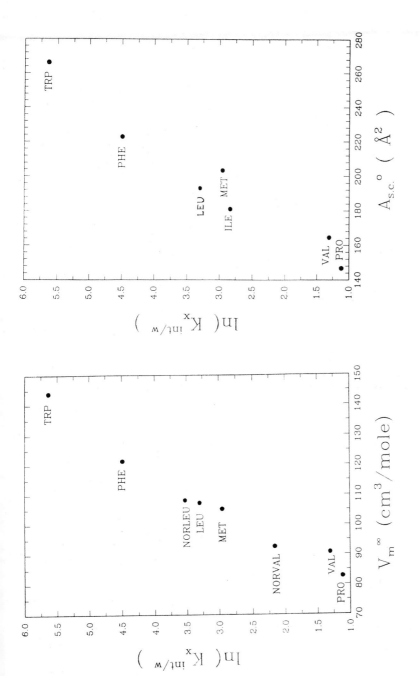

Fig. 10. Correlations of the logarithms of the interfacial partition coefficients of various amino acids with (a) their partial molar volume at infinite dilution , and (b) their "standard" side-chain area, as given by Rose et al.(78).

at higher curvatures. In consequence, adsorbed molecules are squeezed out. For this reason, standardization in the reporting of the results is necessary, and thus all the partition coefficients reported in Figure 10 were obtained at a constant excess phase sodium ion concentration of 0.08 M.

Insight into the thermodynamics of amino acid solubilization in the AOT interface can be obtained by correlating the partition coefficients with a number of meaningful parameters. The best correlations have been obtained with the partial molar volume at infinite dilution (Figure 10(a)) and the standard side-chain area (Figure 10(b)) reported by Rose et al. (78). These parameters are proportional to the van der Waals volume and surface area, respectively, of the molecules (79). This interesting observation is in accord with the literature findings that solubilities of organic compounds in water and partition coefficients of hydrophobic compounds in the water/octanol system correlate well with van der Waals volume or surface area (79-81).

The proportionality of ln $K_{i,x}^{int/w}$ to the van der Waals volume or surface area implies that the main driving force for interfacial association is the removal of the hydrophobic side chains from water, and not some specific interaction with the interface. This is further illustrated by the fact that the presence of polar, hydrogen-bonding groups in the side chains significantly reduces partition coefficients. Thus, serine, cysteine, threonine, glutamine and asparagine do not associate with the interface, while the association constants of tyrosine and para-amino-phenylalanine are much lower than that of phenylalanine (72).

4. SOLUBILIZATION OF PROTEINS IN REVERSED MICELLES

The liquid-liquid extraction of proteins from an aqueous solution to a reversed micellar organic phase is by now a well-established process and needs little introduction. The past decade has seen rapid advances made in going from the initial exploratory experiments (69,82) to small pilot-scale studies (83-85), with the promise of industrial applications in the offing. There are, however, many fundamental questions that need to be addressed before a complete understanding of the protein solubilization process has been obtained. The nature of the protein/reversed micelle complex, in particular protein localization and the size or shape perturbations induced in the reversed micelle by the insertion of the protein, is still poorly understood. The driving forces for solubilization, and how they are influenced by salt and surfactant type and concentration, and by solvent selection and temperature, are not yet totally clear. From a technological standpoint, improvements in extraction selectivity must follow an enhanced understanding of the protein interactions with the reversed micelles.

Most of the characterization of proteins in reversed micelles has been undertaken using micellar solutions obtained via the injection method, and has been reviewed recently by Luisi and coworkers (7,12). It is not advisable to assume that these results carry over directly to the phase transfer situation because of the important differences between the two solution preparation methods. As noted earlier, the electrolyte distributions in the water pools are vastly different in the two cases. In the injection method coions are forced into the water pools together with a desired amount of water, always with the proviso that a transparent reversed micellar solution be obtained. In the phase transfer case, however, coions are excluded from the reversed micelles, and the water uptake is determined solely by thermodynamic considerations, as dictated by salt type and concentration. In these cases, a significant redistribution of counterions between the two phases can occur if a cation other than Na^+ (for the AOT system) is introduced (21). It is therefore evident that the actual size of the protein-filled and the protein-empty reversed micelles, and the trends that the micellar sizes follow when important system parameters are changed, may be quite different in the two cases because of the differences in their thermodynamic descriptions.

A number of important papers have discussed the use of ultracentrifugation (29-31), small-angle neutron scattering (27), quasi-elastic light scattering (28), fluorescence recovery after fringe photobleaching (32), and small-angle X-ray scattering (23) for the determination of sizes of empty and filled reversed micelles. In all these studies, however, the reversed micellar solutions were prepared with the injection method, and the results cannot be extrapolated to the phase transfer situation because of the fundamental differences between the two experiments. The effect of salt type and surfactant concentration should certainly be different in the two cases, and it may well be that the effects of protein type and concentration and salt concentration are also different.

Sheu et al. (25) used SANS to determine filled and empty reversed micellar radii obtained with cytochrome-c using the phase transfer method, reporting that protein-filled micelles were always larger than the empty micelles, irrespective of salt concentration, and that protein concentration in the excess aqueous phase did not affect the micellar sizes to a significant extent. The shell-and-core model of Bonner et al. (30), in which the protein was assumed to be located concentrically within the micelle water pool, was used in the analysis of the SANS spectra. There is now significant evidence, however, that cytochrome-c associates with the AOT micellar interface (23,86,87). This has two important ramifications with regard to the Sheu et al. analysis: (i) the shell-and-core model was not an appropriate description of the system, and (ii) the implicit assumption of conservation of micelle core surface area, as determined by the total surfactant head area, was incorrect because the protein also occupied portions of the interface. Another complicating factor in their study was that KCl was used as the accompanying electrolyte in the extraction process, which introduced the complexity of significant K^+/Na^+ redistribution between the two phases (21), with a consequent modification of the micellar size distributions.

The recent SANS study of Rahaman and Hatton (26) was designed to circumvent the limitations inherent in the Sheu et al. analysis. They used a sodium phosphate buffer as the sole accompanying electrolyte in their study of the solubilization of the hydrophilic protein α-chymotrypsin, which, according to popular opinion (23,86) does not associate with the reversed micellar interface. In this case the results, presented in Figure 11, were less ambiguous. It was found that the protein-filled micelles were smaller than the empty micelles at low salt concentrations but eventually were larger than the empty micelles at high salt concentrations. Furthermore, for low pH the radius of the filled micelles remained almost constant over a wide range of salt concentrations, while for the pH value above the pI of the protein, i.e., for an unfavourably charged protein, the behaviour of the filled micelle size tracked that of the empty micelles closely. It was also found that the protein concentration had essentially no effect on the micelle sizes, in contrast to the theoretical prediction of Caselli et al. (88) for the injection case.

The trends observed experimentally were duplicated in the theoretical calculations of Rahaman and Hatton (26), which showed that the electrostatic free energy of the filled and empty micelles is the primary determinant of protein solubilization behaviour. These results emphasize the important role that electrostatics play in the solubilization process.

Effect of pH, surfactant concentration and salts

The effect of pH on protein solubilization in AOT/reversed micelles is well known and qualitatively and semi-quantitatively (89-91) understood today. The pH determines the net charge and conformational stability of the protein. High positive charge on the protein favors

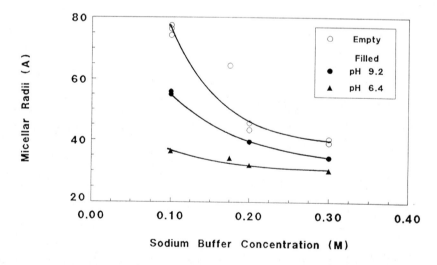

Fig. 11. Radii of protein-filled and empty micelles as functions of external sodium buffer concentration and pH.

increased solubility in an anionic reversed micellar system, although it is recognized today that significant solubilization can be observed even when the protein is negatively charged at low salt concentration.

Surfactant concentration can also have an effect on the solubilization process. With sodium salts in the aqueous phase at a constant concentration, increasing the surfactant concentration yields an increase in the number of micelles of a constant size. Consequently, there is an increase in protein solubilization if the pH and micellar size/protein size ratio permit (92), or no surfactant concentration effect if the micellar size is too small for proteins to be solubilized. If the counterion introduced in the excess aqueous phase is other than sodium, then increasing AOT concentration increases the amount of sodium ions in the system, leading to modified micellar sizes, or w (Figure 4). This can have a profound impact on protein solubilization.

Salt concentration affects the solubilization process in that it determines the equilibrium size of the micelles. Protein salting-out from reversed micelles at increased ionic strength is by now a well-known experimental fact (83,91,93,94). A theoretical explanation of the effect of salt concentration on protein solubilization in reversed micelles has been presented by Bratko et al. (89). This model, based on electrostatic considerations, cannot explain the specific salt effects on protein solubilization observed by Kelley et al. (94,95) and shown in Figure 12. For a given salt concentration, the amount of protein solubilized is greater for

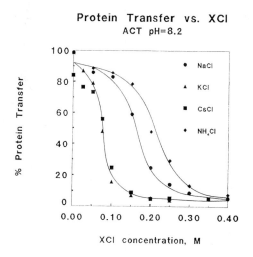

Fig. 12. Effect of salt concentration and cation type on α-chymotrypsin transfer to an AOT/isooctane reversed micellar solution at pH = 8.2. The ordinate is salt concentration over and above a background concentration of 0.2 M Sodium Phosphate buffer.

sodium than for cesium or potassium, results which can be rationalised qualitatively using the arguments put forward below.

The equilibrium size of empty reversed micelles is a significant factor in determining protein solubilization. If the empty micelle size is smaller than the size of the protein, then a significant surfactant redistribution is required for the formation of a micelle large enough to host a protein molecule under the prevailing conditions. This creates an unfavourable entropic effect which increases at smaller equilibrium empty micellar radii. A gradual exclusion of the proteins from the micelles is observed, which becomes a total exclusion at low enough w values. The pH is, of course, the second important factor in the process since it determines protein charge. Increased charge implies increased electrostatic interaction with the micellar interface and hence increased ability for counterbalancing the unfavourable entropic effect created by surfactant redistribution. This means that at higher protein charge we can tolerate progressively lower w values and still achieve significant protein solubilization.

These points are illustrated in Figure 13, where it is evident that higher and higher sodium concentrations were required before the unfavourable entropic effect began to exclude the proteins from the micelles as the pH was lowered (i.e. at higher positive protein charge).

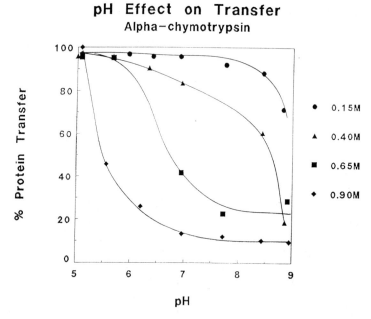

Fig. 13. Illustration of the balance between electrostatic and entropic free energy contributions to protein solubilization in reversed micelles. α-chymotrypsin transfer vs. pH, with Sodium Chloride molar concentration as the parameter.

As noted before (Figure 12), protein partitioning at constant pH and salt strength depended on the salt type. Potassium and cesium induced the formation of smaller empty micelles than were obtained with sodium (Figure 5), and hence protein exclusion began at lower salt concentrations for these ions. Figure 14 shows that if protein transfer is plotted as a function of w the spread in the solubilization curves is reduced significantly, although there are still specific salt effects in evidence at the low pH values. Under these conditions, the water shell surrounding the protein consists of only a few layers of water, and there will be a significant interaction of the hydrated cations in the micelle electrical double layer with the protein. Since the volume exclusion effect is more pronounced with sodium than with potassium it stands to reason that the water shell surrounding the protein be thinner with potassium and that for a given level of protein extraction the water uptake with potassium will be smaller than with sodium. On the other hand, as the protein net charge decreases, i.e., as the pH is increased, the data for all salts at a given pH collapse onto a single curve. The reason for this is thought to be that as the charge interaction becomes less favourable, the protein-filled micelle becomes larger, and the volume exclusion of the hydrated cations in the increased water shell around the protein is not affected by the presence of the protein. Under these circumstances, the solubilization is not driven as much by electrostatics as by the entropy of mixing of the protein in the reversed micellar phase, and the protein concentration based on the solubilized water volume is similar to that of the bulk aqueous solution.

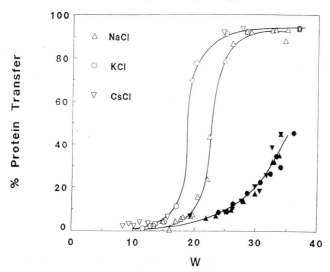

Fig. 14. Fractional protein transfer as a function of water uptake w with sodium, cesium, and potassium chloride. Data obtained at pH = 6.4 (empty symbols), and pH = 9.3 (filled symbols).

Finally, Figure 15 shows the effect of AOT concentration when KCl is used as the accompanying electrolyte. The results can again be rationalized in terms of increased micellar sizes. As the ratio of sodium to potassium ions in the system increases, there is less displacement of the sodium from the reversed micelles by the potassium. The increased volume exclusion effects associated with the larger hydrated sodium ion result in larger reversed micelles, with an accompanying greater capacity for solubilizing proteins. This effect is, of course, in addition to the enhanced capacity expected solely on the basis of mass action kinetics with the increased micellar concentration.

The organic solvent used in the reversed micellar phase can have a significant effect on the structure of the surfactant aggregates that form, particularly with respect to the aggregation numbers for the reversed micelles. There have been little in the way of systematic studies on solvent effects in protein extraction operations, but it is known that proteins can be conveniently recovered from the reversed micellar phase by the addition of more polar, water-immiscible solvents such as ethyl acetate which tend to disrupt the micellar structure (90,96). Temperature effects, too, have received little attention to date, and this is an area that deserves to be emphasized because of the important thermodynamic information that can be gleaned from this type of study.

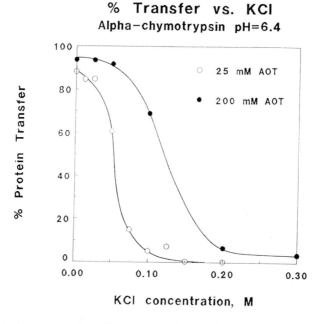

Fig.15. Surfactant concentration effect on protein transfer to a reversed micellar solution with Potassium chloride as an external electrolyte. Illustration of the K^+-Na^+ ratio effect. The ordinate is KCl concentration over and above a background of 0.2 M Sodium Phosphate.

Affinity Partitioning: A New Concept

Protein solubilization in reversed micellar solutions has been proposed as a novel separation method for the separation and purification of proteins from fermentation broths. There are many parameters that can be manipulated to improve the selectivity of the extraction for a targetted protein. It has been demonstrated, for instance, that the separation of a synthetic mixture of proteins (14,90,91) and the selective extraction of an extracellular alkaline protease from an unclarified fermentation broth (85) can be achieved using reversed micelles by manipulating the aquous phase pH and ionic strength.

Recently, Woll et al. (96) showed that it is also possible to exploit the biospecific nature of proteins in their interactions with selected ligands to improve the selectivity of the extraction process (13,96). Octyl-beta-D-glucopyranoside was used as a bioaffinity surfactant to enhance the extraction of concanavalin-A, a carbohydrate-binding protein, into an AOT reversed micellar solution without affecting the simultaneous extraction of ribonuclease-a introduced as a control. This was the first demonstration that additional specificity can be incorporated in protein extraction operations by using a second surfactant species having molecular recognition capabilities. Figure 16 shows the effect of the concentration of the affinity ligand (2 - 10% of the total surfactant in the reversed micellar phase) on the partitioning of concanavalin-A.

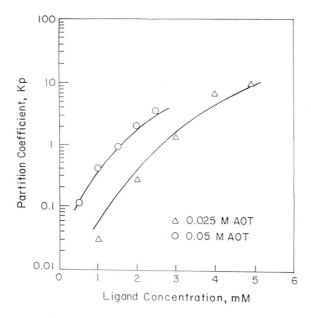

Fig. 16. Concanavalin-A partition coefficient between a reversed micellar and an equilibrium electrolyte solution. Effect of ligand concentration in the organic phase.

296

Correlational Methods and Thermodynamic Theories

The first attempts to model protein solubilization in reversed micelles via the phase transfer method have appeared only recently. A phenomenological model based on simple thermodynamic arguments was developed by Woll and Hatton (96) to correlate protein partitioning data. They showed that close to the protein isoelectric point the partition coefficient should depend on pH and surfactant concentration according to the relation

$$\ln K = A + B pH + (C + D pH) \ln [S] \tag{5}$$

where K is the partition coefficient, [S] the surfactant concentration, and A, B, C and D are constants related to the aggregation number of the micelles, the potential difference between the phases, the slope of the protein titration curve and the isoelectric point of the protein. In Figure 17 experimental solubilization data for the proteins concanavalin-A and ribonuclease-a are shown together with the fitted model curves. While it could be argued that the model contains too many adjustable parameters, it appears to be capable of capturing the important features of the solubilization phenomenon, at least with respect to pH and surfactant concentration variations, and should be considered a useful correlational method for application purposes.

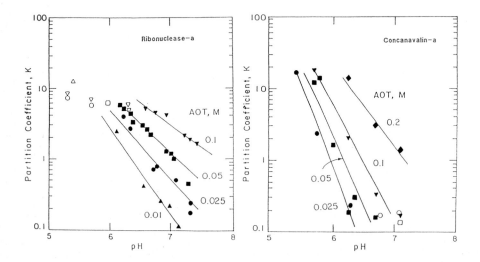

Fig. 17. Illustration of successful fits obtained with equation (5). The cases of (a) Ribonuclease-a, and (b) Concanavalin-a are presented.

More fundamental modelling approaches have also appeared recently, using as the starting point the Bonner et al. shell-and-core model (30). The protein and the micelle are viewed as concentric spheres with charges evenly smeared over their respective surfaces, and the electrostatic interactions between these surfaces are treated in terms of suitable approximations to the distribution of ions in the aqueous shell. Discrimination between proteins is on the basis of size and net charge.

Caselli et al. (88) treated the protein-reversed micelle complex as a series of concentric electric capacitors, developing an expression for the free energy of the reversed micellar phase which was to be minimized to determine radii of empty and filled micelles. Bratko et al. (89) used the cell model to treat the protein in both the excess aqueous phase and the reversed micelle. The excess free energy of the protein-reversed micelle complex was assumed to be of purely electrostatic origin, and was computed according to the cell-model approximation by Marcus (97):

$$A_e = \frac{1}{2} \int_{V_k} \rho \psi dV + kT \sum \int_{V_k} (n_i \ln n_i - \bar{n}_i \ln \bar{n}_i) dV \qquad (6)$$

where ρ is charge density in the micelle, ψ is the electrostatic potential, n_i is the local concentration of ionic species i, and $\bar{n}_i$ its average concentration in the excess aqueous phase. These authors assumed that protein-filled and empty micelles have the same size, independent of salt concentration, and allowed for incomplete surfactant dissociation. The theoretical predictions on the effects of ionic strength on protein transfer were in qualitative agreement with the experimental results of Göklen and Hatton (91). It was claimed that the negative free energies of complexation are due mainly to the increase in translational entropy of the positively charged counterions released from the micelle upon the accommodation of a protein molecule. This interpretation appears to break down when protein solubilization in AOT micelles above the protein's pI is to be considered, however. Moreover, it does not take into consideration the additional screening of interactions between surfactant headgroups on opposing sides of the micelle on insertion of the dielectric in the centre of the micelle.

Rahaman and Hatton (26) have independently developed a similar model for protein solubilization in reversed micellar systems, allowing also for variations in both filled and empty micelle sizes. This work recognizes the importance of the reversed micellar interface, in particular the variation with surface curvature of steric interactions between the surfactant tail segments, and the distribution of surfactant molecules between filled and empty micelles. The electrostatic free energies inside the micelles were calculated using classical charging process arguments, thereby obviating the need for the independent evaluation of the entropy of ions in the system, i.e.,

$$A_e = \int_0^{\sigma_m} \psi_m \, d\sigma + \int_0^{\sigma_p} \psi_p \, d\sigma \tag{7}$$

The contribution to the system free energy due to the mixing of surfactant aggregates in the organic solvent phase was accounted for by an ideal entropy of mixing term. The model was able to describe the effects of protein charge and concentration, solution ionic strength and surfactant concentration on the respective sizes of the filled and empty micelles, as discussed above in the context of SANS measurements of these quantities. While these theoretical modelling approaches have yielded some insight into the importance of various factors in determining the protein solubilization characteristics, there is still much room for improvement. For instance, the important effect that salt type has on protein extraction can only be understood in terms of a more refined electrostatic model such as that proposed by Leodidis and Hatton (21). Water in reversed micelles should also be treated in a more refined way, especially at low overall w or when the distance between the protein surface and the micellar surface in a filled micelle is equal to a few molecular diameters of water. The excess free energy of the protein-reversed micelle complex should include additional contributions from interactions that have hitherto been ignored. Charge fluctuation on the protein surface (98), Van der Waals forces (99) and hydrophobic interactions could be important. The finding that hydrophobic amino acids, like tryptophan, tyrosine, phenylalanine, methionine and leucine strongly associate with the AOT interface (72) can lead to the identification of possible sites of interaction of a protein with the interface.

These sites must either contain a large number of positively charged groups (which can also be investigated with the help of a "Hammer" map (100)), or must contain a large number of hydrophobic amino acids close to each other. Finally, it is questionable whether Luisi's shell-and-core model (30) is a valid representation of the structure of many protein-reversed micellar complexes, particularly for membrane or membrane-associated proteins and for hydrophobic proteins. The picture of two spheres touching or else intersecting one another seems to be a more realistic depiction for such cases; studies are currently under way to investigate these structures (101). Furthermore, it is doubtful whether the smeared charge model for the protein surface is of any use in these geometries, and allowance must be made for a non-homogeneous distribution of charges over the protein surface. The approaches proposed by Kirkwood and Tanford (102) and refined by Matthew et al. (103,104), or the more elaborate model of Russell and Warshel (105,106), could be invoked here. Even for hydrophilic proteins the shell-and-core model can lead to false conclusions, especially at low water contents.

There is no doubt that the concentric sphere model is a useful idealization and first approximation, but it will eventually have to be replaced by a more elaborate picture, one which recognizes the dramatic differences encountered from protein to protein, differences which defy any attempt to treat all proteins in the same generalized fashion.

5. OVERVIEW OF INFORMATION OBTAINED WITH THE PHASE-TRANSFER METHOD

The utility of the phase transfer method in elucidating the problem of solubilization of ions, small guest molecules and proteins in reversed micellar solutions has been reviewed. In the case of ions, significant information can be obtained with regard to the ionic properties that determine solubilization. The basis for understanding salt effects in protein transfer has also been developed. In the case of small guest molecules we have presented some very recent results which indicate that the phase-transfer method will be of great value in the investigation of solute association with the reversed micellar interface. Finally, in the protein extraction case we have presented a number of new results which, we believe, elucidate the importance of various parameters affecting the process. We have also discussed the importance of the formulation of more refined theoretical models.

We believe that the possibilities for obtaining information from the phase transfer method have not yet been exhausted. Careful new experiments focusing on particular factors could lead to an even better understanding of the solubilization properties of reversed micelles. The temperature dependence of the interfacial partition coefficients of amino acids will prove whether interfacial solubilization is a predominantly enthalpic or an entropic phenomenon. Temperature studies of protein partitioning in reversed micellar solutions will also be of great value. Additional size information on protein-filled and empty micelles under a variety of different conditions will provide a sound basis for better understanding of the complex phenomenon of protein solubilization. Use of different solvent/surfactant systems can greatly improve liquid-liquid extraction with reversed micellar solutions.

REFERENCES

1 K. Shinoda and B. Lindman, Langmuir, 3 (1987) 135.
2 P. Ekwall, L. Mandell and K. Fontell, J. Colloid Int. Sci., 33 (1970) 215.
3 B. Tamamushi and N. Watanabe, Colloid Polym. Sci., 258 (1980) 174.
4 A. S. Kertes and H. Gutmann, in: E. Matijevic (Ed), Surface and Colloid Science, 8, Wiley, New York, 1976, p. 193.
5 H-F. Eicke, Topics Cur. Chem., 87 (1980) 85.
6 H-F. Eicke, in: H-F. Eicke and G. D. Parfitt (Eds), Interfacial Phenomena in Apolar Media, M. Dekker, New York, 1987.

300

7 P. L. Luisi and L. J. Magid, CRC Crit. Rev. Biochem., 20 (1986) 409.
8 P. L. Luisi, in: B. E. Straub (Ed), Reverse Micelles, Plenum, New York, 1984.
9 K. Martinek, A. V. Levashov, N. Klyachko, Y. L. Khmelnitski, I. V. Berezin
   Eur. J. Biochem., 155 (1986) 453.
10 K. Martinek, I. V. Berezin, Y. L. Khmelnitski, N. Klyachko and A. V. Levashov,
    Collection Czech. Chem. Commun., 52 (1987) 2589.
11 A. Kitahara, Adv. Colloid Int. Sci, 12 (1980) 109.
12 P. L. Luisi, M. Giomini, M. P. Pileni and B. H. Robinson, Biochim. Biophys. Acta, 947
    (1988) 209.
13 J. M. Woll, A. S. Dillon, R. S. Rahaman and T. A. Hatton, in: R. Burgess (Ed), Protein
    Purification: Micro to Macro, Alan R. Liss, Inc., New York, 1987, pp. 117.
14 T. A. Hatton, in: J. F. Scamehorn and J. H. Harwell (Eds), Surfactant-Based Separations,
    M. Dekker, New York, 1989.
15 K. Kon-No and A. Kitahara, J. Colloid Int. Sci, 37 (1971) 469.
16 S. G. Frank and G. Zografi, J. Colloid Int. Sci, 29 (1969) 27.
17 D. O. Shah, (Ed), Surface Phenomena in Enhanced Oil Recovery, Plenum, New York,
    1981.
18 K. Kon-No and A. Kitahara, J. Colloid Int. Sci., 41 (1972) 47.
19 K. Osseo-Assare and M. E. Keeney, Sep. Sci. Tech., 15 (1980) 999.
20 F. J. Ovejero-Escudero, H. Angelino and G. Casamatta, J. Disp. Sci. Tech., 8 (1987) 89.
21 E. B. Leodidis and T. A. Hatton, Submitted for publication.
22 E. Gulari and C. Vijayalakshmi, Paper presented at the ACS Symposium, Los Angeles,
    1988.
23 M-P. Pileni, T. Zemb and C. Petit, Chem. Phys. Let., 118 (1985) 414.
24 A. M. Howe, C. Toprakcioglu, J. C. Dore and B. H. Robinson, J. Chem. Soc. Faraday
    Trans. 1, 82 (1986) 2411.
25 E. Sheu, K. E. Goklen, T. A. Hatton and S-H. Chen, Biotech. Progress, 2 (1986) 175.
26 R. S. Rahaman and T. A. Hatton, to be published.
27 P. D. I. Fletcher, A. M. Howe, N. M. Perrins, B. H. Robinson, C. Toprakcioglul and J.
    C. Dore, in: K. L. Mittal and B. Lindmann (Eds), Surfactants in Solution, 3, Plenum, New
    York, 1984.
28 D. Chattenay, W. Urbach, A. M. Cazabat, M. Vacher and M. Waks, Biophys. J., 48
    (1985) 893.
29 A. V. Levashov, Y. L. Kmelnitsky, N. L. Klyachko, V. Y. Chernyak and K. Martinek, J.
    Colloid Int. Sci., 88 (1982) 444.
30 F. J. Bonner, R. Wolf and P. L. Luisi, J. Solid-Phase Biochem., 5 (1980) 255.
31 G. G. Zampieri, H. Jackle and P. L. Luisi, J. Phys. Chem. 90 (1986), 1849.
32 D. Chattenay, W. Urbach, C. Nicot, M. Vacher and M. Waks, J. Phys. Chem. 91 (1982)
    2198.
33 K. A. Connors, Binding Constants: The Measurement of Molecular Complex Stability,
    Wiley, New York, 1987.
34 P. A. Winsor, Solvent Properties of Amphiphilic Compounds, Butterworths, London,
    1954.
35 M. L. Robbins and J. Bock, J. Colloid Int. Sci., 124 (1988) 462.
36 H. Kunieda and K. Shinoda, Colloid Int. Sci., 70 (1979) 577.
37 H. Kunieda and K. Shinoda, Colloid Int. Sci., 72 (1980) 601.
38 H. Kunieda and K. Shinoda, Colloid Int. Sci., 118 (1987) 586.
39 O. Ghosh and C. A. Miller, J. Phys. Chem. 91 (1987) 4528.
40 W. C. Tosch, S. C. Jones and A. W. Adamson, J. Colloid Int. Sci., 31 (1969) 297.
41 M. Barthe, J. Biais, M. Bourrel, B. Clin and P. Lalanne, in: P. Bothorel (Ed), Surfactants
    in Solution, Plenum, New York, 1986.
42 P. D. I. Fletcher, J. Chem. Soc. Faraday Trans. 1, 82 (1986) 2651.
43 R. Aveyard, B. P. Binks, S. Clark and J. Mead, J. Chem. Soc. Faraday Trans. 1, 82 (1986)
    125.
44 H-F. Eicke and J. Rehak, Chim. Acta, 59 (1976) 2883.

45　A. W. Adamson, J. Colloid Int. Sci. 29 (1969) 261.
46　S. Levine and K. Robinson, J. Phys. Chem. 76 (1972) 876.
47　S. Mukherjee, C. A. Miller and T. Fort, J. Colloid Int. Sci., 91 (1983) 223.
48　J-F. Jeng and C. A. Miller, Colloids Surf., 28 (1987) 247.
49　E. Ruckenstein and R. Krishnan, J. Colloid Int. Sci., 71 (1979) 321.
50　C. Huh, Soc. Pet. Eng. J., (1983) 829.
51　J. Th. G. Overbeek, G. J. Verhoecks, P. L. DeBruyn and H. N. W. Lekkerkerker, J. Colloid Int. Sci., 119 (1987) 422.
52　F. J. Booth, Chem. Phys. 19 (1951) 391.
53　E. Ruckenstein and D. Schiby, Langmuir, 1 (1985) 612.
54　B. E. Conway, Ionic Hydratiton in Chemistry and Biophysics, Elsevier, Amsterdam, 1981.
55　Y. Gur, I. Ravina and A. J. Babchin, J. Colloid Int. Sci., 64 (1978) 326.
56　R. M. Noyes, J. Amer. Chem. Soc., 84 (1962) 513.
57　G. Lindblom, B. Llindman and L. Mandel, J. Colloid Int. Sci., 34 (1970) 262.
58　M. Wong, J. K. Thomas and T. Nowak, J. Amer. Chem. Soc., 99 (1977) 4730.
59　L. Sepulveda, E. Lissi and F. Quina, Adv. Colloid Int. Sci., 25 (1986) 1.
60　A. Verbeek and F. C. DeSchryver, Langmuir, 3 (1987) 494.
61　K. Vos, C. Laane and A. J. W. G. Visser, Photocheme. Photobiol., 45 (1987) 863.
62　C. Chachaty, Progress in NMR Spectroscopy, 19 (1987) 183.
63　J. H. Fendler, F. Nome and J. Nagyvary, J. Molec. Evol., 6 (1975) 216.
64　Y. Nozaki and C. Tanford, J. Biol. Chem. 246 (1971) 2211.
65　H. B. Bull and K. Breese, Arch. Biochem. Biophys., 161 (1974) 665.
66　C. A. Boicelli, F. Conti, M. Giomini and A. M. Giuliani, Chem. Phys. Let., 89(6) (1982) 490.
67　M. A. J. Rodgers and P. C. Lee, J. Phys. Chem. 68 (1984) 3480.
68　A. Dossena, V. Rizzo, R. Marchelli, G. Casnati and P. L. Luisi, Biochim. Biophys. Acta 446 (1976) 493.
69　P. L. Luisi, F. J. Bonner, A. Pellegrini, P. Wiget and R. Wolf, Helv. Chim. Acta, 62 (1979) 740.
70　K. K. Lee, in: B. Sc. Thesis, MIT, Cambridge, MA, 1986.
71　M. P. Thien, K. K. Lee, K. M. Thompson and T. A. Hatton, Paper presented at Annual Meeting of the AIChE, Miami Beach, Florida, 1986.
72　E. B. Leodidis and T. A. Hatton, to be published.
73　C. A. Bunton and L. Sepulveda, J. Phys. Chem., 83 (1979) 680.
74　C. Hirose and L. Sepulveda, J. Phys. Chem., 85 (1981) 3689.
75　Y. Katz and J. M. Diamond, J. Membr. Biol., 17 (1974) 101.
76　N. K. Anderson, S. S. Davis, M. James and I. Kozima, J. Pharm. Sci., 72 (1983) 443.
77　S. S. Davis, M. J. James and N. M. Anderson, Disc. Faraday Soc., 81 (1986) 313.
78　G. D. Rose, A. R. Geselowitz, G. J. Lesser, R. H. Lee and M. H. Zehfus, Science 229 (1980) 834.
79　R. S. Pearlman, in: W. J. Dunn, J. H. Block and R. S. Pearlman (Eds), Partition Coefficient Determination and Estimation, Pergamon Press, New York, 1986.
80　R. B. Hermann, J. Phys. Chem., 75 (1971) 363.
81　R. B. Hermann, J. Phys. Chem., 76 (1972) 2754.
82　P. Meier, V. E. Imre, M. Fleschar and P. L. Luisi, in: K. L. Mittal and B. Lindman (Eds), Surfactants in Solution, 2, Plenum, New York, 1984.
83　M. Dekker, K. Van't Riet, S. R. Weijers, J. W. A. Baltussen, C. Laane and B. H. Bijsterbosch, Chem. Eng. J., 33 (1986) B27.
84　S. Giovenco, F. Verheggen and C. Laane, Enzyme Microb. Technol., 9 (1987) 470.
85　R. S. Rahaman, J. Y. Chee, J. M. S. Cabrall and T. A. Hatton, Biotech. Progress, Accepted for publication.
86　C. Petit, P. Brochette and M. P. Pileni, J. Phys. Chem., 90 (1986) 6517.
87　P. Brochette, C. Petit and M. P. Pileni, J. Phys. Chem., 92 (1988) 3505.
88　M. Caselli, P. L. Luisi, M. Maestro and R. Roselli, J. Phys. Chem., 92 (1988) 3899.
89　D. Bratko, A. Luzar and S-H. Chen, J. Chem. Phys., 89 (1988) 545.

90 J. M. Woll and T. A. Hatton, Bioprocess Engineering, Accepted for publication.
91 K. E. Goklen and T. A. Hatton, Sep. Sci. Tech. (1987).
92 P. D. I. Fletcher and D. J. Parrott, J. Chem. Soc. Faraday Trans. 1, 84 (1988) 1131.
93 K. E. Goklen and T. A. Hatton, Biotech Progress, 1(1) (1985) 69.
94 B. D. Kelley, E. B. Leodidis, R. S. Rahaman and T. A. Hatton, Paper presented at the ACS 62nd Colloid and Surface Science Symposium, Pennsylvania State University, 1988.
95 B. D. Kelley and T. A. Hatton, To be published.
96 J. M. Woll, T. A. Hatton and M. L. Yarmush, Biotech Progress, Accepted for publication.
97 R. D. Marcus, J. Chem. Phys. 23 (1955) 1057.
98 J. G. Kirkwood and J. B. Schumaker, Proc. Natl. Acad. Sci. 38 (1952) 855.
99 V. A. Parsegian and G. M. Weiss, J. Chem. Phys., 60 (1974) 5080.
100 D. J. Barlow and J. M. Thornton, Biopolymers, 25 (1986) 1717.
101 P. Hurter and T. A. Hatton, Work in progress.
102 C. Tanford and J. G. Kirkwood, J. Amer. Chem. Soc., 79(20) (1957) 5333.
103 J. B. Matthew, G. I. H. Hanania and F. R. N. Gurd, Biochemistry, 18 (1979) 1919.
104 J. B. Matthew, Ann. Rev. Biophys. Chem., 14 (1985) 382.
105 A. Warshel and S. T. Russell, Quart. Rev. Biophys., 12 (1984) 203.
106 S. T. Russell and A. Warshel, J. Molec. Biol., 185 (1985) 389.

PROTEIN PARTITIONING BETWEEN MICROEMULSION PHASES AND CONJUGATE AQUEOUS PHASES

Paul D.I. FLETCHER and David PARROTT

1. INTRODUCTION

Water-in-oil (W/O) microemulsion phases can co-exist at equilibrium with excess phases consisting of virtually pure water. Proteins added to the system partition between the two phases to an extent which depends upon the nature of the stabilising surfactant and the protein together with system variables such as the addition of electrolyte, pH and temperature. In this chapter, we describe the properties of microemulsion two-phase systems, the factors affecting the extent of partitioning of proteins between the phases and the relationship between the partition coefficients and the chemical reactivity of the partitioning species in each phase. Finally, we outline the potential use of these systems for the liquid/liquid extraction and concentration of enzymes.

2. MICROEMULSION PHASE EQUILIBRIA

2.1 Single-Phase Microemulsions

Microemulsion phases are single-phase, thermodynamically-stable mixtures of oil and water, the system being stabilised by a surfactant. The surfactant can be pure or a mixture and may contain a co-surfactant such as a medium chain length alcohol.

Microemulsions can consist of oil droplets dispersed in water (O/W) or water droplets in oil (W/O). These "droplet type" microemulsions normally occur for mixtures containing less than about 20% by volume of the dispersed component. However, microemulsions can also form in mixtures containing comparable volumes of oil and water. These mixtures are apparently continuous with respect to both the oil and the water and are called "bicontinuous" microemulsions.

The structure of microemulsion phases is discussed in detail earlier in this book by Langevin. For the droplet type systems of interest here, the relationship between microemulsion droplet size and mixture composition can be predicted to a good first approximation by simple geometrical arguments. If it is assumed :
(i) that the droplets are spherical and monodisperse,
(ii) that all the surfactant is bound to the oil/water interface and
(iii) that each surfactant molecule occupies a constant (ie. droplet size independent) area a_S at the interface, then equation [1] is obtained.

$$r_c = (3 \, mv_{dc}/a_s) \, [\text{dispersed component}]/[\text{surfactant}] \qquad [1]$$

r_c is the radius of the droplet core (ie. this value does not include the interfacial surfactant shell thickness) and mv_{dc} is the volume of a single molecule of the dispersed component.

Equation [1] is obeyed quite closely for for single-phase, "made-up" microemulsions stabilised by sodium bis(2-ethylhexyl) sulphosuccinate (AOT). Both W/O and O/W systems and glycerol-in-oil microemulsions are satisfactorily described (1-4). For W/O microemulsions, r_c is proportional to the molar ratio of water to AOT and the slope of the line indicates the value of a_s is 0.50 nm^2 (1,2). Changing the water and AOT concentrations together such that their ratio remains constant alters only the droplet concentration but not their size.

The behaviour of the AOT stabilised systems described above agree well with the simple model predictions and hence can be described to a first approximation in terms of spherical, monodisperse droplets. More complex systems such as those containing a mixture of surfactants or a surfactant plus a co-surfactant can be treated similarly but proper account must be made of the different values of the areas occupied by each surface active, adsorbed component. Also, alcohol co-surfactants commonly partition between the interface and the continuous solvent. The resulting variation in interfacial composition must be taken into account in attempting to model the structural behaviour.

Microemulsion types can be classified according to the curvature of the surfactant monolayers surrounding the particles. Droplet type microemulsions can be either positive or negative curvature structures. (A positively curved monolayer is defined here as one in which the surfactant hydrophilic headgroups lie on the exterior surface and vice versa). These are normal micelles and O/W microemulsions or reversed micelles and W/O microemulsions respectively. Bicontinuous microemulsions are zero (or close to zero) curvature structures (5-7). Zero curvature lamellar liquid crystalline phases may be formed in preference to a bicontinuous microemulsion phases for rigid surfactant monolayers.

2.2 Multi-phase Equilibria involving Microemulsions.

Three types of multi-phase equilibria are commonly observed in mixtures of oil, water and a surfactant. These are a two-phase system of O/W with an excess oil phase, a three-phase system of a bicontinuous microemulsion with excess oil and water phases and a W/O microemulsion plus excess water as shown in figure 1. These multi-phase systems are commonly called Winsor I, Winsor III and Winsor II systems respectively after their description by Winsor some decades ago (8). For a particular surfactant, the progression of phase system from Winsor I - III - II may be obtained by progressively changing a system variable such as the electrolyte

excess oil

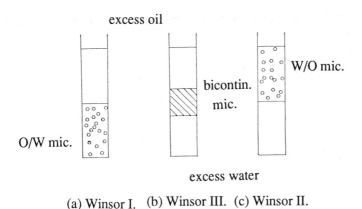

excess water

(a) Winsor I.   (b) Winsor III.   (c) Winsor II.

Fig. 1. Multi-phase equilibria observed in water/oil/surfactant mixtures.

concentration, temperature, co-surfactant concentration, the nature of the oil component (eg. chain length of n-alkane type oils) and others.

At low concentrations the surfactant is present as monomers in either the oil or water phases or distributed between the phases. Above a critical concentration called the critical aggregation concentration (cac) aggregates of the surfactant are formed. These aggregates can be either positive curvature structures formed in the aqueous phase. negative curvature structures formed in the oil phase or zero curvature structures formed in a third, "middle" phase. Which type of aggregate is formed (ie. oil microemulsion droplets in the water phase, water microemulsion droplets in the oil phase or a bicontinuous third phase) can be understood in terms of the tendency of the surfactant monolayer to form curved surfaces (the spontaneous curvature). A surfactant which has a hydrophilic headgroup area larger than the hydrophobic tailgroup area will tend to form positively curved surfaces such that the headgroups form the exterior surface. In this case O/W microemulsion droplets will be formed in the aqueous phase and all the aggregated surfactant will be present in the water phase. The equilibrium concentration of monomeric surfactant (equal to the value of the cac) partitions independently of the aggregated surfactant and may be distributed between both phases. If the surfactant or the conditions are changed such the spontaneous curvature becomes negative then W/O microemulsions droplets are formed in the oil phase. Under conditions where the spontaneous curvature is close to zero, the aggregated surfactant is located within the middle, bicontinuous microemulsion phase of a Winsor III system. It is important to note that the microemulsion droplet size in these multi-phase systems can adjust to an equilibrium size by solubilisation of water or oil to a value that is normally close to the reciprocal of the spontaneous curvature. This is in contrast to single-phase, "made-up" microemulsions where the size is dictated by the phase composition according to the simple geometrical considerations outlined earlier. The droplet size in single-phase, "made-up" systems is only equal to the spontaneous size (= 1/curvature) at the phase boundary where an excess phase of the dispersed component separates.

What factors affect the spontaneous curvature of surfactant monolayers ? Firstly, there is the molecular structure of the surfactant. For example a double tailed surfactant like AOT is expected to show a greater tendency to form reversed curvature structures as compared with a single tailed surfactant such as sodium dodecyl sulphate. However, the molecular structure as visualised using molecular models is not sufficient in attempting to estimate the spontaneous curvature. The "effective" areas of the surfactant head and tail groups depend upon all the intermolecular forces (van der Waals', chain packing, electrostatic) between adjacent molecules within the surfactant monolayer. The effective head and tail group areas, and hence the spontaneous curvature, can be manipulated to a particular value required by adjusting one of a number of possible variables which include the electrolyte concentration, co-surfactant concentration, the nature of the oil component and temperature.

The effect of salt on the spontaneous curvature of charged surfactant monolayers is caused by screening of the repulsive electrostatic forces between the headgroups of adjacent surfactant molecules in the interface. This decreases the effective headgroup area resulting in a decreased (ie. more negative) spontaneous curvature. The effect of changing the alkane oil component can be rationalised in terms of penetration of the tailgroup region of the surfactant monolayer by the oils. Short chain length alkanes penetrate to a greater extent than long chains and hence swell the tail region to a greater extent resulting in an increased negative curvature. Alcohol co-surfactants affect the monolayer curvature by adsorbing at the oil water interface forming a mixed monolayer with the surfactant. Depending on the alcohol chain length, the tail region can be swelled more or less than the head region. Thus short chain alcohols generally increase the positive curvature whereas long chain alcohols decrease it. Alcohols containing four or so carbon atoms have little effect. The effect of temperature is different for ionic and nonionic surfactants. For ionic surfactants, increased temperature causes increased positive curvature. This effect is probably due to increased counter ion dissociation. For nonionic surfactants, increased temperature is thought to cause dehydration of the headgroups thereby decreasing the effective size of the head causing decreased curvature. The factors affecting spontaneous curvature are summarised in table 1.

Variable	Effect on curvature	Cause
Increase [salt]	more negative	screened repulsion between headgroups
Increase oil chain length	more positive	decreased penetration of surfactant tail region
Addition of short chain alcohol	more positive	alcohol swells surfactant head region more than tail
Addition of long alcohol	more negative	alcohol swells surfactant chain tail region more than head
Increased temp. (ionic surfactant)	more positive	increased surfactant counterion dissociation
Increased temp. (nonionic surfactant)	more negative	headgroup size reduced by dehydration

----

Table 1. Factors affecting the spontaneous curvature of surfactant monolayers.

2.3 Multi-phase Behaviour of AOT

Much of the interest in these multi-phase microemulsion systems stems from the fact that ultra-low oil/water interfacial tensions occur when the surfactant spontaneous curvature is close to zero (9). Hence these systems have potential applications in enhanced oil recovery. They also have potential, however, as an alternative to conventional solvents for the extraction and separation of solute mixtures. Of particular interest here is the use of Winsor II systems for the extraction or separation of proteins. One of the best characterised surfactant multi-phase systems is that containing AOT (4,10-14). Many of the general principles of protein partitioning will be illustrated using results obtained in this system. For this reason, the Winsor II behaviour of this surfactant is discussed in more detail.

AOT is a double tailed surfactant and might be expected favour reversed curvature structures. However, when AOT is allowed to distribute between n-heptane and water at 25°C, all the AOT resides in the water phase. On addition of NaCl at concentrations greater than 0.05 mol dm^{-3} all the aggregated AOT partitions to the oil phase. The concentration of monomeric AOT (equal to the cac) is in the range 1 - 0.1 mmol dm^{-3} (depending on the salt concentration) and is contained entirely within the water phase. At NaCl concentrations around 0.05 mol dm^{-3} the aggregated surfactant is contained within a third, middle phase. This phase is not a bicontinuous microemulsion but is a lamellar liquid crystal in the case of AOT. Hence, addition of NaCl causes a progressive change in the spontaneous curvature of the AOT monolayer. It starts at high and positive (small normal micelles), decreases (forming larger O/W microemulsion droplets) passes through zero (when the zero curvature

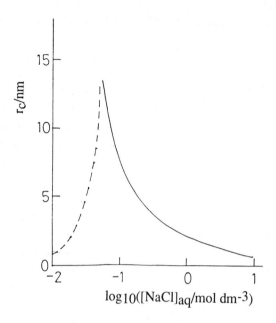

Fig. 2. The variation of microemulsion droplet core radii (r_c) with NaCl concentration for AOT/n-heptane/brine mixtures at 25°C. The dashed line refers to O/W microemulsion droplets in equilibrium with excess oil. The full line refers to W/O microemulsion droplets in equilibrium with excess brine phases.

lamellar phase is formed), passes to small and negative (corresponding to the formation of W/O microemulsion droplets) and finally becomes very large and negative (when small reversed micellles are formed). The equilibrium droplet sizes in the Winsor I and II systems obtained as a function of salt (4,14) are shown in figure 2. It can be seen that the droplet type (ie. O/W or W/O) and size can be systematically varied by adjusting the salt concentration in accordance with the general principles outlined above. For the particular case of AOT stabilised W/O microemulsions in equilibrium with excess water, the phase boundary is not affected by the AOT concentration. Hence, if the NaCl concentration is kept constant the same size W/O micromeulsion droplets are produced at equilibrium independent of the AOT concentration (13). Doubling the AOT concentration produces, at equilibrium, twice the concentration of water droplets of the same size. This point is important in understanding the partitioning of an added solute between the two phases.

3. PARTITIONING OF PROTEINS IN WINSOR II MICROEMULSION SYSTEMS

3.1 General Behaviour

Many different proteins can be solubilised within the water droplets of W/O microemulsions with retention of their native activity (for reviews see 15-18). The pioneering work of Luisi et al.(19-21) demonstrated that it is possible to extract proteins from an aqueous phase into a W/O microemulsion phase. Furthermore, by changing the conditions it is possible to back-extract the proteins into the water phase. Different proteins partition to greatly differing extents and hence these systems offer some potential for the separatiion of protein mixtures.

Globular proteins generally have radii in the range 2 - 20 nm and hence are of comparable size with microemulsion droplets. Proteins are oil-insoluble and are therefore located within the dispersed water droplets in a W/O microemulsion phase. The microenvironment of the protein within the droplet remains essentially "water-like" as the protein does not "see" the oil. Indeed, the hydration of chymotrypsin, as measured using a quasi-elastic neutron scattering technique, is virtually identical in bulk water and in AOT stabilised microemulsion water droplets of 3.5 nm radius (22). However, since the sizes of the protein and the water droplets are generally comparable, proteins hosted within microemulsion water droplets are neccessarily in close proximity to, or may be embedded within, the surfactant monolayer stabilising the droplet. The transfer of a protein from an aqueous phase to a microemulsion phase involves tranfer from bulk water to micro-dispersed water with relatively little change in the microenvironment experienced by the protein. The essential difference in the two situations is the proximity of the surfactant interface in the micromeulsion phase. The extent of protein partitioning between the phases is expected to be determined by the interactions of the protein with the surfactant monolayer coating the droplets.

Experimentally, protein partitioning is determined by adding the protein to the oil/water mixture and vigorously shaking or stirring. The mixture is then allowed to separate to two clear phases and the protein concentration in each phase measured spectrophotometrically. Both the achievement of equilibrium partitioning and the phase separation can be slow depending upon the particular system used and the conditions. It is good practice to check that the same result is obtained when the protein is added initially to the oil phase containing surfactant and when it is added initially to the water phase.

3.2 Defining the Partition Coefficient

The partitioning of low molecular weight solutes between AOT stabilised W/O microemulsions and conjugate aqueous phases has been reported (14). As discussed in reference (14), the partition coefficient in this system may be expressed in a number of ways. For the case of proteins it is assumed these species are located exclusively within the dispersed water droplets. The microemulsion water droplets are assumed to behave as a separate phase (a "pseudophase") within the microemulsion phase. The partition coefficient of the protein $P(protein)_{dw/aq}$ is then defined as

$$P(protein)_{dw/aq} = [protein]_{dw}/[protein]_{aq} \tag{2}$$

where $[protein]_{dw}$ is the concentration of protein in the microemulsion phase expressed per unit volume of dispersed water within the microemulsion phase and $[protein]_{aq}$ is the aqueous phase concentration of protein. The value of $[protein]_{dw}$ is calculated from the measured overall concentration of protein in the microemulsion ($[protein]_{ov}$ = protein concentration per total volume of microemulsion) and the volume fraction of water in the microemulsion phase (O_W).

$$[protein]_{dw} = [protein]_{ov}/O_W \tag{3}$$

The validity of this "pseudo-phase" definition of the partition coefficient can be checked by determining the concentration ratio between water and microemulsion phases in which the volume fraction of water is changed. (This is analagous to changing the phase volume ratio in a partitioning experiment using conventional solvents.) An example of this is shown in figure 3 for chymotrypsin partitioning between an aqeous phase containing 0.5 mol dm^{-3} NaCl and water-in-heptane microemulsions stabilised by AOT (23). For this system, the water droplet radius is 2.9 nm independent of the AOT concentration. Altering the AOT concentration produces a proportional change in the droplet concentration and hence the volume fraction of water in the microemulsion phase is proportional to the AOT concentration. The solid lines in figure 3 are calculated using a value of $P(chymotrypsin)_{dw/aq}$ of 44. The partition coefficient calculated in this way is independent of the surfactant concentration and thus represents a more meaningful measure of the protein partitioning between the bulk water phase and the dispersed water pseudophase than the ratio of the overall concentrations in each phase determined at an arbitrary surfactant concentration.

3.3 Effects of salt concentration and pH

The variation of $P_{dw/aq}$ for chymotrypsin partitioning between W/O microemulsions stabilised by AOT and conjugate aqueous phases containing NaCl (23) is shown in figure 4. Since the microemulsion droplet size varies with the aqueous phase concentration of NaCl both these variables are shown on the graph. Note that each value of $P_{dw/aq}$ represents the average determined at a number of AOT concentrations from plots similar to figure 3. It is clear that in this example, where the droplet size depends upon the salt concentration, the effects of electrolyte concentration and droplet size cannot be determined independently.

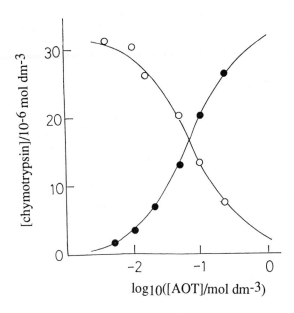

Fig. 3. Concentration of chymotrypsin in the lower aqueous phase (○) and in the upper microemulsion phase (●) as a function of the AOT concentration in the upper phase. The conditions were $[NaCl]_{aq} = 0.5$ mol dm^{-3} (corresponding to a droplet radii of 2.9 nm). The solid lines are calculated from eqn. [2] using a value of $P_{dw/aq}$ of 44.

Chymotrypsin is a globular protein of MW 25,000 dalton and has an approximate radius of 2.2 nm. The isoelectric point is 8.2±0.3. Chymotrypsin is solubilised within AOT stabilised W/O microemulsion droplets with no detectable change in the droplet size (22). This is true except when the droplet size before solubilisation is smaller than the protein size in which case the droplet swells to accommodate the protein by exchange of material with "empty" droplets.

At low salt concentrations (corresponding to large droplets), the protein partitioning is sensitive to pH. Strong partitioning to the microemulsion phase occurs when the protein is positively charged. The value of $P_{dw/aq}$ drops by 2 - 3 orders of magnitude as the pH is increased above the isoelectric point rendering the protein negatively charged. The effect of changing pH is decreased at high NaCl concentrations (corresponding to small water droplets). This behaviour is expected on the basis that electrostatic interactions between the protein and the charged AOT interface dominate the partitioning. However, the partitioning remains in favour of the microemulsion dispersed water even when the protein has a negative charge. (If the protein had an equal affinity for the bulk and dispersed water, a value of

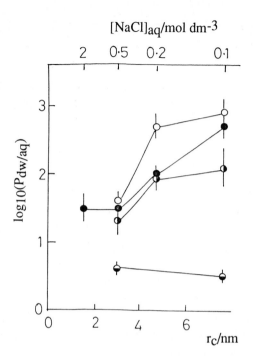

Fig. 4. Variation of $P_{dw/aq}$ for chymotrypsin as a function of microemulsion droplet radius (and corresponding aqueous phase salt concentration) for pH 6.1 (○), pH 7.2 (●), pH 8.2 (◑) and pH 9.8 (◓).

$P_{dw/aq} = 1$ would be obtained.) At high salt concentrations, the lines for the different pH values appear to converge to a value of $\log_{10}(P_{dw/aq})$ of approximately 1.3. At this point it may be assumed that all electrostatic interactions are small and hence the data indicate that an additional non-electrostatic interaction favouring partitioning to the droplets is present. This interaction may be a hydrophobic interaction of the protein with the oil/water interface.

Pepsin is a globular protein with a MW of 35,000 dalton and an isoelectric point of approximately 3. the partitioning behaviour of this protein in the AOT system is very similar to that of chymotrypsin when due allowance is made for the different isoelectric point (23).

The partitioning of these proteins within this particular surfactant two-phase system is adequately described by the pseudophase model description. This corresponds to the proteins occupying the microemulsion droplets with little change in droplet size. A more complex situation can occur for different systems. Instead of proteins occupying a microemulsion droplet with little change in the numbers of water and surfactant molecules in the aggregate, the protein can "create" an aggregate of different stoichiometry around itself. This is likely to occur when the "empty" microemulsion droplets (or reversed micelles) are smaller than the protein or when there is a particularly strong interaction between the protein and the surfactant monolayer.

An example of this behaviour is observed in the case of lysozyme (23) (MW. 14,400 and pI 11.1) which partitions very strongly to AOT W/O microemulsion phases over the pH range 7.2 - 9.8 (as expected from the high positive charge on the protein at the pH values tested). The simple behaviour predicted by equation [2] is not found for this protein. Over the pH range 7.2 - 9.8 no lysozyme is detected in the bulk aqueous phase. At AOT concentrations less than that required for monolayer coverage of the lysozyme, the lysozyme accumulates in an opaque, semi-solid layer at the interfacial region between the oil and water phases. At higher AOT concentrations the lysozyme is dispersed in the oil phase. Whereas chymotrypsin is solubilised within AOT W/O microemulsion droplets with little change in the conformation and hydration of the protein and of the microemulsion droplet size (this is true for droplets of initial radius larger than the protein), the droplet size and protein conformation is affected by solubilisation of lysozyme.

Four possibilities exist when protein is added to a Winsor II system. The protein can remain in the water phase or be solubilised within the microemulsion droplets either with or without a change in the aggregate stoichiometry (ie. the numbers of water and surfactant molecules contained within the aggregate before and after solubilisation of the protein). Lastly, the protein can form an insoluble (and probably denatured) complex with the surfactant. Which possibility, or combination of possibilities occurs depends on the properties of the particular Winsor II system used and the strength of the interaction between the protein and the surfactant. Under conditions where the microemulsion droplet size before protein solubilisation is larger than the protein size and the electrostatic interaction between the protein and the surfactant monolayer is not too attractive (ie. the protein is of the same charge as the surfactant or close to its pI value) then the partitioning can reasonably be described in terms of the bulk water and unperturbed microemulsion droplets. This is the case for chymotrypsin and pepsin partitioning between aqueous phases and AOT stabilised W/O microemulsion phases at salt concentrations such that large droplets (ie. larger than the protein size) are formed and at pH values such that the protein/surfactant monolayer is not too attractive (23). In these circumstances the extent of partitioning varies with pH and salt addition in a manner consistent with electrostatic interactions dominating the process. It should be remembered, however, that the effects due to salt addition and to droplet size variation cannot be separated.

Similar trends with pH and salt concentration have been reported for the partitioning of ribonuclease-a, cytochrome-c, lysozyme and other proteins (19-21, 24-28).

Under conditions where there is a strong attractive electrostatic interaction between the protein and the surfactant, precipitation of a protein/surfactant complex is commonly seen. This has been reported for trypsin, chymotrypsin, elastase and chymotrypsinogen in the AOT system at pH values well below the pI values of the proteins (26).

3.4 Effects of other variables

In addition to the work using AOT as surfactant, the partitioning of proteins to apolar phases containing trioctylmethylammonium chloride (TOMAC) has been studied (19, 29-31). TOMAC is a tri-tailed surfactant and hence would be expected to show a tendency to form highly negatively curved surfaces. Although little characterisation of this system has been carried out, it has been observed that only 2 - 3 water molecules per surfactant are solubilised in this system (19). Hence the aggregates present are very small reversed micelles rather than the microemulsion droplets found with AOT. Solubilisation of proteins in these aggregates must involve the formation of a larger aggregate around the protein. Amylase (pI 5.2) will transfer from water to apolar solutions of TOMAC but only when the pH is close to 10 (29-31). It appears that a rather high negative charge on the protein and hence a very strong attractive electrostatic interaction is neccesary for tranfer in this case. Furthermore, the extent of tranfer decreases rapidly as the pH is increased further. It is unclear whether this decrease is associated with precipitate formation as might be expected from the previous discussion. Chymotrypsin, trypsin and pepsin can be similarly transferred at high pH values. When the data of Luisi et al. for chymotrypsin (19) is expressed as a value of $P_{dw/aq}$, a value of approximately 10^3 is obtained at a pH of 10 with an aqueous phase containing zero added electrolyte. This value, at 2 pH units higher than the pI, can be compared with the value at 2 pH units lower than the pI in the anionic AOT system. There a value of 10^3 is obtained at low salt concentrations corresponding to a large droplet radius. The similarity of these two values (at approximately equal charge difference between the protein and surfactant monolayer), imply that the relative sizes of the unfilled surfactant aggregate and the partitioning protein is unimportant in determining the partitioning. The major factor determining the extent of partitioning appears to be the electrostatic interaction between the protein and the surfactant monolayer.

The importance of the electrostatic interaction is further illustrated by the work of Hatton et al (26). They have systematically measured the extent of partitioning of amino acids to an AOT stabilised microemulsion as a function of the net charge on the partitioning species (controlled by varying pH). Extraction to the microemulsion stabilised by the negatively charged AOT only occurs when the amino acid bears a net positive charge. However, the extent of partitioning also depends on the relative hydrophobicities of the amino acids with the more hydrophobic residues showing the greatest affinity for the microemulsion phase. Hence, although electrostatic interactions appear to dominate the partitioning, these results indicate the presence of a hydrophobic interaction in some systems.

In addition to the relatively non-specific electrostatic and hydrophobic interactions operating in these systems, it is also possible to introduce protein specific interactions. Hatton et al. (27) were able to enhance the extraction of concanavilin-A (con-A) into AOT stabilised microemulsion phases by "doping" the microemulsion phase with the biological surfactant octyl-beta-D-glucopyranoside. There is a specific binding interaction between the con-A and carbohydrate headgroup of the biological surfactant. Low concentrations of the biological surfactant (approx. one molecule per microemulsion droplet) are sufficient to enhance the extraction ten fold.

3.5 Summary of factors controlling partitioning

To summarise, the protein partition coefficient between a bulk aqueous phase and a dispersed water pseudophase of a conjugate W/O microemulsion phase is determined primarily by the electrostatic interaction between the protein and the surfactant monolayer. This can be modulated over three orders of magnitude by altering the pH and electrolyte concentration. A further attractive interaction between the protein and the surfactant monolayer is probably a hydrophobic interaction and is of secondary importance. The extent of partitioning can be further altered by changing the volume fraction of dispersed water pseudophase either by changing the relative volumes of each phase or by altering the surfactant concentration. Under conditions where the protein interacts strongly with the surfactant, precipitation of a surfactant/protein complex can occur in preference to solubilisation within the microemulsion phase.

4. RELATIONSHIP BETWEEN PARTITIONING AND ENZYME ACTIVITY IN MICROEMULSIONS

Enzymes are generally catalytically-active in W/O microemulsions (15-17). However, their catalytic activity is normally altered relative to that observed in bulk water. Since the partition coefficient between two phases is an expression of the free energy of transfer between the phases, the reactivity in one phase is related to the reactivity in the other phase through a combination of the partition coefficient of all species participating in the reaction. This relationship has been investigated in detail for the chymotrypsin catalysed hydrolysis of N-glutaryl-L-phenylalanine-p-nitroanilide (GPNA) in AOT W/O microemulsions (23). The kinetic behaviour of this reaction is described by the Michaelis-Menten scheme in both bulk water and in W/O microemulsions (32). In single phase microemulsions containing large droplets, the turnover number k_{cat} is equal to the value observed in bulk water at the same pH and temperature but increases by a factor of five when the droplet size is reduced.

The Michaelis constant K_M (equal to the dissociation constant of the enzyme-substrate complex, ie. [enzyme][substrate]/[enzyme-substrate complex]) is increased by a factor of 100 relative to water in large droplet microemulsions and increases slightly further for small droplets. Note that the value of K_M in the microemulsions is expressed per volume of dispersed water (not per volume of total microemulsion) in order to allow a direct comparison with bulk water values. Hence it can be seen that the dominant effect of the microemulsion upon the reactivity of chymotrypsin is one of reduced binding affinity between the enzyme and substrate. The virtually unchanged value of k_{cat} and spectral measurements (33) indicate the protein conformation in the microemulsion is very similar to that in water.

One possible explanation for the effect on K_M is that the chymotrypsin and the GPNA are located within the microemulsion droplets in different environments (eg. the protein may be located within the water core and the GPNA within the surfactant monolayer). An alternative explanation is that the enzyme undergoes inhibition by the surfactant. Both explanations would lead to a reduced binding affinity and hence increased K_M. The difference between the explanations is that the first involves an effect on the substrate whereas the second involves only the enzyme. The partitioning behaviour in the two-phase system can yield information concerning which effect is dominant.

In a two-phase system the dissociation constant in the dispersed water pseudo-phase of the W/O microemulsion ($K_M(dw)$) is related to the dissociation constant in the aqueous phase ($K_M(aq)$) by equation [4], involving the partition coefficients of all species involved in the binding equilibrium :

$$K_M(dw) = K_M(aq)P_{dw/aq}(E)P_{dw/aq}(GPNA)/P_{dw/aq}(ES) \qquad [4]$$

where E is the enzyme and ES is the enzyme-substrate complex. If values of $K_M(aq)$ at the salt and AOT monomer concentrations corresponding to those of the conjugate aqueous phases and the partition coefficients of all species are known, then the values of $K_M(dw)$ in the corresponding microemulsion phases may be calculated. However, the partition coefficient of ES cannot be measured because the latter is a reactive species. We can proceed further by making one of two limiting assumptions. If it is assumed that the ES species partitions similarly to E (ie. $P_{dw/aq}(E) = P_{dw/aq}(ES)$) since the enzyme partitioning might be expected to be relatively insensitive to substrate binding, then

$$K_M(dw) = K_M(aq)P_{dw/aq}(GPNA) \qquad [5]$$

Alternatively, if it is assumed that the ES partition coefficient is given by the geometric mean of the enzyme and substrate partition coefficients, then

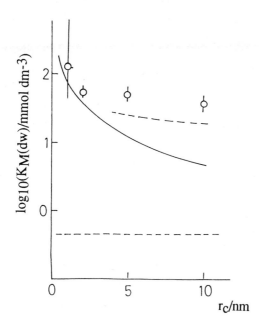

Fig. 5. Comparison of experimental data for $K_M(dw)$ as a function of droplet radii (circles) with that calculated from measured $K_M(aq)$ values and partition coefficient data using equation [5] (solid line) and equation [6] (upper dashed line). The lower dashed line indicates the value in bulk water containing no NaCl.

$$K_M(dw) = K_M(aq)[P_{dw/aq}(E)P_{dw/aq}(GPNA)]^{1/2} \qquad [6]$$

The values of $K_M(dw)$ calculated using measured values of $K_M(aq)$ and $P_{dw/aq}(GPNA)$ are compared with experimental values in figure 5 for a range of microemulsion droplet sizes (corresponding to a range of salt concentrations in the conjugate aqueous phase). The reasonable agreement found is improved if proper account is taken of the monomeric AOT present in the bulk water phase (23). The large difference in K_M values observed in the microemulsion as compared with bulk water is mainly determined by an effect of the microemulsion upon the GPNA (as evidenced by the large values of $P_{dw/aq}$ for this species (23)) rather than the chymotrypsin.

The turnover number k_{cat} can be treated similarly. The magnitude of k_{cat} is determined by the difference in free energy between the species ES in the ground state and in the transition state of the reaction. Hence any change in k_{cat} in the

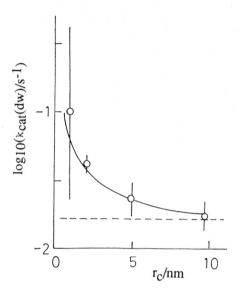

Fig. 6. Comparison of experimental data for $k_{cat}(dw)$ as a function of droplet radii (circles) with that measured in the corresponding bulk aqueous phase containing the appropriate concentration of NaCl (solid line). The dashed line indicates the value in bulk water containing no NaCl.

microemulsion as compared with its value in water is determined by the free energies of transfer of the two species between bulk water and the microemulsion dispersed water droplets. Obviously the partition coefficient (and hence the free energy of transfer) of the ES(transition state) cannot be experimentally measured. If it is assumed that the ground and transition state ES species partition similarly then it is predicted that the value of k_{cat} in the microemulsion phase is identical with the value measured in the conjugate bulk water phase (ie. $k_{cat}(dw) = k_{cat}(aq)$). Note that $k_{cat}(aq)$ refers to the value measured in the particular aqueous phase composition corresponding to that found in equilibrium with the particular microemulsion phase (ie. of the correct surfactant monomer and salt concentrations). This prediction is tested in figure 6 for the case of the chymotrypsin catalysed hydrolysis of GPNA in AOT stabilised W/O microemulsions. Excellent agreement is found implying that the free energies of transfer of the ground and transition states are similar in this case.

Although a large body of data on enzyme reactivity in microemulsions exists, the above example is the only one which has been quantitatively related to the activity in water through this type of analysis. It would be of interest to examine further examples where the enzymic reactivity is altered to a greater extent than chymotrypsin or where the partitioning involves solubilisation of the protein with a large change in the microemulsion droplet stoichiometry.

5. LIQUID-LIQUID EXTRACTION OF PROTEINS

It has been demonstrated that protein partitioning between microemulsion and aqueous phases is sensitive to the nature of the protein and the experimental conditions. Using an understanding of the partitioning behaviour, proteins can be successfully separated from a mixture or concentrated from a fermentation broth. The work of the group in Wageningen is particularly noteworthy in this respect. They have shown that amylase can be concentrated from a fermentation broth in a process involving continuous forward and back extraction between water and iso-octane solutions of TOMAC using two mixer/settler units. They have achieved concentration of the enzyme by a factor of 11.5 with a recovery of enzyme activity of 75% (31).

Although promising laboratory results have been obtained, the exploitation of Winsor II systems for the large scale extraction of proteins requires further work in assessing their economic viability and scale up. Even at this stage of development a number of problems associated with Winsor II systems remain to be overcome. These include :

1. Microemulsion phase boundaries are generally very sensitive to composition, temperature and the presence of impurities in the surfactant. These factors must be carefully controlled in order to achieve reproducible extraction.

2. As pointed out in the previous discussion, the formation of microemulsions is always associated with very low oil/water interfacial tensions. Hence, when these systems are shaken a very finely dispersed macro-emulsion is produced. Depending on the stabilising surfactant, the phase separation to give two phases at equilibrium can take as long as several days. Such a long separation time is a serious drawback in a liquid-liquid extraction system.

3. Enzyme inactivation can be a problem in microemulsion phases. For the example of chymotrypsin partitioning to AOT stabilised microemulsions, the amount of enzyme inactivated during extraction to a microemulsion and back to a water phase is three times that of a control kept for the same period of time (23).

At the present time relatively few Winsor II systems have been investigated for their potential in liquid-liquid extraction of proteins. A large number of different systems have been described in the literature concerned with surfactant phase behaviour and enhanced oil recovery (see for example (33)). Investigation of these systems and others using biological surfactants (35) may provide a route to overcoming some of these problems. The microenvironment experienced by proteins when solubilised within W/O microemulsion droplets may mimic to some extent the natural environment of proteins associated with cell membranes. In principle, it may be possibile to "tailor" the surfactant aggregate (choosing from the wide range of systems available) to provide the optimum microenvironment for a particular protein to be extracted. This possibility makes microemulsions an interesting alternative to more established methods such as those employing two-phase incompatible polymer solutions (36); particularly for trans-membrane proteins.

6. CONCLUSIONS

Equilibrium two-phase systems consisting of a W/O microemlsion together with an aqueous phase (Winsor II systems) can be established for a large variety of surfactants. The properties of the system such as droplet size and the composition and charge of the surfactant monolayer can be systematically varied by controlling parameters including the electrolyte and co-surfactant concentrations, the nature of the oil and surfactant and the temperature. The extent of partitioning of proteins between the two phases of a Winsor II system is controlled mainly by the electrostatic interaction between the protein and the surfactant. There is also a relatively minor interaction which is thought to be hydrophobic in nature. The electrostatic interaction can be varied through controlling the electrolyte concentration and pH. By changing these conditions, proteins can be extracted to or stripped from microemulsion phases. When the conditions are adjusted such that the electrostatic interaction is very strongly attractive, it is generally found that precipitation of a surfactant/protein complex (with resultant protein denaturation) results. Winsor II systems can be successfully used for the liquid-liquid extraction and concentration of proteins but further work is required in a number of areas before the full potential of these systems can be realistically assessed.

We thank the SERC for support of this work through provision of a studentship for D.P.

REFERENCES

1. J.D. Nicholson and J.H.R. Clarke in: K. Mittal and B. Lindman (Eds.), Surfactants in Solution, Plenum, New York, 1984, vol. 3, pp. 1663.
2. B.H. Robinson, C. Toprakcioglu, J.C. Dore and P. Chieux, J. Chem. Soc. Faraday Trans. 1, 80 (1984) 13.
3. P.D.I. Fletcher, M.F. Galal and B.H. Robinson, J. Chem. Soc. Faraday Trans. 1, 80 (1984) 3307-3314.
4. P.D.I. Fletcher, J. Chem. Soc. Faraday Trans. 1, 83 (1987) 1493-1506.
5. D. Langevin, Physica Scripta, T13 (1986) 252-258.
6. A. De Geyar and J. Tabony, Chem. Phys. Letts., 113 (1985) 83
7. D. Langevin, Mol. Cryst. Liq. Cryst., 138 (1986) 259-305.
8. P.A. Winsor, Trans. Faraday Soc., 44 (1948) 376-398.
9. R. Aveyard, Chem. and Ind., July (1987) 474-478.

10. H. Kunieda and K. Shinoda, J. Colloid Interface Sci., 75, (1980), 601.

11. R. Aveyard, B.P. Binks and J. Mead, J. Chem. Soc. Faraday Trans. 1, 82, (1986) 1755.

12. R. Aveyard, B.P. Binks, S. Clark and J. Mead, J. Chem. Soc. Faraday trans. 1, 82 (1986) 125.

13. P.D.I. Fletcher, Chem. Phys. Letts., 141 (1987) 357.

14. P.D.I. Fletcher, J. Chem. Soc. Faraday Trans. 1, 82 (1986) 2651-2664.

15. P.L. Luisi, Angew. Chem., 24 (1985) 439.

16. K. Martinek, A.V. Levashov, N. Klyachko, Y.L. Khmelnitski and I.V. Berezin, Eur. J. Biochem., 155 (1986) 453-468.

17. P.L. Luisi, M. Giomini, M.P. Pileni and B.H. Robinson, Biochim. Biophys. Acta, 947, (1988) 209-246.

18 K. Martinek, A.V. Levashov, N. Klyachko, Y.L. Khmelnitski and I.V. Berezin, Biocatalysis, 1, (1987) 9-15.

19. P.L. Luisi, F.J. Bonner, A. Pellegrini, P. Wiget and R. Wolf, Helv. Chim. Acta, 62 (1979) 740-753.

20. P.L. Luisi, V.E. Imre, H. Jaeckle and A. Pande, in: D.D. Breimer and P. Speiser (Eds.), Topics in Pharmaceutical Sciences, Elsevier, Amsterdam, 1983, pp. 243-255.

21. P. Meier, V.E. Imre, M. Fleschar and P.L. Luisi, in: K. Mittal and B. Lindman (Eds.), Surfactants in Solution, Plenum, New York, 1984, vol. 2, pp. 999-1012.

22. P.D.I. Fletcher, B.H. Robinson and J. Tabony, Biochim. Biophys. Acta, 954 (1988) 27-36.

23. P.D.I. Fletcher and D. Parrott, J. Chem. Soc. Faraday Trans. 1, 84 (1988) 1131 -1144.

24. K.E. Goklen and T.A. Hatton, Biotechnol. Prog., 1 (1985) 69

25. E. Sheu, K.E. Goklen, T.A. Hatton and S-H. Chen, Biotechnol. Prog., 2 (1986) 175-186.

26. T.A. Hatton, in: W.L. Hinze and D.W. Armstrong (Eds.), The Use of Ordered Media in Chemical Separations, ACS Symposium Series 342 (1987) 170-183.

27. J.M. Woll. A.S. Dillon, R.S. Rahaman and T.A. Hatton in: R. Burgess (Ed.), UCLA Symp. Mol. Cell Biol. New Ser. (Protein Purification), 68 (1987) 117-130.

28. K.E. Goklen and T.A. Hatton, Separation Sci. and Technol., 22 (1987) 831-841.

29. K. Van't Riet and M. Dekker, Proc. 3rd Eur. Congr. Biotechnol, Verlag Chemie, Vol. 3 (1984) pp. 541-544.

30. M. Dekker, K. Van't Riet, S.R. Weijers, J.W.A. Baltussen, C. Laane and B.H. Bijsterbosch, Chem. Eng. J., 33 (1986) B27-B33.

31. M. Dekker, J.W.A. Baltussen, K. Van't Riet, B.H. Bjisterbosch, C. Laane and R.

Hilhorst in: C. Laane, J. Tramper and M.D. Lilly (Eds.), Biocatalysis in Organic media, Proceedings of an International Symposium held at Wageningen, 7-10 Dec. 1986, Elsevier, Amsterdam, (1987) pp. 285-288.

32. P.D.I. Fletcher, R.B. Freedman, J. Mead, C. Oldfield and B.H. Robinson, Colloids and Surfaces, 10 (1984) 193-203.

33. S. Barbaric and P.L. Luisi, J. Amer. Chem. Soc., 103 (1981) 4239.

34. K. Shinoda, H. Kunieda, T. Arai and H. Saijo, J. Phys. Chem., 88 (1984) 5126 -5129.

35. N. Kosaric, W.L. Cairns and N.C.C. Gray (Eds.), Biosurfactants and Biotechnology, Surfactant Science Series Vol. 25, Dekker, New York, (1987).

36. P.A. Albertsson, Partition of Cell Particles and Macromolecules, Wiley, New York, (1986).

APPLICATIONS OF ENZYME CONTAINING REVERSED MICELLES

R. HILHORST

1. INTRODUCTION

Several fields of application have been suggested for protein containing reversed micelles. The conversion of apolar compounds by enzymes entrapped in the aqueous core of reversed micelles has drawn most attention. These aggregates have a very dynamic character because they can exchange the contents of their waterpools by a collision process. Apolar compounds can diffuse from the continuous organic phase into the interphase. Enzymes in the water pool can convert both polar and apolar compounds. This enables conversions that are otherwise difficult to perform.

Protein extraction is a second application of reversed micelles. Several approaches have been described, ranging from extraction of proteins from the solid state, extraction of proteins from an aqueous solution to extraction of intracellular proteins from intact bacteria.

A third field of application is the use of reversed micelles for analytical purposes. The presence of an organic phase allows detection procedures that are difficult to perform in aqueous media, e.g. the assay of apolar compounds.

Finally, application of reversed micelles as drug delivery systems has been suggested.

In this chapter, the state of the art in these fields of application will be reviewed and the potentials and drawbacks related to the fulfillment of those potentials will be discussed.

2. CONVERSION OF APOLAR COMPOUNDS

The trend to use enzymes in preparative organic synthesis of poorly water-soluble chemicals, has provoked considerable interest in the potential of reversed micelles as media for the conversion of apolar compounds. As compared to other systems that have been described for the conversion of apolar compounds, e.g. water containing a substantial percentage of organic solvent (1,2), two phase systems (3,4) and organic solvents containing a low percentage of water (5), reversed micelles have several advantages (6,7).

TABLE 1

Long term enzymatic reactions in reversed micellar media.

Enzyme	EC-number	Medium composition	Reaction catalysed	Reference
hydrogenase	1.12.2.1	CTAB/octane/chloroform	2 thiophenol $\xrightarrow{\text{light}}$ H$_2$ + (thiophenol)$_2$	10
hydrogenase	1.12.2.1			
lipoamide dehydrogenase	1.6.4.3	CTAB/octane/hexanol	H$_2$ + progesteron → 20β-hydroxyprogesteron	31
20β-hydroxysteroid dehydrogenase	1.1.1.53			
cholesterol oxidase	1.1.3.6	CTAB/cyclohexane/butanol Triton X100/cyclohexane/butanol CTAB/heptane/octanol hexane/isopropanol	cholesterol + O$_2$ → cholestenon + H$_2$O$_2$	32,33 34 30
alcoholdehydrogenase	1.1.1.1	SDS/cyclohexane/butanol CTAB/cyclohexane/butanol	butanol + cinnamaldehyde → butanal + cinnamylalcohol	35
alcoholdehydrogenase		AOT/cyclohexane	ethanol + cyclohexanon → ethanal + cyclohexanol	36
lipase	3.1.1.3	AOT/heptane	glycerol + oleic acid → glycerol mono-, di- or triolate	37
lipase		AOT/isooctane	olive oil → glycerol + fatty acids	38
tryptophanase	4.1.99.1	trioctylmethylammoniumchloride/ Brij 56/cyclohexane/hexanol	indole + serine → tryptophane	39
enoate reductase	1.3.1.31	CTAB/octane/aliphatic alcohols	tributylamine + 3-phenylpropenoic acid $\xrightarrow{\text{light}}$ oxidized tributylamine + phenylpropanoic acid	40,41
α-chymotrypsin	3.4.21.1	AOT/hexanol	N-acetyl-L-tyrosine ethylester + hexanol → N-acetyl-L-tyrosine ethanol	42

In general, the presence of organic solvents has a deleterious effect on enzyme activity and stability but, when encapsulated in reversed micelles, the enzyme is shielded from the continuous organic phase by a monolayer of surfactant, and exhibits a stability that can be equal to or even higher than the stability in aqueous solution (8-11). Enzyme activity is also influenced by the microenvironment. For several enzymes an enhancement in catalytic activity has been reported (8,12-16), which has been attributed to an increased conformational rigidity of the enzyme (8), to reduction of substrate inhibition (9), to a decreased polarity of the water pool resulting in a change in the pK of groups at the active site (17) and to stabilization of the transition state by interaction with the surfactant head groups (18). The changes in microenvironment can also lead to changes in substrate specificity, due to differences in partitioning (15,19,20).

Understanding of factors regulating enzyme activity in reversed micelles will enable the design of microemulsions tuned to the requirements of an individual reaction. Compared to other systems for the conversion of apolar compounds, reversed micelles possess an enormous interfacial area, that can be as high as 100 m^2/ml. For this reason reaction rates are not limited by diffusion processes. Because rapid exchange takes place between the water pools of reversed micelles (21,22), enzymes can convert water soluble compounds, but they can also react with apolar compounds, dissolved in the interphase or in the organic phase. This makes reactions involving polar and apolar compounds possible. In order to achieve high rates of conversion of apolar compounds, the concentration of these compounds in the vicinity of the enzyme must be high. A system for the conversion of apolar compounds can be optimized using the following guidelines for an optimal activity (16,18)

a)   The substrate concentration in the interphase must be optimized by matching the polarities of the substrate and the surfactant and cosurfactant, and

b)   the substrate concentration in the organic phase must be lowered by making the polarity difference between substrate and continuous phase as large as possible.

Reversed micellar media are very versatile, because they can be composed in many different ways.

Sodium-di-ethylhexyl-sulphosuccinate (Aerosol OT) is the surfactant most often used for the preparation of reversed micelles. Also cetyltrimethylammonium bromide, trioctylmethylammonium chloride, Triton X-100, Tween and Brij have been used as surfactants in combination with a variety of organic solvents, and, in some cases, with an aliphatic alcohol as cosurfactant.

The need for a surfactant to stabilize the system limits the application of reversed micellar media in industry, as most of the surfactants presently applied are not allowed in the preparation of drugs or food additives. Therefore a trend in reversed micellar research is the use of biologically produced surfactants. Some reports have appeared on the use of lecithin for the preparation of reversed micelles (23-25). Another, promising approach is the use of detergentless microemulsions. They consist of hexane, isopropanol and water (26). When mixed in the appropriate proportions, droplets are formed that can host enzymes, without loss of enzyme activity. Trypsin (27,28), chymotrypsin (27), laccase (29) and cholesteroloxidase (30) have been shown to be active in this medium. Enzyme and product recovery in detergentless microemulsions are performed by the addition of excess volumes of aqueous phase and hexane. This leads to phase separation, the organic phase containing the product, which can be recovered by evaporation of hexane, and the aqueous phase containing the biocatalyst, that can be reused (30).

Product recovery from surfactant containing microemulsions has not yet received a lot of attention. Procedures have been reported (31,41) where the surfactant is precipitated from the organic solvent by the addition of acetonitrile or 2-methylbutane, while the product remained in solution. The residual surfactant concentration in the remaining solution however has not been investigated. If surfactant containing reversed micellar media are to be used for the conversion of apolar compounds, the problem of product recovery without surfactant contamination has to be tackled. Not only may the surfactant pose problems, also the presence of organic solvents in large quantities may be a hazard, because of their flammability and explosiveness. Long chain alkanes are preferable, but are more viscous than hexane and octane that are mostly used on lab scale.

A prerequisite for industrial application of enzymes is that they show a high activity, that should be retained for a long period of time. Most literature data on enzyme activity deal with enzymes belonging to the class of hydrolases (6) and with initial enzyme activity and storage stability. Table 1 lists enzymatic conversions in reversed micelles performed for longer periods of time. In some cases, quantitative conversion was the aim of the study, in others operational stability was studied. A major proportion of the enzymes listed in this table belongs to the class of the oxidoreductases. For these enzymes to be active over long periods of time, a cofactor regenerating system must be included. The systems that have been applied in reversed micellar media are shown in Fig. 1. Reducing equivalents can be recycled by the same enzyme that performs the desired conversion (35,36) (Fig. 1A). Such systems require high concentrations of both substrates to drive the reaction. The fact that the reaction cannot go to completion makes a difficult separation procedure for both substrates and products necessary.

Reducing equivalents can also be produced photochemically (Fig.1B). When the electron donor is located in the organic phase and the acceptor in the

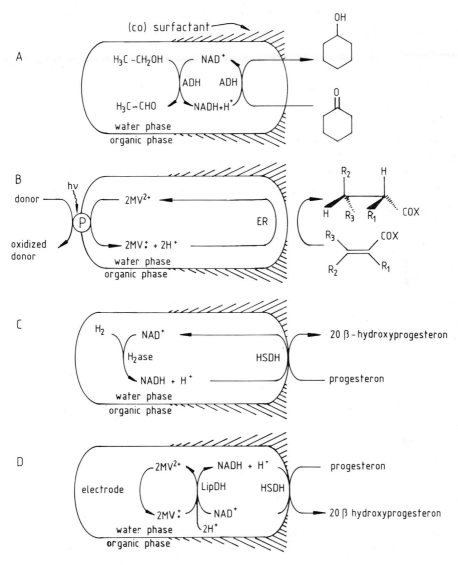

Figure 1. Methods for the production of reducing equivalents in reversed micellar media. A. Use of ADH for cofactor regeneration and conversion. B. Photochemical production of reducing equivalents. C. Cofactor regeneration with hydrogen gas and hydrogenase. D.Electrochemical cofactor regeneration. Abbrevations: ADH: alcohol dehydrogenase; P: photosensitizer; MV2+: methyl viologen; ER: enoate reductase; H2ase : hydrogenase; HSDH: hydroxysteroid dehydrogenase; LipDH: lipoamide dehydrogenase.

waterpool, vectorial electron transport is achieved. The yield of this photochemical reaction depends on the location of the photosensitizer and is highest when the sensitizer is located in the reversed micellar interphase. Spatial separation of donor and acceptor is necessary to obtain charge separation (41).

Hydrogenase can utilize hydrogen gas to reduce methyl viologen, that can be used directly by the production enzyme (40), or be converted into NAD(P)H in a second enzymatic step (31). NADH can be formed directly from H_2 and NAD^+ by the hydrogenase from *Alcaligenes eutrophus* (40). Hydrogen gas as the ultimate electron donor has the advantages that it can be supplied continuously in large quantities during the reaction, that it dissolves very well in organic solvents, that it is cheap and that it is consumed totally in the reaction leaving no byproducts.

An alternative way for the production of reducing equivalents is the electrochemical reduction of methyl viologen (40). Advantages of this method are that electricity is cheap, leaves no byproducts in the medium and the reducing power can be set at will by the potentiostat. A drawback is that this is not a very easy method because of the low conductivity of organic media. Furthermore, as the electrode only produces electrons, the medium may become depleted for protons, resulting in a pH-shift to higher values.

An approach that has not been described yet, would be a combined oxidation-reduction reaction that yields two valuable products.

Whereas a considerable amount of research has been spent on cofactor regeneration, engineering aspects of the use of reversed micellar media have been neglected till now. For a process to be most cost-effective, it must be run as a continuous process. Two papers report preliminary results on reactor design for reversed micelles. Doddema et al. (34) studied the conversion of cholesterol by cholesterol oxidase entrapped in reversed micelles. This process was carried out in a stirred tank reactor combined with a plug flow reactor, with a total volume of 80 ml. The product was separated by leading the effluent product stream through a hollow fiber unit. Luthi and Luisi (42) used a hollow fiber as reactor for the α−chymotrypsin catalysed synthesis of a dipeptide, that was continuously separated from the reaction medium. A third paper has tentatively proposed a reactor design, but no attempts have been made to test it in reality (39).

When a continuous process is not possible, costs of the enzyme can be reduced by its repeated use. Methods for the recovery of enzymes from reversed micellar media are relatively well investigated and will be discussed in the next section.

In conclusion it can be said that reversed micelles offer several advantages for the enzymatic conversion of apolar compounds over other systems that have been described. Some advantages are: they are thermodynamically stable, they are easily prepared, they allow a high enzyme activity and stability, enable cofactor regeneration, their composition is easily adapted, and no diffusion limitation occurs. For incorporation in an industrial process these advantages do not suffice by themselves, for, apart from having to compete with already existing processes that are well integrated into the total production chain, a new process must not conflict with restrictions on the use and disposal of organic solvents or surfactants, imposed by law, or with already existing patent rights, and not require a high capital investment.

3. PROTEIN EXTRACTION USING REVERSED MICELLES

Now that advances in biotechnology such as genetic engineering, facilitate the production of proteins, new large scale techniques must be developed for downstream processing. Selective protein extraction from fermenter liquids by reversed micelles is a promising new recovery technique. Reversed micelles can also be applied for protein extraction from the solid state (44,45) and recently they were shown to be effective in the isolation and partial purification of intracellular enzymes from whole cells (46).

3.1 Liquid-liquid extraction

At the end of the seventies, it was discovered that not only amino acids (47) but also proteins (48) could be transferred from an aqueous phase into an organic solvent with the aid of a quaternary ammonium salt. This transport was accompanied by the transport of small amounts of water, suggesting that under those conditions reversed micelles were formed. From those studies it became clear that electrostatic interactions between surfactant molecules and amino acids and proteins are very important. Amino acids were only transferred into the organic phase when they possessed a charge opposite to the surfactant molecule (47). Similar observations were made for proteins (48). The existence of electrostatic interactions between protein and surfactant head groups is not in agreement with the water shell model, that depicts an enzyme in the center of a reversed micelle, surrounded by a layer of water (49). From spectroscopic studies evidence has been obtained that an interaction between protein and surfactant head groups exists (50-52).

For proteins having a radius smaller than the micellar radius, the electrostatic interaction seems to drive the solubilisation of proteins in reversed micelles (53). In Fig. 2A uptake of cytochrome c, lysozyme and RNAse by AOT reversed micelles begins at the isoelectric point, and is 100% at lower pH values. For larger proteins different patterns are obtained (Fig. 2B), that show a sharp drop at lower pH values. For α-amylase, with a M_r of 55 000, a very narrow uptake profile was reported (54) (Fig. 3). Note that in this case a positively charged surfactant was used, and that uptake now takes place above the isoelectric point. It has been shown that addition of salt influences the uptake

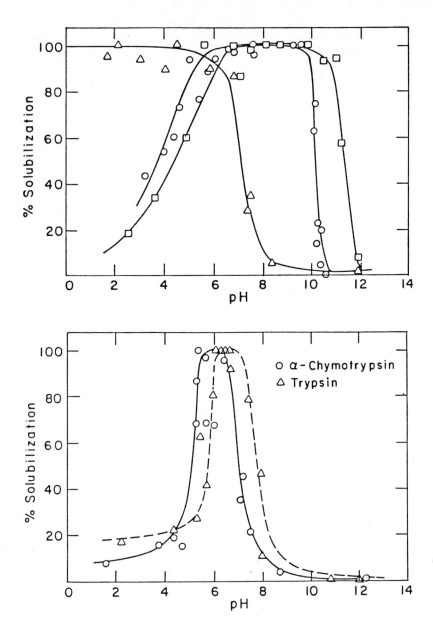

Figure 2. Effect of pH of a 0.1 M KCl solution on protein transfer to a 50 mM AOT/isooctane solution.
A: lysozyme (□), cytochrome c (o) and ribonuclease A (Δ).
B: a–chymotrypsin (o) and trypsin (Δ) (from reference 57).

profile; for α-amylase for instance, the uptake profile shifted to higher pH-values, but still 100% of transfer was obtained (61). Not for all proteins the partition behaviour over the aqueous phase and micellar phase is influenced in the same way (Fig. 4). Also the type of salt used is important in determining the partition (49).

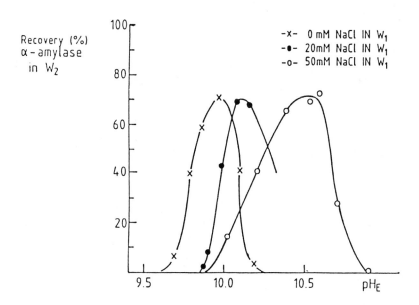

Figure 3. Effect of pH and ionic strength of a 35 mM EDA solution on α–amylase transfer to a 10 mM TOMAC/isooctane solution (from reference 61).

From literature data it becomes clear that protein charge with respect to surfactant charge, protein size, and composition of the aqueous phase (pH, ionic strength, types of ions present) have been subjects of investigation, but also type of solvent, size of reversed micelle, type of surfactant, presence of a cosur-factant and temperature play a role (55).

With the data yet available, it cannot be predicted how variation of each of those factors will affect the partitioning over the two phases.

Till now only uptake of proteins from an aqueous solution into reversed micelles has been discussed. When a protein containing reversed micellar solution is brought into contact with an aqueous phase of a pH that changes the sign of the overall protein charge or with a solution of a high ionic strength, conditions that favour the expulsion of the protein from the reversed micelles are created. When a reversed micellar medium is in contact with two aqueous phases of different composition, protein transport from one phase to the other via the organic solvent is observed (48). This process is diffusion controlled, so it takes some days before equilibrium is reached. Recently an experimental set-up was described where the volume of the organic phase was reduced to one droplet so as to create a liquid membrane (56). Here too, equilibrium was reached very slowly and transfer percentages were low.

For this method of protein extraction to be attractive for industrial application, several conditions must be fulfilled: the process has to be selective, preferentially it has to be run as a continuous process, with high rates of mass

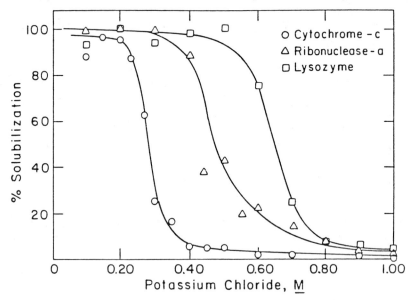

Figure 4. Effect of ionic strength on the solubilisation of cytochrome c (o), ribonuclease A (Δ) and lysozyme (□) in a 50 mM AOT/isooctane solution (from reference 57).

transfer and it has to yield a high percentage of active enzyme. Göklen and Hatton (57) have successfully employed this liquid-liquid extraction method for the separation of a mixture of ribonuclease, cytochrome c and lysozyme. Recently (58) the same group reported the extraction of an extracellular alkaline protease from a fermentation broth. Some 36% of enzyme activity was recovered, in two extraction stages and the specific activity increased 2.2 fold.

An alternative way to increase the specificity of the extraction process is the use of a surface active affinity ligand (58). Addition of low concentrations of an affinity ligand was shown to enhance the partitioning of concanavalin A into reversed micelles. Dekker et al. (59) described a system for the continuous extraction of α–amylase using trioctylmethylammonium chloride in isooctane as medium. Using the mixer-settler unit depicted in Fig. 5, they recovered 45% of

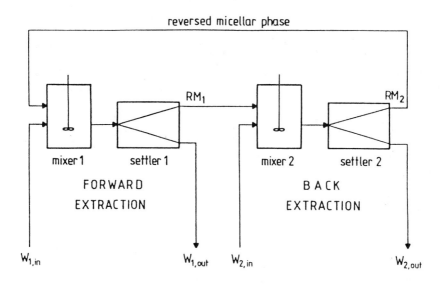

Figure 5. Flow sheet of a combined forward and back extraction, for two mixer-settler units with the reversed micellar phase circulating between the two extraction units (from reference 59).

the α-amylase activity, which was 8-fold concentrated. Inactivation of α–amylase in the first aqueous phase by surfactant molecules diffusing into this phase was found to be mainly responsible for this loss. Decreasing the residence time in this phase by increasing the rate of mass transfer (achieved by increasing the stirrer speed and by the addition of a non ionic surfactant) and/or increasing the distribution coefficient, considerably improved the performance of the system, for 85% of the α–amylase activity, was recovered, concentrated 17-fold (60-62).

Replacement of the mixer settler unit by a membrane extraction making use of hollow fibers has been investigated (63,64). The use of a membrane to separate an aqueous phase from a micellar phase brings some special problems: because of the low interfacial tension between the two phases, permeation of the membrane is difficult to avoid, and because of the amphiphilic character of surfactants, they easily adsorb onto the membrane, regardless of whether this is hydrophobic or hydrophilic in nature.

3.2 Protein extraction from the solid state

Protein extraction is performed by contacting a reversed micellar solution with a protein powder. In contrast to liquid-liquid extraction, where a micellar phase is in equilibrium with an aqueous phase, only a micellar phase is present, so the water content of the micelles can be adjusted at will.

The first reports on protein uptake from the solid state were on the uptake of myelin basic protein by AOT-reversed micelles (44,45). Solubilization was found to be maximal at low w_0 values, where the micellar size is less than the protein dimensions. Luisi et al. (6, 65, 66) reported a similar effect for lysozyme uptake, but at higher water contents of the reversed micelles an increase in protein solubilization was observed. It was suggested that protein uptake is governed by two different mechanisms. At high w_0, the core is sufficiently large to accommodate a protein molecule. At low w_0, the micellar size must increase and energy is derived from electrostatic interactions between the protein and the surfactant molecules. This model is supported by the observation that the ionisation state of the protein is very important. The ionisation state is determined by the pH and composition of the solution that the protein was dissolved in prior to precipitation or freeze drying. In this case too, proteins were only taken up if their overall charge was opposite to the charge of the surfactant.

Extraction of crude sunflower meal, resulted in the selective uptake of small proteins. Only when the small proteins were depleted, larger ones were taken up. Increasing the size of the waterpool also resulted in the uptake of larger proteins.

A patent application for this method of protein extraction has been filed (67).

3.3 Extraction of intracellular proteins

This process is based on the observation that bacterial cells, upon injection into reversed micellar media, containing charged surfactants, are rapidly desintegrated by the medium. For *Azotobacter vinelandii* it has been shown that the contents of the cell are liberated into the medium and that some proteins are taken up in the water pools of the reversed micelles, composed of cetyltrimethylammonium bromide in octane with hexanol as cosurfactant. The proteins can be recovered from the reversed micelles by addition of a solution containing a high KBr concentration. Using this procedure, isocitrate dehydrogenase and β−hydroxy butyrate dehydrogenase activities can be recovered completely, while simultaneously purified sixfold (Table 2). The water content of the system can be used as a selection criterion. Glucose-6-phosphate dehydrogenase could not be recovered, probably because it is too large to be incorporated into reversed micelles. Here, as for extraction from the solid state, small proteins seem to be solubilized preferentially (68).

Table 2. Isolation of intracellular enzymes from intact cells of <u>A.vinelandii</u> using reversed micelles. CFE is cell free extract. Taken from ref. 46.

Enzyme	Conditions	Total protein	Total activity	Specific activity	Recovery (protein)	Recovery (activity)	Purification factor
		µg	m.units	U mg^{-1}	%	%	X
Isocitrate	CFE	225	262	1.2	100	100	1
dehydrogenase	Wo = 5	49	172	3.5	23	65	2.8
	Wo = 15	41	296	7.2	18	113	6.2
	Wo = 25	21	83	4.0	9	31	3.4
β-hydroxybuty-	CFE	225	30	0.13	100	100	1
rate dehydro-	Wo = 5	49	25	0.15	23	85	3.7
genase	Wo = 15	41	33	0.80	18	110	6.1
	Wo = 25	21	21	1.0	9	69	7.6
Glucose-6-phos-	CFE	225	36	0.16	100	100	1
phate dehydro-	Wo = 5	49	0	0	23	0	0
genase	Wo = 15	41	0	0	18	0	0
	Wo = 25	21	0	0	9	0	0

At a first glance, this report is contradictory to the observation that living micro organisms can be solubilized in reversed micelles (69,70). For those experiments Tween was used as surfactant, and isopropylpalmitate as organic solvent. Bacterial cells are in general less sensitive to Tween surfactants compared to ionic surfactants (71). In aqueous solutions Tween is often used to increase the solubility of sparingly soluble compounds (72). Isopropylpalmitate is a solvent with a high log P value, so less likely to affect bacterial viability adversely as compared to octane and hexanol (73,74). Finally, it must be noted that different microorganisms were used, that might greatly differ in their susceptibility to organic solvents.

4. ANALYTICAL PROCESSES IN REVERSED MICELLES.

For the analytical application of reversed micelles, no problems are encountered connected with enzyme or product recovery or scaling up. For this reason, this might become the first field where reversed micelles are employed commercially.

As enzymes in reversed micelles are able to convert poorly water-soluble compounds, enzymatic detection procedures can be extended to the class of water insoluble substances. The analysis can be preceeded by an extraction step, where apolar compounds are extracted from a very dilute aqueous phase into an organic solvent, that subsequently can be used for the preparation of reversed micelles. By manipulating the polarity of the micellar interphase, the partitioning behaviour of substrates can be influenced, resulting in a higher interfacial concentration and consequently a higher enzyme activity and improved sensitivity. As shown by Kurganov et al. (75) not only enzyme activity, but also inhibition of enzymes by rather apolar compounds can be studied in reversed micelles and used as analytical tool.

Reversed micelles have been shown to possess some advantages for the analysis of water soluble compounds. The bioluminescence assay for the determination of the water soluble ATP for instance, is much more sensitive in reversed micelles than in aqueous solution (76,77). Only a very small sample volume is needed for the preparation of reversed micelles and, because the substrate is located in the water pools, the concentration experienced by the enzyme is comparable to the concentration in the sample. Furthermore, in reversed micelles, the intensity of the emitted light remains constent until the ATP is exhausted, whereas in aqueous solution, the signal fades out within a minute. Performing this ATP assay in reversed micelles increases the sensitivity ten to hundred fold (77).

Visser and Santema (78) showed the potential of reversed micelles for analytical procedures for hydrogen peroxide producing enzymes. This compound gives, upon reaction with luminol, rise to chemiluminescence, a reaction that does not take place in aqueous solutions below pH 9, unless a catalyst or cooxidant is present. These requirements make the coupling between an enzymatic assay and the chemiluminescence in aqueous solution virtually impossible.

In reversed micelles, the chemiluminescence reaction takes place below pH9, probably catalysed by the surfactant head groups, and under these conditions a coupling between the two reactions is possible, as was shown for glucose oxidase (78).

Recently, in a detailed study of the chemiluminescence reactions in reversed micelles, it was reported that the yield has increased by 3-3.5 orders of magnitude (79). Here no coupling was made with an enzymatic reaction.

A different, unexplored field with potential for applications is antigen-antibody interactions in reversed micelles. Eremin et al. (80) showed that those interactions are not hampered by enclosure in reversed micelles, a fact that might extend the use of antibodies for analytical purposes to organic solvents.

5. MEDICAL APPLICATIONS

Medical science has developed interest in the use of enzymes as drugs. This poses problems, not yet encountered. After oral administration of enzymes, they are digested in the intestinal tract or excreted. When administered intravenously, they cause immunological reactions. Such effects may be circumvented by masking the antigenic determinants of enzymes, by encapsulation of enzymes, by targetting such capsules to the inflicted organ, etc.

Hardened reversed micelles have been proposed as drug delivery systems (81,82). Varying the surfactant properties might make targetting possible, and encapsulation of the drug in such a capsule may retard its release, thereby reducing the required frequency of administration.

Kabanov et al. (83,84) modified trypsin and α–chymotrypsin in reversed micelles with stearoylchloride. Whereas in aqueous solution between six and twelve amino acid residues per protein were alkylated, in reversed micelles only one or two stearoyl chains were attacked. This modification procedure for enzymes in reversed micelles might prove to be useful for the pharmaceutical industry, because enzymes thus hydrophobisized can interact with membranes.

These procedures use reversed micelles. As pointed out before only few surfactants are approved by law for incorporation in drugs and if approved, often only very low concentrations are tolerated. This limits the choice of micellar media that can potentially be used for medical applications and biocompatible surfactants must be used preferentially. These considerations prompted the group of Luisi to direct its attention to natural surfactants. They reported that with very low amounts of water, phospholipids can gellify considerable volumes of organic solvents (85,86). Such systems might have potential as slow release systems.

6. CONCLUSIONS

Possibilities for the application of reversed micelles have been envisaged in the fields of conversions of apolar compounds, protein extraction, analytical chemistry and pharmacy. Although the research effort put into these areas is rapidly increasing, there is still a lack of understanding of many aspects of enzyme behaviour in reversed micelles. For pharmaceuticals and products for the food and beverage industry, the compounds that can be used during preparation are severely restricted by law, so media must be composed with components that are approved by law.

Because analytical methods require neither large scale procedures nor enzyme recovery, application of reversed micelles can first be expected in this field. For large scale applications, procedures must be developed and scaled up for cofactor regeneration, biocatalyst recovery, product recovery, medium recycling, reactor design etc. Finally, a process based on reversed micelles must be incorporated in a production chain.

7. REFERENCES

1. L.G. Butler, Enzyme Microb. Technol. 1 (1979) 253-259.
2. M. Reslow, P. Adlercreutz and B. Mattiasson, Appl. Microbiol. Biotechnol. 26 (1987) 1-8.
3. G. Carrea, Trends Biotechnol. 2 (1984) 102-106.

338

4. P.J. Halling, Biotech. Adv. 5 (1987) 47-84.
5. A.M. Klibanov, Chemtech. 16 (1986) 354-359.
6. P.L. Luisi and C. Laane, Trends Biotechnol. 4 (1986), 153-161.
7. K. Martinek, A.V. Levashov, N.L. Klyachko, Yu.L. Khmelnitsky and I.V. Berezin, Eur. J. Biochem. 155 (1986) 453-468.
8. S. Barbaric and P.L. Luisi, J. Am. Chem. Soc. 103 (1981) 4239-4244.
9. K. Martinek, A.V. Levashov, N.L. Klyachko, V.I. Pantin and I.V. Berezin, Biochim. Biophys. Acta 657 (1981) 277-294.
10. R. Hilhorst, C. Laane and C. Veeger, Proc. Natl. Acad. Sci. USA 79 (1982) 3927-3930.
11. A.N. Eremin and D.I. Metelitsa, Biokhimiya 51 (1986) 1612-1623.
12. F.M. Menger and K. Yamada, J. Am. Chem. Soc. 101 (1979) 6731-6734.
13. R. Wolf and P.L. Luisi, Biochem. Biophys. Res. Commun. 89 (1979) 209-217.
14. R.C. Srivastava, D.B. Madamwar and V.V. Vyas, Biotechnol. Bioeng. 29 (1987) 901-902.
15. K. Martinek, A.V. Levashov, Yu.L. Khmelnitsky, N.L. Klyachko and I.V. Berezin, Science 218 (1982) 889-891.
16. R. Hilhorst, R. Spruijt, C. Laane and C. Veeger, Eur.J.Biochem. 144 (1984) 459-466.
17. C. Grandi, R.E. Smith and P.L. Luisi, J.Biol.Chem. 256 (1981) 837-843.
18. C. Laane, R. Hilhorst and C. Veeger, Methods in Enzymol. 136 (1987) 216-229.
19. K. Martinek, Yu.L. Khmelnitsky, A.V. Levashov and I.V. Berezin, Dokl.Akad.NaukSSSR 263 (1981) 737-741.
20. P.D.I. Fletcher and B.H. Robinson, J. Chem. Soc. Faraday Trans. I, 81 (1985) 2667-2679.
21. R. Zana and J. Lang, in: K.L. Mittal and E.J. Fendler (Eds.), Solution Behaviour of Surfactants. Theoretical and Applied Aspects, Plenum Press, New York, 1982, pp. 1195-1206.
22. P.D.I. Fletcher and B.H. Robinson, Ber. Bunsenges. Phys. Chem. 85 (1981) 863-867.
23. A.V. Levashov, N.L. Klyachko, A.V. Pshezhetskii, I.V. Berezin, N.G. Kotrikadze, B.A. Lomsadze and K. Martinek, Dokl.Akad.Nauk SSSR 289 (1986) 1271-1273.
24. L. Magid, P. Walde, G. Zampieri, E. Battistel, Q. Peng, E. Trotta, M. Maestro and P.L. Luisi, Colloids and Surfaces 30 (1988) 193-207.
25. R.L. Misiorowski and M.A. Wells, Biochemistry 13 (1974) 4921-4927.
26. G.D. Smith, C.E. Donelan and R.E. Barden, J.Colloid Interface Sci. 60 (1977) 488-496.

27. Yu.L. Khmelnitsky, I.N. Zharinova, I.V. Berezin, A.V. Levashov and K. Martinek, Annals N.Y. Acad. Sci. 501 (1987) 161-164.
28. Yu.L. Khmelnitsky, A. van Hoek, C. Veeger and A.J.W.G. Visser, J. Phys. Chem. (1988) in press.
29. Yu.L. Khmelnitsky, personal communications.
30. Yu.L. Khmelnitsky, R. Hilhorst and C. Veeger, Eur. J. Biochem. 176 (1988) 265-271.
31. R. Hilhorst, C. Laane and C. Veeger, FEBS Letters 159 (1983) 225-228.
32. K.M. Lee and J.-F. Biellmann, Bioorganic Chemistry 14 (1986) , 262-273.
33. K.M. Lee and J.-F. Biellmann, NouveauJ.deChimie 10 (1986) 675.
34. H.J. Doddema, J.P. v.d. Lugt, A. Lambers, J.K. Liou, H.J. Grande and C. Laane, Annals N.Y. Acad. Sci. 501 (1987) 178-182.
35. J.P. Samana, K.M. Lee and J.-F. Biellmann, Eur.J.Biochem.163 (1987) 609-617.
36. K.M. Larsson, P. Adlercreutz and B. Mattiasson, Eur.J.Biochem. 166 (1987) 157-161.
37. P.D.I. Fletcher, R.B. Freedman, B.H. Robinson, G.D. Rees and R. Schomacher, Biochim. Biophys. Acta 912 (1987) 278-282.
38. D. Han and J.S. Rhee, Biotechnol. Bioeng. 28 (1986) 1250-1255.
39. D.K. Eggers and H.W. Blanch, Bioprocess Engineering 3 (1988) 83-91.
40. C. Laane and R. Verhaert, Isr. J.Chem. 28 (1987/88) 17-22.
41. R.M.D. Verhaert, T.J. Schaafsma, C. Laane, R. Hilhorst and C. Veeger, Photochem. Photobiol., 49 (1989) 209-216.
42. E. Testet, M. Baboulene, V. Speziale and A. Lattes, J. Chem.Tech.Biotechnol. 41 (1988) 69-74.
43. P. Luthi and P.L. Luisi, J. Am. Chem. Soc. 106 (1984) 7285-7286.
44. A. Delahodde, M. Vacher, C. Nicot and M. Waks, FEBS Letters 172 (1984) 343-347.
45. C. Nicot, M. Vacher, M. Vincent, J. Gallay and M. Waks, Biochemistry 24 (1985) 7024-7032.
46. S. Giovenco, F. Verheggen and C. Laane, Enz.Microb.Technol.9 (1987) 470-473.
47. J.P. Behr and J.M. Lehn, J. Am. Chem. Soc. 95 (1973) 6108-6110.
48. P.L. Luisi, F.J. Bonner, A. Pellegrini, P. Wiget and R. Wolf, Helv.Chim.Acta 62 (1979) 740-753.
49. P. Meier, V.E. Imre, M. Fleschar and P.L. Luisi, in: K.L. Mittal and B. Lindman (eds.), Surfactants in Solution, Plenum Press, New York, Volume 2 (1984) pp. 999-1012.
50. C. Petit, P. Brochette and M.P. Pileni, J.Phys.Chem. 90 (1986), 6517-6521.

340

51. K. Vos, C. Laane, S.R. Weijers, A. van Hoek, C. Veeger and A.J.W.G.Visser, Eur.J.Biochem. 169 (1987) 259-268.
52. K. Vos, C. Laane, C. Veeger and A.J.W.G. Visser, Eur.J.Biochem. 169 (1987) 275-282.
53. K.E. Göklen and T.A. Hatton, Sep. Sci. Technol. 22 (1987) 831-841.
54. K. van 't Riet and M. Dekker in: Proc. 3rd Eur. Congr. Biotechnol., Verlag Chemie, Weinheim, vol. 3 (1984) pp. 541-544.
55. K.L. Kadam, Enz. Microb. Technol. 8 (1986) 266-273.
56. D.W. Armstrong and W. Li, Anal.Chem. 60 (1988) 88-90.
57. K.E. Göklen and A. Hatton, Separation Science and Technology 22 (1987) 831-841.
58. J.M. Woll, A.S. Dillon, R.S. Rahaman and T.A. Hatton in: Protein Purification: Micro to Macro, Alan R. Liss, New York (1987) pp. 117-130.
59. M. Dekker, K. van 't Riet, S.R. Weijers, J.W.A. Baltussen, C. Laane and B.H. Bijsterbosch, Chem. Eng. J. 33 (1986) B27-B33.
60. M. Dekker, J.W.A. Baltussen, K. van 't Riet, B.H. Bijsterbosch, C. Laane and R. Hilhorst, in: C. Laane, J. Tramper and M.D. Lilly (Eds.) Biocatalysis in Organic Media, Elsevier, Amsterdam (1987) pp. 285-288.
61. M. Dekker, K. van 't Riet, J.W.A. Baltussen, B.H. Bijsterbosch, R. Hilhorst and C. Laane, Proc. 4th Eur. Congr. Biotechnol., Volume 2, Elsevier, Amsterdam (1987) pp. 507-511.
62. M. Dekker, K. van 't Riet, B.H. Bijsterbosch, R.B.G. Wolbert and R. Hilhorst, A.I.Ch.E. Journal (1988) in press.
63. M. Dekker, K. van 't Riet, J.M.G.M. Wijnans, J.W.A. Baltussen, B.H. Bijsterbosch and C. Laane, Proceedings ICoM Tokio (1987) 793-794.
64. L. Dahuron and E.L. Cussler, A.I. Ch. E.J. 34 (1988), 130-136.
65. M.E. Leser, G. Wei, P.L. Luisi and M. Maestro, Biochem. Biophys.Res.Commun. 135 (1986) 629-635.
66. M. Leser, G. Wei, P. Luthi, G. Haering, A. Hochkoeppler, E. Blochliger and P.L. Luisi, J. de Chim.Phys. 84 (1985) 1113-1118.
67. M. Leser and P.L. Luisi, personal communications.
68. S. Giovenco, C. Laane and R. Hilhorst, Proc. 4th European Congress on Biotechnol. Volume 2, Elsevier, Amsterdam 1987, pp. 503-506.
69. G. Haering, P.L. Luisi and F. Meussdoerffer, Biochem. Biophys. Res. Commun. 127 (1985) 911-915.
70. G. Haering, A. Pessina, F. Meussdoerffer, S. Hochkoeppler and P.L. Luisi, Annals, N.Y. Acad. Sci. 506 (1987) 337-334.
71. S. Boeren, personal communications.
72. S. Boeren and C. Laane, Biotechnol. Bioeng. 29 (1987) 305-309.
73. C. Laane, S. Boeren and K. Vos, Trends Biotechnol. 3 (1985) 251-252.

74. C. Laane, S. Boeren, K. Vos and C. Veeger, Biotechnol.Bioeng. 30 (1987) 81-87.
75. B.I. Kurganov, L.G. Tsetlin, E.A. Malakhova, N.A. Chebotareva, V.Z. Lankin, A.V. Levashov, G.D. Glebova, V.M. Berezovskii, K. Martinek and I.V. Berezin, Dokl.Akad.Nauk SSSR 282 (1985) 1263-1267.
76. E.I. Belyaeva, L.Yu. Brovko, N.N. Ugarova, N.L. Klyachko, A.V. Levashov, K. Martinek and I.V. Berezin, Dokl.Akad.Nauk SSSR 273 (1983) 494-497.
77. N.L. Klyachko, M.Yu. Rubtsova, A.V. Levashov, E.M. Gavrilova, A.M. Egorov, K. Martinek and I.V. Berezin, Annals N.Y. Acad. Sci. 501 (1987) 267-270.
78. A.J.W.G. Visser and J.S. Santema, in: L.J. Krika, G.H.G. Torpe and T.P. Whitebread (Eds.) Analytical Applications of Bioluminescence and Chemiluminescence, Academic Press, London 1984, 559-563.
79. H. Hoshino and W.L. Hinze, Anal.Chem. 59 (1987) 496-504.
80. A.N. Eremin, M.I. Savenkova, and D.I. Metelitsa, Bioorg.Khim. 12 (1986) 606-612.
81. P. Speiser, in: P.L. Luisi and B.E. Straub (Eds.) Reversed micelles. Biological and Technological Relevance of Amphiphilic structures in apolar media, Plenum Press, New York (1984) pp. 339-346.
82. E.G. Abakumova, A.V. Levashov, I.V. Berezin and K. Martinek, Dokl. Akad. Nauk SSSR 283 (1985) 136-139.
83. A.V. Kabanov, A.V. Levashov and K. Martinek, Vestn.Mosk.Univers. 41 (1986) 591-594.
84. A.V. Kabanov, A.V. Levashov and K. Martinek, Annals. N.Y. Acad. Sci. 501 (1987) 63-66.
85. G. Haering and P.L. Luisi, J.Phys.Chem. 90 (1986) 5892-5895.
86. R. Scartazzini and P.L. Luisi, J.Phys.Chem. 92 (1988) 829-833.

REVERSE MICELLES AS A MODEL FOR THE STUDY OF MEMBRANE PROTEINS
AT MYELIN INTERLAMELLAR AQUEOUS SPACES.

C. NICOT and M. WAKS

1 INTRODUCTION

Although the existence and the possible role of reverse
micelles in biology remain a topic of research and discussion
(1), such a mode of organization can be observed in a highly
specialized membrane: myelin of central nervous system. The
multilamellar sheath originates from the rotation of the
oligodendroglial cell process around axons greatly increasing the
rate of nerve conduction. The subsequent compaction of myelin
leads to the tight apposition of bilayer leaflets. At both
apposed cytoplasmic and extracellular membrane surfaces, aqueous
spaces sandwiched between two lipid layers, shelter perimembrane
proteins or the polar segments of transmembrane proteins. If one
considers only those lipid layers adjacent to the aqueous spaces,
the protein or peptide chains are surrounded by an inverted
bilayer environment (Fig. 1). This observation prompted us a few
years ago to study the reactivity, the behavior and the structure
of myelin proteins in reverse micelles. In these organized
microassemblies the anisotropic and amphipathic nature of
biological membranes is preserved to a certain extent, as well as
the functionally important interface between surfactant head
groups and sequestred water. Such an environment includes the
charge density gradient, the head group packing, the availability
of hydrogen-bond forming groups, as well as the local proton
activity, the polarity and the dielectric constant of the
interfacial region (2).

A comparison of the organization of reverse micelles of
sodium bis (2-ethylhexyl) sulfosuccinate, water and isooctane
with the myelin aqueous spaces reveals a number of similarities.
In both structures the dimension of the protein-containing
aqueous spaces is of a comparable order of magnitude (30-50 Å)
(3). In both structures the affinity of polar heads for water is

the dominant force opposing close apposition of surfaces (4). The presence of salt in the aqueous content of native myelin corresponds to the presence of sodium counterions in the water pool of reverse micelles (5). Furthermore, Gent et al., (6) have observed by dielectric studies the presence in native myelin of highly organized water of "ice-like" character comparable to the bound water in the water pool of reverse micelles and displaying the reduced mobility of biological interfacial water (7). In this chapter we will attempt to describe and discuss in some detail the specific features in this membrane-mimetic system, of the two major structural myelin proteins which are believed to be involved in the maintenance of the integrity of myelin. They both acquire distinctive physicochemical and structural properties upon incorporation into reverse micelles.

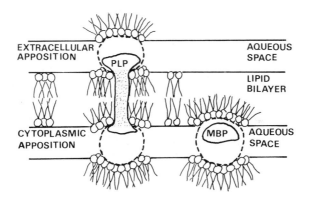

FIGURE 1.
    Schematic representation of the myelin basic protein (MBP) and of the Folch-Pi proteolipid (PLP) in a myelin aqueous space. Dashed circles underline the analogy with the structure of reverse micelles. Adapted from Ref. [19].

2 THE MYELIN PROTEINS
    A unique set of proteins constitutes about 80% of the total protein of the central nervous system myelin (8): they include the Folch-Pi proteolipid protein (PLP) and the myelin basic protein (MBP). Both proteins have been completely sequenced: the latter by Eylar (9) and the former by Stoffel et al.,(10) and Laursen et al.,(11). They are both basic: pI = 10.6 for MBP and

9.2 for PLP respectively. Their molecular weights (MW) are different: 30.000 daltons for PLP, 18.500 daltons for MBP. But the most important feature is their localisation in different domains of the myelin lamellae (Fig.1). MBP a water-soluble perimembrane protein, is located between the cytoplasmic apposition of the membrane, whereas the Folch-Pi proteolipid protein, an exceptionally hydrophobic protein, displays several membrane spanning α-helices and charged domains protruding into both the extracellular and cytoplasmic aqueous spaces. Some additional points deserve attention: MBP, a single tryptophan protein, is particularly suitable for fluorescence spectroscopic measurements. The absence of disulfide and thiol residues confers on the protein its markedly flexible conformation. In contrast, PLP displays 4 or 5 disulfide bridges and about 4 sulfhydryl residues which are crucial for the preservation of its native conformation (12). Finally, the proteolipid contains 2% (by weight) of covalently bound fatty acids (13) and is more stable as a lipid-protein complex.

3 SOLUBILIZATION OF MYELIN PROTEINS IN REVERSE MICELLES

Myelin proteins in the aqueous core of reverse micelles display unusual solubility properties. They can be related in part to the specific amino-acid sequences of the peptide chains and to the existence in water of physicochemical forces probably similar to those responsible for maintaining the packing of the intact myelin membrane.

When the solubilization of MBP in reverse micelles is studied from absorption measurements at 280 nm, as a function of the water content of the system (w = [H_2O] / [AOT]), at a constant concentration of AOT (50mM), it is obvious that the protein is totally insoluble in the absence of water. The solubility curve, increases sharply with increasing amounts of water up to a maximum value for w = 5.6, then decreases rapidly until a plateau is reached at a value of 11.2 (Fig.2 A). In the very narrow range of maximal solubilization where the mean aggregation number (i.e. the number of molecules of surfactant per micelle) is in the vicinity of 100 (14), the saturating overall concentration of MBP reaches a value of about 0.55 millimolar; however in the waterpool of reverse micelles, at 0.5% v/v water content, this represents a protein concentration 200 fold higher (0.11 M). A more realistic calculation, however, should take into account the increase in the volume of the water pool after protein incorporation (15).

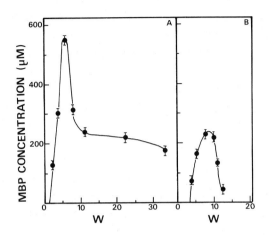

FIGURE 2.
    Solubilization curves of MBP   AOT /isooctane/water reverse micelles. The protein concentration in the overall solution is calculated from absorbance measurements at 278 nm and plotted versus w , the molar [H$_2$0] / [surfactant] ratio. In curve A the surfactant is 50 mM sodium bis (2-ethylhexyl) sulfosuccinate (AOT), in curve B, 100 mM tetraethyleneglycol mono-n-dodecyl ether (Nikkol). Adapted from Ref. [15].

    The very sharp profile of the solubility curve of MBP at low water to surfactant molar ratios around w = 6.0, is unusual (Fig. 2A). Optimal solubilization of other water-soluble proteins requires in general higher w values (16); this holds also for a basic peptide from adrenocorticotropin ACTH (1-24) as well (unpublished results). Although important, as will be seen below, the net charge of the peptide chain is not the sole origin of the observed phenomenon. It may reflect, instead, a specific interaction of the protein with the water entrapped in the micelles, the unusual physical properties of which at low w values result from the interaction with the ionic head groups of the surfactant. In AOT reverse micelles the presence of bound water (6 or 7 water molecules per AOT molecule), as opposed to free water, has been reported by various spectroscopic techniques (3, 5, 17). Therefore, it may be safely assumed that at optimal solubilization of MBP in AOT (w = 5.6), the amount of free water present is negligible. The solubility curve indicates, then, a preferential solvation of MBP by bound water. A change of conformation of the peptide chain due to interactions with this type of solvent and with nearby polar head groups may also favor solubility in the bound water phase. Evidence bearing on such a conformational change is described below.

The solubilization of MBP in reverse micelles of a non-ionic surfactant, tetraethylene glycol mono-n-dodecyl ether (Nikkol BL 4S4) was also investigated. Figure 2 B shows that the maximal solubilization is shifted toward a higher w value of 10.0. Moreover, the maximal overall concentration of solubilized MBP in micelles made of 100mM Nikkol reaches about half the value obtained with AOT micelles.

In contrast to MBP in AOT, the solubility of the Folch-Pi proteolipid, which is very low at 50mM AOT, increases with AOT concentration, leveling off at a relatively high (300 mM) value, (18). These results suggest the involvement of more than one micelle per protein molecule in the solubilization process. In this connection, we have noted that after the insertion of the proteolipid into reverse micelles, the surfactant concentration can be decreased by successive dilutions with isooctane, for example from 300 to a 50 mM concentration, with only a partial loss of protein by precipitation (unpublished results). This may indicate that at high AOT concentration, the protein is stabilized by interactions with protein-free micelles.

The solubilization of the water-insoluble Folch-Pi proteolipid was also studied as a function of the water content of the micellar system, at optimal (300 mM) AOT concentration. It is evident that the proteolipid is also totally insoluble in the absence of water, or at very low w values. Solubilization starts around w = 3.0, increases with the water content up to a w $\simeq$ 7.0 and then decreases for higher values of the water-to-surfactant mole ratio. The maximum overall concentration of solubilized proteolipid obtained is 150μM. In the water pool of reverse micelles at low w = 7, it represents a protein concentration 25 fold higher. However, this is 30 times less than that for MBP under optimal experimental conditions. In contrast, all the attempts to incorporate a completely delipidated protein into reverse micelles failed at all w values studied, from 5.6 to 16.8 (19).

The lipid requirements for the incorporation of the lipid-protein complex into the system were therefore investigated after a careful delipidation carried out by successive isooctane precipitations. Quantitative analyses by HPLC of protein-bound lipids reveal the persistence of a lipid-protein complex of stable relative composition after several isooctane precipitations and displaying optimal micellar solubilization. This complex is made of 6 ± 1.0 moles of acid lipid, including phosphatidylserine (PS ≃ 2.0); sulfatide (≃ 3.7) and the phosphatidylinositol (PI ≃ 0.6).

What might be the role of lipids (including the covalently bound fatty acids) in the incorporation of the proteolipid into reverse micelles? They apparently stabilize a conformation which confers on the protein the optimal solubility observed. Whether this conformation is also promoted in part by a microenvironment bearing a resemblance to that of native myelin, is at present more difficult to ascertain.

4 CONFORMATIONAL STUDIES.

Both myelin proteins share a significant property: after insertion into reverse micelles they undergo significant conformational changes. This effect is certainly due to their structural flexibility and to the transfer of the macromolecules from a homogeneous solvent to an ordered, anisotropic, membrane-like environment. The result of these conformational adjustments demonstrate that for membrane proteins or peptides active at interfaces, reverse micelles constitute a more appropriate membrane-mimetic milieu than the widely used organic solvent mixtures.

4-1. Structural changes of MBP.

Upon incorporation of MBP in reverse micelles, there is an obvious change in structure from a flexible, extended chain to a more ordered α-helical structure, as indicated by circular dichroism spectra (Fig.3). Such an increase in structural order was predicted from the sequence analysis by Martenson (20).

The circular dichroism spectrum of MBP measured in aqueous and in micellar solutions is illustrated in Fig.3. A shift from

348

198 to 205 nm is observed in reverse micelles, in addition to
striking differences in ellipticities at the negative extrema.
Fig.3 (insert) shows the calculated difference circular dichroism
spectrum between aqueous and micellar solutions of MBP. The
spectrum is characterized by a negative extremum at 222 nm,
indicating that the most probable structure generated consists of
α-helices. However, although the difference spectrum lacks a
distinctive second maximum near 208 nm, the presence of β
pleated sheet and β turns is probable (15). The conformation of
MBP was unaffected by surfactant concentration in the 25 to 100
mM range. No measurable differences were observed at increasing
water-to-surfactant molar ratios, i.e. for w values ranging from
2.0 to 22.4. It was concluded that the conformation of MBP was
not dependent either on the water content of the micellar system
or on the surfactant concentration, indicating the lack of
interaction between the polypeptide chain and the surfactant
monomers (15).

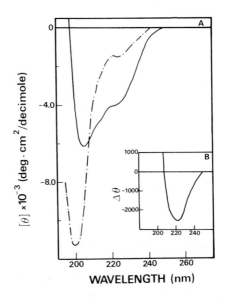

FIGURE 3
    Circular dichroism spectra of MBP:
    a) Far ultraviolet circular dichroism spectrum of MBP in
aqueous (⌐—·) and micellar (⟜—⟶) solutions of 50 mM AOT / isooctane
/ water, at w 5.6.   b) Calculated difference spectrum between
aqueous and micellar solutions.   Adapted from Ref. [15].

When measured in reverse micelles of Nikkol, a non ionic surfactant, at w = 5.6 the circular dichroism of MBP was changed relative to aqueous solution, but less so than in AOT. Calculations of the secondary structure yield a value of approximatively 18 ± 2 % α-helix for MBP in AOT and only 10% in Nikkol reverse micelles (15). Thus MBP adopts in reverse micelles a more folded, compact structure compatible with the model suggested by Shapira et al. (21).

4-2. Structural changes of the proteolipid.

The conformation of the water-insoluble Folch-Pi proteolipid in reverse micelles cannot be compared with that of the aqueous form of the lipid-depleted protein. We have indeed shown that delipidated protein cannot be incorporated into reverse micelles, the complete removal of lipids inducing a perturbation which cannot be reversed by readdition of lipids (19). This is similar to the finding that tightly bound lipids are essential in preserving the functional integrity of several membrane proteins (22).

As for many such proteins or peptides, structural measurements have been previously carried out in organic solvents, although it is now clear that the latter artificially increase the amount of ordered structural content. For PLP, Cockle et al., (23) have reported 95% α-helix in 2-chloroethanol, whereas in reverse micelles of AOT (in the 50 to 300 mMolar range) and at a w value of 5.6, Delahodde et al., (18) have found 55% α-helix. In fact, only the latter helix content is compatible with recent models of the proteolipid obtained from sequence and proteolytic studies. These models have 4 or 5 helices extending through the bilayer (10, 11).

4-3. Origin of the conformational changes.

The conformational changes occurring upon uptake of myelin proteins by reverse micelles arise from at least two major types of mechanism: electrostatic interactions and the state of hydrogen-bonding in entrapped water. The first originates from the existence of complementary charges between the protein basic side-chains and the sulfonates of the AOT polar head groups. In the case of MBP this is suggested strongly by the observation that the final structure is totally unaffected by the amount of

water present. Moreover by replacing AOT by a non ionic
surfactant, the electrostatic interactions are weakened with the
unfolding of MBP as a consequence.

These results are confirmed by fluorescence studies of MBP
in reverse micelles, for upon excitation at 290 nm a blue shift
of the sole Trp is observed from 350 nm in aqueous solution to
335 nm in reverse micelles. This wavelength remains unchanged
with increasing amounts of water and indicates a shielding from
it. Similar behavior has been also observed in the proteolipid,
although the interpretation of the results is rendered more
difficult by the presence of 4 tryptophan residues. The influence
of electrostatic interactions is supported also by the
fluorescence emission maximum (λ = 341 nm) of MBP in Nikkol
micelles at w = 5.6. Thus, there is a correlation between the
absence of charged head groups, an increased tryptophan
accessibility to solvent and a loss of α-helix of MBP in Nikkol.
The firm anchoring of MBP at the micellar interface thus
represents the predominant mechanism responsible for the
conformational change observed (15).

The physical properties of entrapped water constitute
another important factor responsible for changes in the periodic
structure of myelin proteins. At low water content,
intramolecular bonding of the protein will be favored as a
consequence of the disruption of the three dimensional hydrogen-
bonded network characteristic of bulk water; at high w values
peptide backbone-water interactions will be the more prominent.
Such observations have been first reported by Seno et al.,(24)
with basic homopolyaminoacids. It is thus clear that water
relations will affect the macromolecular conformation in part by
modifying the hydrogen bonding potential.

Gallay et al.,(25) have investigated the respective
importance of charges and of the state of micellar water in the
conformational changes of synthetic adrenocorticotropin peptides
ACTH (1-24) and (5-10), at various values and in comparison with
the small molecule N-acetyl-tryptophanamide (NATA). For NATA the
wavelength of maximum fluorescence emission varies continuously
as a function of w . It indicates a progressively higher polar
environment of the chromophore and/or changes in the physical
properties of entrapped water, such as dielectric constant and
viscosity. It has to be emphasized that, even at the highest

water-to-surfactant molar ratio measured (w = 22.4), the maximum
wavelength of fluorescence emission (340 nm) is still to the blue
of a fully exposed indole derivative (350 nm). The behavior of
the basic ACTH (1-24) (6 positive charges) differs from that of
the short ACTH (5-10) (2 positive charges). The latter peptide
behaves as NATA, with continuously varying wavelength of
maximum fluorescence emission from 336 nm to 345 nm. ACTH (1-24),
in contrast, undergoes a variation from 334 nm at w = 2.0 to
338 nm at w = 7, where all the water is still tightly bound. The
maximum emission wavelength then remains invariant between
w = 7 and 22.4, indicating that the environment of the
chromophore is shielded from additional free water due to the
attractive interactions between the peptide and the micelle inner
wall.

At the same time ACTH (1-24) undergoes a dramatic
conformational change from β-pleated sheet to unfolded structure,
as a function of the presence of unbound water, in which
interaction of the peptide chain with mobile water dipoles is
possible (Fig.4).

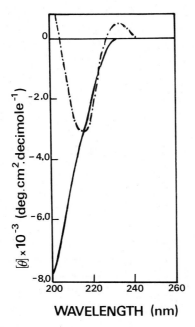

WAVELENGTH (nm)

FIGURE 4
    Far ultraviolet circular dichroism spectra of ACTH (1-24) in
micellar solutions at 25 mM AOT and at w values 3.5 (—·—) and 22.4
(———). Adapted from Ref. [25].

These examples point to the respective roles of the attractive charges as compared to the properties of immobilized water in stabilizing the protein conformation. In conclusion, electrostatic interactions may not prevent an extensive unfolding of short polypeptide chains as in ACTH (1-24), whereas in larger proteins such as MBP a more subtle balance between the stabilizing forces must exist. For example initially the electrostatic forces may induce the conformational change while the intramolecular hydrogen bonds could stabilize and subsequently orient it.

5 ORGANIZATION OF MYELIN PROTEINS WITHIN REVERSE MICELLES.

5-1. Reactivity of protein side-chains within micellar water.

In the interlamellar aqueous spaces MBP and the water soluble segments of the proteolipid interact both with interfacial water and the charges of the membrane surfaces. Thus, an important question in the understanding of myelin protein-lipid interactions is the localization of peptide residues or segments coming in contact with the polar head groups.

Our results with MBP encapsulated into reverse micelles show that many segments of the polypeptide chain are in contact with water. For example, all the 13 lysine ϵ-amino groups were found accessible to the water soluble fluorescent reagent o-phthalaldehyde. The binding of hemin (26) by micellar solutions of the protein indicates that the residues adjacent to the binding site (in the vicinity of phe-phe 44-45) are also accessible to aqueous solvents. These results argue against extensive contacts or penetration into the interface although strong head group interactions are not precluded (15).

The sole tryptophan residue of MBP, which is completely accessible to oxidation by N-bromosuccinimide in aqueous solutions (15), becomes inaccessible to the reagent upon insertion of MBP into reverse micelles. In contrast, the model tryptophan compound (NATA) or the basic ACTH (1-24) tested in identical experimental conditions, were fully reactive with N-bromosuccinimide (25), indicating that in MBP the tryptophan is shielded from water.

For the proteolipid, segments of the protein, for example those including most SH groups, are exposed to the aqueous core of reverse micelles. This also holds for all the ϵ-amino groups and for at least 2 of the 4 tryptophans, which can be oxidized by N-bromosuccinimide (19). It indicates that these side chains remain accessible to aqueous solutions of reagents. In fact titration of such groups can be used experimentally to delineate the topography of the protein in contact with the aqueous core of reverse micelles or with the apolar organic solvent. The internal dynamics of protein molecules brought about by time resolved fluorescence techniques also yields interesting information about the organisation of the macromolecule within micelles. In this respect, MBP with its sole Trp residue is an ideal candidate for such studies as compared to the proteolipid with four tryptophans.

5-2. <u>Internal dynamics of MBP in reverse micelles</u>.

As mentioned above, the tryptophan environment within the protein becomes more apolar and/or more constrained on the nanosecond time scale as indicated by the blue-shift of the emission maximum. Steady-state and time-resolved fluorescence anisotropy measurements demonstrate that these changes are accompanied by severe restrictions of the tryptophan rotational motion upon incorporation of MBP into reverse micelles (15) in striking contrast to its motion in water. Furthermore, the rotational motion of the fluorophore becomes faster with increased w values. It appears that increasing the water content of micelles does not measurably modify the overall protein conformation but only the internal protein dynamics.

Gallay et al. (25) have studied in detail the Trp internal dynamics and overall peptide rotational diffusion of the basic peptide ACTH (1-24) and of the shorter peptide ACTH (5-10) in comparison with neutral glucagon. The results give on a more limited scale a clue to what happens with myelin proteins. Upon transfer of all these peptides from aqueous solution into reverse micelles, a severe slowing down of the peptide internal dynamics and a strong limitation of the angle of Trp rotation are observed. The amplitude of these effects is more important when the Trp residue is close to the surfactant/water interface as in Trp-9 in ACTH (1-24). The variation of the long correlation time of the Trp fluorescence anisotropy decay as a function of the water content of reverse micelles is depicted in Fig.5. The overall tumbling of all peptides is slowed in reverse micelles as compared to water. The amplitude of this effect is a function of both the micellar water content and the peptide charge.

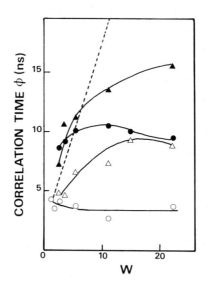

FIGURE 5
Variation of the long correlation time in nanoseconds of the Trp fluorescence anisotropy decay in peptides inserted into reverse micelles as a function of w , (▲) ACTH (1-24), (●) glucagon, (△) ACTH (5-10), (○) NATA. (‹--›) Variation of the rotational correlation time of the micelles calculated from the Stokes- Einstein relation. Adapted from Ref.[25].

For the overall rotational motion of a rigid body entrapped in reverse micelles, two limiting cases can be considered according to their size (27). At low w values, corresponding to small-size micelles, the microscopic viscosity is considerably higher than in bulk water (28). The solute is not allowed to rotate freely within the micelle, and the long correlation time of the anisotropy decay corresponds to the rotational diffusion of the micelle itself. At high w values, the rotational correlation time of the micelles is larger since their volumes increase. However, the microscopic viscosity inside the micelle decreases at the same time, and therefore tumbling of the solute within the micellar water pool starts to occur. In this case, the long correlation time is a combination of both rotational processes: the rotational diffusion of the peptide within the micelle and the rotational diffusion of the micelle itself.

The strength of coupling between the two rotational motions is governed not only by the interactions between the solute and the surfactant, but also by interactions with embedded water. In this respect, ACTH (1-24) interacts more strongly than glucagon and ACTH (5-10) with the AOT sulfonate group and therefore displays a stronger coupling. The influence of the peptide-water interactions on the coupling mechanism is suggested by the fact that despite their differences in molecular weight, ACTH (5-10) (MW: 830) and glucagon (MW: 3.500) display similar overall tumbling at high w values. It can be concluded that the electrostatic interactions of the side chains, hydrogen bonding with water and probably also with the surfactant head groups are crucial for the localisation and the conformational distribution of the peptides and proteins with respect to the micellar interface.

6 REORGANIZATION OF THE MICELLES IN PRESENCE OF MYELIN PROTEINS.

The structural reorganization of reverse micelles after myelin protein uptake has been investigated in detail by Chatenay et al.,(29,30) and Binks et al.,(31) using non perturbing physical techniques such as quasi elastic light scattering (QELS) and fluorescence recovery after fringe pattern photobleaching (FRAPP). The physical picture accompanying the uptake of proteins varies according to the protein and to its location in the myelin interlamellar aqueous spaces. Thus, a different equilibrium situation is established for a perimembrane protein as MBP or for a transmembrane protein such as the Folch-Pi proteolipid.

For MBP, the solubilization of the protein is accompanied by two types of perturbations. At low w values ($\approx$ 6), an increase from 29 to 43 Å of the micellar apparent hydrodynamic radius (R_H^{app}) is observed. This is consistent with purely geometrical model derived from the water-shell model (32), where one protein molecule is enclosed within a larger micelle built up from several smaller initially protein-free micelles. The number of such micelles required for the formation of a protein-filled one was calculated to be around 3.3. The resulting redistribution of surfactant and water at the molecular level probably originates from a competition between the surfactant and the protein for interfacial water layers. This result can be juxtaposed with the unusual solubility of MBP in reverse micelles (Fig.2) and probably mirrors some of the properties of the aqueous myelin spaces where the protein interacts with both interfacial water and charged membrane surfaces. At a high water content (w = 22.4) a more subtle perturbation takes place. Here the hydrodynamic radius (51 Å), does not measurably vary upon MBP incorporation. However it appears that intermicellar interactions increase the apparent R_H as a function of AOT concentration in the 25-100 mM range. The presence of the protein thus induces attractive interactions. This type of interactions may play a role in maintaining the compaction and the integrity of the myelin sheath (30).

A much more intricate picture emerges from observations of the Folch-Pi proteolipid in reverse micelles (31). From solubility curves we know that the integral membrane protein needs a rather high surfactant concentration (200-300 mM) for optimal solubilization. This is indicative of a new situation in which the water-shell model does not apply even at a low water content. At a finite AOT concentration (200 mM), QELS experiments carried out as a function of protein concentration reveal the presence of two apparent hydrodynamic radii differing 10 fold in size. The smallest apparent radius (52 Å) corresponding to that of protein-free micelles at the same AOT concentration remains almost constant throughout the studied proteolipid concentration range. The apparent radius of the larger one increases linearly with protein concentration (from 2.5 to 15 μM) (Fig.6). Since QELS is a global measurement of protein-free and protein-containing micelles, FRAPP measurements have been carried out in order to obtain R_H values for both the protein-free and protein-containing micelles after extrapolation to zero AOT concentration. The extrapolated value is 44 Å while 29 Å was found for the "empty" micelles. Since large size aggregates were not observed, we must assume that the population of these aggregates which probably represent protein bridges between micelles, is too small to be observed by FRAPP experiments.

If we compare the R_H of the proteolipid-micelle complex (44 Å) with that of PBM-containing micelles (43 Å) we observe that it is identical for two myelin proteins of significantly different molecular weights (30.000 and 18.500 daltons respectively). This is understandable if we compare the Stokes radius of PBM (28 Å) with that of other proteins in aqueous solvents. For example, for ovalbumin a 45.000 dalton protein a value of 28 Å is given, while for myoglobin a 17.000 dalton protein, it is 19 Å only. This is most probably due to the exceedingly high amount of water bound to MBP. Martenson has indeed reported that upon hydration, the volume of the protein doubles from 22.000 A³ to 47.000 A³ (33). In contrast, it has been shown that the proteolipid is weakly hydrated (unpublished results) although differences in its Stokes radius and molecular volume are not known.

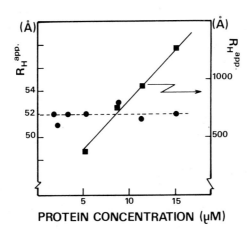

FIGURE 6
    Apparent hydrodynamic radii ($R_H^{app.}$) as a function of proteolipid concentration, at [AOT] = 200 mM and w = 7. The values are deduced from QELS experiments. ● corresponds to the left hand ordinate (low R_H^{app} values), ■ to the right hand (high R_H^{app} values). Adapted from Ref. [31].

From the above results it is difficult to determine, at present, the topography of the Folch-Pi proteolipid in reverse micelles. The chemical reactivity of the protein side-chains described above permits us to conclude that the thiol and ε-amino groups as well as some tryptophans belong to the hydrophilic domains of the protein encapsulated in the aqueous core of the reverse micelles. The apolar α-helices penetrating the membrane bilayer would then interact with isooctane. Such a model has been proposed by Ramakrishnan et al., (34) for rhodopsin-phospholipid complexes in heptane.

CONCLUSION

Myelin proteins incorporated into reverse micelles exhibit a set of distinctive properties not observed previously in homogeneous detergent or organic solvents. The information obtained by this new approach suggests that the mode of organization of these macromolecules in the membrane-mimetic system reflects for each of them the nature of the environment experienced in the aqueous spaces of intact myelin. Thus, for example, while both proteins require a very low amount of water for optimal solubilization in the system, this result does not necessary imply a similarity in the mechanism of interaction with the interfaces. For MBP, it may indicate a high affinity for interfacial water, while for the proteolipid it may arise from a subtle balance between the dual nature of the interactions of the lipid-protein complex with water, a definite amount being necessary for the phospholipid charges and also for the limited surface charges of the protein. As we have shown (27), the specific protein amino-acid sequence seems to play a prominent role in the mode of organization of myelin proteins within reverse micelles. In myelin basic protein where discrete hydrophobic and polar regions alternate (8), the macromolecule interacts with both the charged polar head groups of the surfactant and interfacial water, while the hydrophobic domains are removed from water in the folding of the protein. For the proteolipid, the situation appears much more complex because of the existence both of the charged domains interacting with water and of distinct hydrophobic helices that interact with the apolar solvent and probably also with protein-free micelles. Supporting these ideas is the fact that high surfactant concentrations promote intermicellar interactions. The growth in aggregate size with protein concentration which does not occur in the same way with MBP (29), appears thus to arise from intermicellar protein cross links described for rhodopsin by Montal (1).

Reverse micelles have thus opened the way for the study of myelin proteins in a membrane-mimetic environment. It is hoped

that further investigations will elucidate some of the
fundamental questions concerning membrane proteins and proteins
in general.

ACKNOWLEDGEMENTS

The authors are grateful to Dr. P. Kahn for his critics and
comments. They thank M. Vacher for skilfull editorial help. This
work was supported by grants from CNRS and INSERM.

REFERENCES

1   M. Montal in P.L. Luisi and B.E. Straub (Eds), Reverse
     Micelles. New York: Plenum Press, 1984, pp.221-229.
2   P.L. Luisi and L.J. Magid, CRC Crit.Rev.Biochem., 20, (1986)
     409-474.
3   A. Maitra, J.Phys.Chem., (1984) 5122-5125.
4   D.A. Kirschner, A.L. Ganser and D.L.D. Caspar in P. Morell
     (Ed), Myelin. New York: Plenum , 1984, pp.74-84.
5   K.F. Thompson and L.M.Gierasch, J.Am.Chem.Soc., 106, (1984)
     3648-3652.
6   W.L.G. Gent, E.M.Grant and S.W. Tucker, Biopolymers, 9,
     (1970) 124-126.
7   I.D. Kuntz and W, Kauzmann, Adv.Protein.Chem., 28, (1974)
     239-345.
8   M.B. Lees and S.W. Brostoff, in P.Morell (Ed), Myelin. New
     York: Plenum, 1984, pp.197-209.
9   E.H. Eylar, Proc.Natl.Acad.Sci. USA, 67, (1970), 1425-1431.
10   W. Stoffel, H. Hillen, W. Schröder and R. Deutzmann, Hoppe
     Seylers Z.Physiol.Chem., 364, (1983), 1455-1466.
11   R.A. Laursen, M. Samiullah and M.B. Lees, Proc. Natl. Acad.
     Sci. USA , 81, (1984), 2912-2916.
12   M. Vacher, M. Waks and C. Nicot, Biochem.J., 218, (1984),
     197-202.
13   P. Stoffyn and J. Folch-Pi, Biochem.Biophys.Res.Commun., 44,
     (1971), 157-161.
14   C. Grandi, R.E. Smith, P.L. Luisi, J.Biol.Chem., 256, (1981)
     837-843.
15   C. Nicot, M.Vacher, M.Vincent, J.Gallay and M.Waks,
     Biochemistry, 24, (1985), 7024-7032.
16   P.L. Luisi and R. Wolf, in P. Mittal and J.M. Fendler (Eds),
     Solution behavior of Surfactants. New York: Plenum, 1982,
     pp.887-905.
17   M. Wong, J.K. Thomas and T. Nowak, J.Am.Chem.Soc., 99,
     (1977), 4730-4736.
18   A. Delahodde, M. Vacher, C. Nicot and M. Waks, FEBS Lett.,
     172, (1984), 343-347.
19   M. Vacher, M. Waks and C. Nicot, submitted for publication.
20   R.E. Martenson, J.Neurochem., 36, (1981), 1543-1560.
21   R.Shapira, K.D.Wilkinson and G.Shapira, J.Neurochem., 50,
     (1988), 649-654.
22   Rivnay and G. Fischer Biochemistry, 25, (1986), 5686-5693.
23   S.A. Cockle, R.M. Epand, J.M. Boggs and M.A. Moscarello,
     Biochemistry, 17, (1978), 624-629.

360

24   M. Seno, H. Moritoni, Y. Kuroyanagi, K. Iwamoto and G.
      Ebert, Colloid Polymer Sci., 262, (1984), 727-733.
25   J. Gallay, M. Vincent, C. Nicot and M. Waks, Biochemistry,
      26, (1987), 5738-5747.
26   M. Vacher, C. Nicot, M. Pflumm, J. Luchins and S. Beychok,
      Arch.Biochem.Biophys., 231, (1984), 86-94.
27   E. Keh and B. Valeur, J.Colloid Interface Sci., 79, (1981),
      465-478.
28   K. Tsuji, J. Sunamoto and J.H. Fendler, Bull.Chem.Soc.Jpn.,
      56, (1983), 2889-2893.
29   D. Chatenay, W. Urbach,   A.M. Cazabat, M. Vacher and M.
      Waks, Biophys.J., 48, (1985), 893-898.
30   D. Chatenay, W. Urbach, C. Nicot, M. Vacher and M. Waks,
      J.Phys.Chem., 91, (1987), 2198-2201.
31   B.P. Binks, D. Chatenay, C. Nicot, W. Urbach and M. Waks,
      Submitted for publication.
32   F.J. Bonner, R. Wolf and P.L. Luisi, J.Solid-Phase Biochem.,
      5, (1980), 255-268.
33   R.E. Martenson, J.Biol.Chem., 253, (1978), 8887-8893.
34   V.R. Ramakrishnan, A. Darszon and M. Montal, J.Biol.Chem.,
      258, (1983), 4857-4860.

INVERSE MICROLATEXES : MECHANISM OF FORMATION AND CHARACTERIZATION

F. CANDAU and M. CARVER

1. INTRODUCTION

The use of micelles as sites for organic reactions is well established. These submicroscopic aggregates of surfactants are effective phase-transfer agents and can catalyze bimolecular reactions by bringing reactants together (1,2). Owing to their particular environment, micelles alter reaction rates and control the stereochemical course of reactions. They are used in chiral recognition and in membrane mimetic chemistry (3). Studies of photoprocesses, in these simple organized assemblies has also led to a better understanding of similar processes in biological systems (4).

Free-radical polymerization in colloidal dispersions, e.g. emulsions, is a standard technique for the production of polymers. Isolation of single free radicals in loci of small dimensions affords a means of attaining simultaneoustly high molecular weights and high reaction rates. However, only in the past decade have thermodynamically stable micellar systems been employed in the production of polymers. While a number of reports of the polymerization of oil-soluble monomers in direct micellar systems have recently been published (5-8) it is polymerization in reverse micelles and microemulsions which appears to hold the most promise for commercial applications, owing to the stability of the polymer latexes thus formed, as well as the high molecular weight and water solubility of the polymers themselves (9,10).

In this paper, we discuss some aspects of the inverse microemulsion polymerization process, which can be directly correlated to the structural characteristic of the final latexes. An analysis is made on the role of monomer on the interfacial properties of the micellar systems. These properties govern the stability of the systems, a key parameter for industrial applications.

2. STRUCTURAL AND INTERFACIAL PROPERTIES OF REVERSE MICELLES IN THE PRESENCE OF MONOMER

Nearly all the polymerizations in inverse micellar systems reported thus far have taken place in oil/sodium bis(2-ethylhexyl)sulfosuccinate (AOT)/H_2O + monomer) systems, where the oil is toluene (9-14) benzene (15) heptane (10) or decane (16) and the monomer is acrylamide (AM). The polymerization of methacrylic acid in inverse micelles of dodecylbenzenulfonic acid in benzene has also been reported (17).

The toluene/AOT/H2O systems which have been the most thoroughly studied, display some interesting properties. The solutions are transparent and thermodynamically stable, and contain inverse micelles with diameters on the order of 5 to 10 nm, as measured by quasielastic light scattering (QELS) (9-11). In the absence of AM, the micelles display approximately hard sphere behavior, but when AM is added, the interaction parameters determined by both static and quasi-elastic light scattering and viscometry measurements are indicative of strong interparticular attractive forces (10,11). These results confirm the partitioning of AM between the aqueous core of the droplets and the interfacial shell. It was shown in these systems that AM acts as a cosurfactant, thus increasing the micellar solubilization capacity (10) (see also chapter 14). This extends significantly the domain of the phase diagram accessible to polymerization reactions in inverse micellar systems.

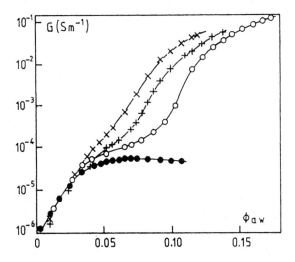

Figure 1. Specific conductivity of inverse micellar systems as a function of the volume fraction of (AM + H2O) in the disperse phase. AM/H2O (weight ratio) = (•), 0.00 ; (o), 0.24 ; (+), 0.40 ; (x), 0.67. (AOT/Toluene (weight ratio) = 0.177).

The effect of AM on the interfacial properties of the systems could also be inferred from conductivity experiments (18). Figure 1 shows the variation of the electrical conductivity of toluene micellar systems containing various acrylamide to water ratios, versus their volume fractions. Addition of AM affects strongly the conductivity of the solutions, and the effect increases as the ratio of AM to water increases. The rapid rise in conductivity in the AM-containing systems was attributed to the phenomenon of percolation. A similar behavior has already been observed for inverse microemulsions containing alcohols as surfactants (19-21). For these labile systems, it is commonly assumed that the interfaces must temporarily open to form water-channels between the particles, allowing the current to flow. It is apparent from Figure 1 that addition of acrylamide can convert a nonpercolating (water only) system to a percolating system.

The effect of AM on the percolative behavior of the micellar systems is directly related to its cosurfactant properties. The presence of monomer molecules in between the AOT molecules increases the flexibility and fluidity of the interface, resulting in a change of its curvature. In systems which are percolating, this structure apparently has an important effect upon the formation of polymer latex particles, as discussed below.

Under certain conditions, the radius of curvature can become so large that the globular configuration transforms toward a bicontinuous structure characterized by randomly interconnected oily and aqueous domains. This transition can be induced by addition of ionic monomers or/and electrolytes to microemulsions stabilized by non ionic surfactants. Studies relative to the polymerization of various water-soluble monomers in non ionic bicontinuous microemulsions have been reported (22-24). It was shown that at the very early stages of the polymerization, the bicontinuous structure breaks down and the reaction proceeds inside globules. The resultant product is a clear and stable microlatex consisting of high molecular weight polymers entrapped within water-swollen particles. It must be noted that bicontinuous microemulsions allow the incorporation of high monomer contents (up to 25%) which make them promising for industrial applications.

3. FORMATION AND CHARACTERIZATION OF LATEX PARTICLES

Any fundamental or applied research on free radical polymerization in micelles or microemulsions requires a careful investigation of the properties of the latexes and the molecular characteristics of the polymers formed. The former comprise stability, fluidity, transparency, size and size distribution of the latex particles. The latter include polymer molecular weight and polydispersity index, electrolyte character and, in case of copolymers, monomer sequence distribution. All these parameters are closely correlated with the mechanism of formation of latex particles and the corresponding kinetics. Therefore, a thorough understanding of the mechanism in inverse micellar systems would allow optimization of reaction conditions and component concentrations. A detailed kinetic study on the effect of initiator, surfactant and monomer concentrations

364

and nature of the solvent has shown the impact of these variables on the polymerization rate and polymer molecular weight (14,15,25).

To describe the building up of the polymer latex particles in micellar systems, it is necessary to determine both particle size and number as a function of the extent of polymerization. This can be achieved by means of quasi-elastic light scattering (QELS) and electron microscopy (EM) experiments. By QELS, one measures the water-swollen polymer particle diameter, by EM the "dry" polymer particle diameter. Both techniques require to work with dilute samples, which implies that the structure and the particle size do not change with dilution. In particular, one thousand-fold dilution and the drying process used in EM can introduce some artifacts or a flattening of the particles resulting in an increase in size (18).

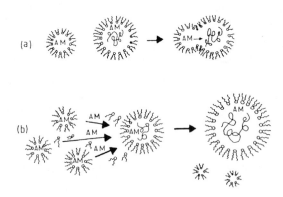

Figure 2. Scheme of the particle growth mechanism (a) by collisions between particles ; (b) by monomer diffusion through the toluenic phase (Reproduced with permission from [ref. 10]. Copyright 1984. Academic Press).

Studies performed on toluene/AOT/(H$_2$O + AM) systems have shown that the diameter of the final latex particle is somewhat dependent upon the initial composition of the system but is in all cases of the order of 40 to 50 nm, that is much larger than that of the initial micelles (5 to 10 nm) (10,11). These results indicate that nucleated polymer particles grow by the addition of monomer from other micelles, either by fusion with neighboring micelles, or by diffusion of the monomer through the continuous phase (figure 2). As a result, the final microlatexes consist of two populations in equilibrium, spherical polymer particles (d ~ 40 nm) with a narrow size distribution and very small AOT micelles (d ~ 3 nm) which arise from the excess of surfactant caused by the particle growth. The microlatexes are transparent or slightly bluish and show no settling over months. They are very fluid (a few cp) and behave like hard-spheres, without strong interactions between the particles. The interparticular forces are indeed reduced after polymerization, as most of the polymer is strongly collapsed within the particles.

The polymers produced and entrapped within the latex particles have ultra-high molecular weights (10^6 - 10^7). The production of high molecular weights required in most applications is achieved by a photochemical process rather than by a thermal initiation (26). Combined measurements of particle diameter and molecular weight of polymer lead to the conclusion that the number of polymer chains per latex particle is quite low, often averaging only one or two molecules per particle (14). These values are far below what is usually found in emulsion polymerization where a number of thousands of chains is commonly observed (27). This unusual finding cannot be interpreted from classical theories established for polymerization in dispersed systems. In order to gain some insight into the processes occurring in these systems, a study on particle nucleation (i.e. formation of polymer particles) was undertaken by means of the previously mentioned techniques. Electron photo-micrographs show that the diameters of the "dry" polymer particles obtained for samples at various degrees of conversion remain approximately constant and of the order of 30-35 nm while the particle number grows steadily. These results show that the systems undergo new particle nucleation at all stages of the reaction. This is in contrast to the somewhat reminiscent technique of emulsion polymerization, where new particles are nucleated only at the beginning of the reaction, then continue to grow by addition of new polymer chains throughout the remainder of the polymerization (27). Unlike the emulsion systems, inverse micellar systems have a larger number of micelles throughout the polymerization, thus the initiating radicals formed in the continuous phase are always more likely to encounter previously unnucleated micelles than already nucleated polymer particles.

QELS experiments provide additional information on the particle nucleation process. Different behaviors were observed depending on whether the starting systems were percolating or nonpercolating. Figure 3 shows the variation of the hydrodynamic radius R_H of the polymer particle as a function of the degree of conversion p for two typical systems (18).

For nonpercolating systems, the hydrodynamic radius first exhibits a sharp rise from the micellar size to a value which remains roughly constant up to 100% conversion, in good agreement with the results obtained from EM.

In percolating systems, the initial rise is much larger, since R_H reaches values as high as 90 nm. When the polymerization proceeds, the size of the polymer particle decreases drastically to reach at high p a value close to that obtained for nonpercolating systems. The different behavior observed in QELS

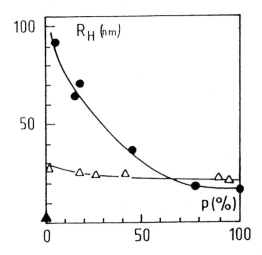

Figure 3. Hydrodynamic radius as a function of the degree of conversion for (●) an initially percolating system and (△) an initially nonpercolating system.

and EM experiments was attributed to the network structure in the percolating systems. As shown in figure 4, a particle initiated at some point along the percolating chain could grow along the chain, forming a "string of beads" -a series of individual micelles held together by the polymer chain. Rather than coalescing to form a single large spherical particle, curvature constrains in these AM-depleted micelles would cause them to maintain their individuality, thus expanding the polymer coil. When the particles are dried for microscopy, they collapse to a much smaller size, as mentioned above. The expanded particles would only be formed while the micellar systems are percolating, during the first 10 to 15% of the polymerization, according to conductivity measurements of the latex solutions. Afterwards, particle nucleation would proceed in the same manner as for nonpercolating systems.

4. APPLICATIONS OF MICROLATEXES AND POLYMERS

A main incentive to the development of polymerization of water-soluble monomers in heterogeneous media comes from the possibility of obtaining ultra-high molecular weight polymers. The polymerization reaction is most often carried out in inverse emulsion. The resultant inverse latex has lower viscosity, higher polymer concentrations and provides better heat transfer than the polymer prepared in homogeneous solution. For use in those applications where the water-soluble polymers are effective, the latex is inverted - by adding excess water and some inverting agent - to an oil in water emulsion and the polymer is

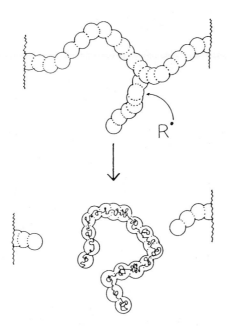

Figure 4. Possible mechanisms for the formation of the large particles at low conversion in percolating systems (R : initiator radical.).

released to the continuous aqueous phase where it easily dissolves. This behavior differs from that observed with solid powder polymer, which forms a gel structure or aggregates when dissolved in water for use (28). However, certain problems are encountered in the production of water-soluble polymers in w/o emulsions which limit the applicability of the process. For example, the inverse latex has usually a poor colloidal stability due to the fact that electrostatic stabilization is not effective in non aqueous media (29). This results in the formation of a sediment which may be hard to redisperse. Gravity can also accelerate flocculation in these systems characterized by a large difference in density between the dispersed and continuous phases. Another difficult problem is the excessive amount of coagulum formed during the polymerization reaction. After removal of the coagulum, the effective amount of polymer formed is reduced and production costs are increased.

Polymerization in inverse micelles and microemulsions overcomes most of these problems and provides a novel technique for the production of water-soluble polymers. The main advantages of the process are the following :

- low viscosity, optical transparency and high stability of the microlatex
- no coagulum formation
- low dimensions and narrow particle size distribution. The small particle size prevents flocculation since gravity forces are reduced

- ultra-high molecular weight polymers. The larger number of micelles contained in microemulsions compared to that in emulsions contribute to the formation of high degrees of polymerization (DP $\propto$ N)

- the microlatex is self-inverting, so that no additional surfactant is needed to promote its inversion.

With respect to the economical aspect, the main drawback of the process is the rather expensive formulation (high surfactant concentration). This is partially balanced by the very high rate of polymerization due to the great number of micelles, loci of the polymerization. Total conversions are usually achieved in a few minutes compared with hours in the usual process.

Most of the applications described for water-soluble polymers prepared in water-oil emulsions can in principle be extended to the inverse microemulsion polymerization process and several patents have recently been issued (30-33). The list of applications given below, although not exhaustive, illustrates some possibilities of the process.

Inverse microlatexes can be used after dilution to water to form thickened solutions for improving the production of oil fields. They have advantages witrh respect to conventional latexes, as a result of their lower particle size, their lower degree of polydispersity and their great stability. They result in a better scavenging of the oil formation and thus in a more efficient oil recovery. The method consists of injecting polymer solutions into a formation, through one injection well, circulating them through the formation and recovering the displaced hydrocarbons through one production well. Test conducted in laboratory have proved their efficiency (31). In other techniques of oil production, inverse microlatexes can be used for ground consolidation, manufacture of drilling muds, prevention of water inflows during the bringing in production of oil wells, and as completion or fracturation fluids.

Most uses in the paper manufacture, water treatment and mining fields are based on the ability of water-soluble polymers such as polyacrylamides to flocculate solids in aqueous suspensions (34). Small mineral or pigment particles settle very slowly and are difficult to eliminate or recover. Addition of charged polyacrylamides permits them to agglomerate. The ultra-high molecular weight polymers produced in the microemulsion process can be very effective in connecting together the small particles through bridging or charge neutralization. Moreover, classical polyacrylamide emulsions are subjected to rapid changes in temperature (in winter time) which cause them to coagulate rather than remain finely dispersed particles. This reduces drastically their usefulness as flocculants. Inverse microlatexes exhibit excellent freeze-thaw properties and contain finely divided polymer particles (~ one polymer chain in a low size particle) which should insure a higher activity.

Other applications include surface coatings, adhesives, photographic emulsions, lubricating and cleaning drains, retention aid in paper making and food processing. Finally, the low viscosity and good stability of microlatexes can be useful for assembling glass fibers. These are the principal uses of water-soluble polymers but many others have been described in the patent literature.

To conclude this brief survey, polymerization in reverse micelles or microemulsions is a rapidly developing field which combines challenging scientific problems with practical applications. This novel technique has been shown to have a unique mechanism characterized by continuous particle nucleation and resulting in one polymer chain per latex particle. This differentiates it from both homogeneous solution and conventional inverse emulsion polymerizations and deserves further investigations.

REFERENCES

1. C.A. Bunton, Pure Appl. chem, 49 (1977) 969-979.
2. J.H. Fendler and E.J. Fendler in "Catalysis in Micellar and Macromolecular Systems" Acad. Pres. New-York, 1975.
3. J.H. Fendler in "Membrane Mimetic Chemistry", Wiley-Interscience, New-York, 1982.
4. J.C. Scaiano, E.B. Abuin and L.C. Stewart, J. Am. Chem. Soc., 104 (1982) 5673-5679.
5. C. Schauber, Thesis, Mulhouse University, 1979.
6. S.S. Atik and K.J. Thomas, J. Am. Chem. Soc. 103 (1981) 4279-4280.
7. P. Lianos, J. Phys. Chem. 86 (1982) 1935-1937.
8. F. Candau in : H. Mark, N. Kikales, C.G. Overberger, G. Menges (Eds), Encyclopedia of Polymer Science and Engineering, 2nd Ed, Wiley, New-York, 1987, 9 pp. 718-724 ; and references therein.
9. Y.S. Leong and F. Candau, J. Phys. Chem. 86 (1982) 2269-2271.
10. F. Candau, Y.S. Leong, G. Pouyet and S. Candau, J. Colloid. Interface Sci., 101 (1984) 167-183.
11. Y.S. Leong, S. Candau and F. Candau in : K.L. Mittal and B. Lindman (Eds) , Surfactants in Solution, Plenum Press, New-York, 3 (1984) 1897-1909.
12. C. Holtzscherer, S. Candau and F. Candau in : K.L. Mittal and P. Bothorel (Eds), Surfactants in Solution, Plenum Press, New-York, 6 (1986) 1473-1481.
13. F. Candau, Y.S. Leong, G. Pouyet and S. Candau in : V. Degiorgio and M. Corti (Eds), Physics of amphiphiles : Micelles , Vesicles and Microemulsions, Amsterdam, 1985, pp. 830-841.
14. F. Candau, Y.S. Leong and R.M. Fitch, J. Polymer Sci., Polymer Chem. Ed., 23 (1985) 193-214.
15. M.T. Carver, F. Candau and R.M. Fitch, J. Polymer Sci. Polymer Chem. Ed., (in press).
16. J.P. Fouasier, D.J. Lougnot, I. Zuchowicz, Eur. Polymer J., 22 (1986) 933-938.
17. K. Arai, Y. Maseki, Y. Ogiwara, Makromol. Chem. Rapid Commun., 7 (1986) 656-660.
18. M.T. Carver, E. Hirsch, J.C. Wittmann, R.M. Fitch and F. Candau (submitted for publication).
19. M. Lagues, R. Ober and C. Taupin, J. Physique, Lett. 39 (1978) 487-491.
20. A.M. Cazabat, D. Chatenay, D. Langevin and J. Meunier, Far. Disc. Chem. Soc., 76 (1983) 291-303.
21. B. Lagourette, J. Peyrelasse, C. Boned and M. Clausse, Nature, 5726 (1979) 60-62.

22. F. Candau, Z. Zekhnini and J.P. Durand, J. Colloid Interface Sci., 114 (1986) 398-408.
23. C. Holtzscherer and F. Candau, J. Colloid Interface Sci. (in press).
24. P. Buchert, Thesis, Louis Pasteur University, Strasbourg, (1988).
25. M.T. Carver, U. Dreyer, R. Knoesel, F. Candau and R.M. Fitch, J. Polymer Sci., Polymer Chem. Ed. (in pres).
26. F. Candau, Z. Zekhnini, F. Heatley and E. Franta, Colloid Polym. Sci., 264 (1986) 676-682.
27. See for example : G. Odian : Principles of Polymerization, 2nd Ed. Wiley, New York (1981).
28. J.W. Vanderhoff, F.V. Distefano, M.S. El Aasser, R.O'Leary, O.M. Shaffer and D.L. Visioli, J. Disp. Sci. and Techn. 5, (1984) 323-363.
29. D.W.J. Osmond and F.A. Waite in : K.E.J. Barrett (Ed), Dispersion Polymerization in Organic Media, Wiley, London, 1975, pp. 9-45.
30. F. Candau, Y.S. Leong, N. Kohler and F. Dawans, French Patent (to CNRS-IFP) 2 524 895, 1984.
31. J.P. Durand, D. Nicolas, N. Kohler, F. Dawans and F. Candau. French Patents (to IFP) 2 565 623 and 2 565 592, 1987.
32. J.P. Durand, D. Nicolas and F. Candau. French Patent (to IFP) 2 567 525, 1987.
33. F. Candau and P. Buckert, French Patent (to Soc. Chim. Charb.) 87 08925, 1987.
34. W.M. Thomas and D.W. Wang in : H. Mark, N. Bikales , C.G. Overberger and G. Menges (Eds) Encyclopedia of Polymer Science and Engineering, Wiley, New-York (1985) pp. 169-211.

SUBJECT INDEX

7OR